10/93

THE
SUCCESSFUL
ENGINEER

PERSONAL AND PROFESSIONAL
SKILLS—A SOURCEBOOK

p.172 Forms of business ownership.

(p.140 Depreciation

Chap 8 Dry but thorough discussion. Ins &
(p.311) exps.

p.351 British engrs. adopted first code of ethics --
U.S. engrs. followed.

p.380 PWRR -- possibly introduce in
Chap. 3 under writing.

THE SUCCESSFUL ENGINEER

PERSONAL AND PROFESSIONAL SKILLS—A SOURCEBOOK

J. Campbell Martin

Professor Emeritus of
Electrical and Computer Engineering Department
Clemson University

McGRAW-HILL, INC.

New York   St. Louis   San Francisco   Auckland   Bogotá
Caracas   Lisbon   London   Madrid   Mexico   Milan   Montreal
New Delhi   Paris   San Juan   Singapore   Sydney   Tokyo   Toronto

This book was set in Times Roman by The Clarinda Company.
The editors were B. J. Clark and Eleanor Castellano;
the production supervisor was Denise L. Puryear.
The cover was designed by Rafael Hernandez.
R. R. Donnelley & Sons Company was printer and binder.

THE SUCCESSFUL ENGINEER
Personal and Professional Skills—A Sourcebook

1 2 3 4 5 6 7 8 9 0   DOH   DOH   9 0 9 8 7 6 5 4 3 2

ISBN 0-07-040725-8

Library of Congress Cataloging-in-Publication Data

Martin, J. Campbell.
    The successful engineer: personal and professional skills—a
sourcebook / J. Campbell Martin.
            p.   cm.
    Includes bibliographical references and index.
    ISBN 0-07-040725-8
    1. Engineers.   2. Engineering—Management.   3. Engineering
economy.   I. Title.
TA157.M325   1993
620'.0023—dc20   92-30123

ABOUT
THE AUTHOR

J. CAMPBELL MARTIN'S education includes a B.S. degree from Clemson University, an M.S. degree from the Massachusetts Institute of Technology, and a Ph.D. from North Carolina State University. All of his degrees are in electrical engineering, with a minor in nuclear engineering at North Carolina State. For several summers he continued his studies in modern control theory at MIT and Stanford University.

Professor Martin has been on the faculty at Clemson since 1948, teaching in the Electrical and Computer Engineering Department. His classes have included engineering economics and a required senior course from which this text was developed.

In addition to his teaching, Dr. Martin has done research at Clemson for Hughes Aircraft in the area of microwave frequency conversion, and for the National Aeronautics and Space Administration (NASA) utilizing digital filters for stellar references in space navigation systems and for aircraft instrument landing systems. This work has been reported in some 20 technical papers and reports.

Professor Martin has been broadening the scope of his work. Twice as a visiting professor in the Engineering Economics Department at Stanford University, he studied, taught, and supervised doctoral tutorials. He recently taught a course in Investments Analysis in Clemson's Department of Financial Management.

Dr. Martin is a member of Tau Beta Pi, Phi Kappa Phi, and the Institute of Electrical and Electronics Engineers (IEEE).

Dedicated to the late William K. Linvill,
my most nearly perfect professor at MIT,
and friend at Stanford.
His influence permeates this text.

CONTENTS

FOREWORD

Every year thousands of engineers graduate from universities and enter business, industry, and government. Most are competent in their technical areas. Yet, 5 to 10 years after graduation many find themselves with floundering careers, unrealized goals, and often unsatisfactory personal lives.[*] This book clearly presents the diverse, personal, and nontechnical aspects of the profession that engineers report they need to know. Its design also facilitates[†] self-study by the graduate engineer.

The engineering approach is the underpinning of each section of the book. Mathematical models are normally used for engineering projects; however, in the nonphysical analysis of Chapters 1 through 4, and in the second part of Chapter 5, nonmathematical cause-and-effect relations are used.

The essence of the engineering approach is using models to make proper decisions. The fortunes of our lives largely depend on our decisions. Often they are made without realizing it, for no decision is, of course, a decision not to make one and to let "happenstance" rule. We might call it decision by default. As Aristotle said, "The unconsidered life is not worth living." The considered decision is always better. It can be emotional or rational. I am convinced that rational decisions always have the higher probability of being correct. Furthermore, all rational decisions require a model, that is, one's concept of reality—intuitive, physical, or mathematical. Naturally, the better the model is, the greater the probability will be of making the correct decision. Note that the correct decision does not ensure the desired outcome, but it has the highest probability of giving the desired results. Decision analysis is presented in Chapter 2.

This book builds on the engineer's background in mathematics and physics, and training in logical analysis. The first step in all sections is to model the nontechnical aspects prevalent in the engineer's daily interaction with individuals, society, and projects. All efforts are directed to reducing the complexities of life to their fundamentals. For easy reference, the chapters are written independently, and suppporting material (in other chapters) has been diligently cross-referenced. A detailed index facilitates retrieval of information.

Chapter 1 will help you recognize your personal characteristics and develop those which give harmony to interpersonal relations. Your potentials are

---

[*]Harris Poll. E E careers: New directions but old issues. *IEEE Spectrum*, June 1984, 53.

[†]A successful engineer primarily uses words to be effective in the areas presented here. Throughout the book techniques are used to develop that competence, building on word fundamentals, that is, their roots. It is interesting how many sophisticated terms are merely simple words in another language, such as facilitate, which is from the French word facile, or "easy".

greater when your activities are matched to your capabilities. They can be realized most easily through the use of time management. Matching and self-management (through self-understanding) are addressed in this chapter. You first review important difficulties which affect your success in assimilating information received orally. In the Appendix, you review factors which significantly influence your success in engineering. Also included are conditions for developing the creativity we all have, and the self-discipline and assertiveness successful engineers must have.

A great portion of an engineer's life is involved with managers or management activities. The essentials of management are presented in Chapter 3. Its four phases draw heavily on the remaining chapters of this text.

Probably one of the most important capability influencing an engineer's success is communication. Chapter 4 gives easy guidelines by which engineers can produce creditable reports, oral and written. Confidence gained in using these guidelines should release the engineer's creativity for producing polished papers and oral presentations.

In the first part of Chapter 5 techniques or methods are presented for managing a project efficiently. They include CPM (critical path method), PERT (program evaluation review technique), and PDM (precedence diagraming methods). PERT and CPM are sometimes used synonymously. PERT, however, considers the random time durations of the activities which make up a project. It is used to determine the probability of a project's completion time and cost. CPM and PERT cannot conveniently model project activities which can be interrupted. PDM, on the other hand, is developed to consider projects which have this type of activity.

One of the main uses of effective communication is to sell a proposal, an idea, or a new concept. In the second part of Chapter 5 concepts and models introduced in Chapters 1 and 4 are extended to analyze the "buyer-seller" interactions. The principles can be used in a forceful but ethical manner.

Most engineers work either in private enterprise or for a federal or local government. In both areas, benefits or profits versus costs are very significant. The concepts and techniques introduced on Chapter 6 are used to make economic evaluations of projects and facilitate their quantitative comparisons. Although this material is significant in engineering practice, understanding it also prepares the student for the required engineering economics portion of the Engineering in Training (EIT) examination.

Additionally, the successful engineer needs to build an estate. Chapter 7 reviews life insurance and presents financial subjects needed for prudent investing in stocks and bonds, and the advantages of a foreign bank account.

Finally, the very important legal and ethical aspects of engineering are covered in Chapters 8 and 9. These both establish and guard engineering as a profession.

Jack C. McCormac
Alumni Professor, Emeritus
Civil Engineering
Clemson University

PREFACE

e, s, s, see p. 102

My bk does not explicitly do this,

This book is intended to enhance the personal growth of engineers, and is designed for easy self-study. The text provides what many graduates wish had been available in their undergraduate studies. Its use in a senior- or junior-level engineering course will give the student a significant advantage: guidance and knowledge for developing the characteristics industry seeks in engineering graduates. The scientific or engineering approach is applied to the personal and nontechnical aspects of life—aspects which every professional encounters, and successful ones have mastered.

Building on the engineer's technical background, the text covers the following areas:

Listening and time management: self-concept, interpersonal relations, assertiveness, and decision analysis (personal and professional)

Management, group relations

Technical communications: oral and written

Proposal presentation and project planning: CPM (critical path method), PERT (program evaluation review techniques), and PDM (precedence diagraming method)

Engineering economics: adequate for the required portion of the Engineering in Training (EIT) examination

Financial planning: includes fundamentals for prudent investing in stocks and bonds, and reviews the values of a foreign bank account

Legal and ethical aspects of engineering

This text was developed from notes used for 10 years in a required senior course in the Electrical and Computer Engineering Department of Clemson University. A major portion of those notes was compiled while I was a visiting professor in the Engineering Economic Systems Department of Stanford University.

I hope that you will derive the tremendous satisfaction that I have gained from the material of this book. If not all, almost all I learned from significant others: my family, professors, and especially my colleagues. Please share your experiences with me.

I am indebted to many colleagues at Clemson, with whom I have discussed these ideas, especially Professor Jack McCormac, Professor Claire O. Caskey, and Associate Dean A. W. Bennett. I also thank Professors William Baron (Civil Engineering), John K. Butler (Management), Joel S. Greenstein (Industrial Engineering), and John R. Sullivan (Mathematics) for their critical reviews, respectively, of Chapters 7, 3, 6, and 5, and Mr. William A. Utic, superintendent, Pacific Gas and Electric, Emeryville, California for his critical analysis of Chapters 1 and 9. Professor Albert Duke (Electrical and Computer Engineering) graciously reviewed the entire manuscript. The following reviewers have been very helpful through their keen insights: John Biddle, California State Polytechnic University; William LeBold, Purdue University; William Ledbetter, Clemson University; Abraham J. Rokach, Hypermedia Systems, Inc.; and Joseph F. Shelley, Trenton State University.

I am especially appreciative to my McGraw-Hill family: B. J. Clark, Executive Editor for Engineering, College Division, Meredith Hart, his Editorial Assistant, and Judy Pietrobono, B. J.'s former Assistant, for their encouraging support; to Eleanor Castellano, Senior Editing Supervisor, for her patience and understanding; to Leslie Anne Weber, Copy Editor, for her close scrutiny; to Kathy Bendo, Research Department and Sue Bell, Marketing Manager, for their help in "getting" an appealing text to you. Most of all, I am indebted to my closest colleague, my wife, Willie Anne. She has been my continuous guide and editor.

J. Campbell Martin

THE SUCCESSFUL ENGINEER

PERSONAL AND PROFESSIONAL SKILLS—A SOURCE BOOK

Courtesy Fluor Daniel, Inc., Greenville, South Carolina. Photo by Fred Martin, Jr.

HUMAN BEHAVIOR: UNDERSTANDING YOURSELF AND OTHERS

CHAPTER OUTLINE

Dr. William Menninger found that 60 to 80 percent of the discharges in industry were due to social incompetence. The Carnegie Institute of Technology noted that in fields such as engineering, 85 percent of one's financial success is due to "human engineering," that is, personality and the ability to interact pleasantly with colleagues [1]. This chapter concerns our special characteristics as engineers. It also deals with understanding these characteristics and the conditions for promoting personal growth. Generally, Chapter 1 provides guidance for self-management toward personal and professional goals. Chapter 2 gives the essence of decision analysis, and Chapter 3, the essence of management (of others). Both Chapters 1 and 3 are based on the fundamentals of psychology, the first of the individual and the other of the organization. Thus, they are complementary.

Most goals in life are easier to achieve when we have the proper information. Frequently that information comes to us orally, often in daily conversation. Thus, effective listening is among the most important tools for achieving our goals. Listening is of paramount importance to our job. It can make or break our success in selling a proposal or in completing a project on time.

1-1 THE DYNAMICS OF LISTENING

Listening is a difficult and complex process for gaining information. It is difficult because so much noise and trivia are usually present in this channel of information. A rational model is important in achieving the best "information to noise ratio." The process is complex because two interfaces are involved: one, sound to brain; then, brain to perception. The first is degraded by external distractions; the second, by internal ones (discussed below). Not yielding to them is the key to effective listening, that is, to assimilating the information. Attentive listening is work! As discussed later, it requires ego control. We humans tend to avoid both, work and ego control. We make excuses for doing otherwise. The economists say, "There is no free lunch." Engineers put it differently: "Every joy has its cost." Engineering minimizes that price by planning and making decisions based on models. These models are developed from fundamentals which often include human characteristics.

Avoiding bad listening habits will improve the effectiveness of oral communication. It will help the speaker as well as the listener.

Listening requires energy; an attentive listener looks alive and maintains good eye contact. Again, most important is avoiding distractions, both external and internal.

External

These are mainly audible and visual:

1. Noises, low-voiced conversations, and music are all audible distractions. They give us "good excuses" for not paying full attention to the speaker. There is no easy way to overcome these distractions, but by effort we must continually re-guide our thinking to the material being presented. As in everything else, we improve with practice. Make planned and regular efforts to concentrate on difficult presentations. Start with only 5- or 10-min sessions. Even shorter ones are effective, if we use them as a

means to develop our concentration span and add 1 or 2 min each time. You can also start with less difficult material. Definitely, it takes effort, as it does for runners who extend their distance only a short amount each day.

2. All of the above comments about audible distractions apply to visual ones as well. They include the speaker's physical appearance, the room's decor, and even a view from the window. All of these can be distracting channels. For instance, the physical appearance of speakers or their mannerisms give us information about them. The verbal presentation, however, must be given top priority and concentration if we are to assimilate the message. Its content is by far the most important.

Internal

These distractions depend on prejudices, developed since childhood. They are difficult to change, but we can be better listeners after recognizing them. Analogies help us understand their effects.

White light contains all colors, but when passed through blue glass, all other colors are filtered out. This filtering is a resonant effect used to give special receptivity to certain stimuli. Radios and televisions use tuned or resonant circuits to select desired signals. Our past experiences and general prejudices cause us to have emotional (resonant) filters. They cause us to ignore certain information and give preferential response to other. Thus, words such as *gay, Republican* or *Democrat, Protestant, Jew* or *Catholic* can cause us to either overreact to or discount information. Psychologically they are ego involvements. They can also cause us to dwell on a subject and waste excess time, as reviewed below.

Another internal distraction is preexposure to the subject material. A competent professor designs successive lectures to be confluent (flow together); thus, at the beginning of each, certain aspects of the previous lecture will be reviewed. From personal experience, at the beginning of a class I have often thought, "Gee, I already know this material," and start thinking about other things. In only a short time the lecturer is covering new material, and I am disadvantaged while trying to catch up.

We frequently distract ourselves by too quickly thinking that we already know "all about it," or from too quickly deciding that the material is uninteresting. Another tendency is to decide that the subject is unimportant if we find it difficult to understand.

The result of all these distractions can cause a downward spiral. Regardless of the reason, poor listening leads to poor understanding, which makes listening even more difficult. In turn the poor understanding makes assimilation of subsequent material more difficult. Soon we are totally lost. A question directed to the speaker could give us assistance and time to catch up. But unless we exert the effort to shut out distractions, through the hard work of concentration, the final assimilation will be meager. As noted previously, planned and periodic practice in listening to difficult material is needed.

Concentration is also difficult because of a special phenomenon: the greater speed of thinking over speaking. The average speech rate is approximately 125 words per minute, while our thought speeds vary between 300 to 1200 words per minute. The excess time can lull us into focusing on the distractions previously noted. It is especially important for us to use the spare time to generalize on the speaker's presentation. Predict by trying to anticipate the speaker's development; and then correct, summarize, and encode the material. One writer has compared recalling information to talking to ourselves. The ease and accuracy of the retrieval of information are proportional to how well the material is encoded. Two common examples are the verses regarding Columbus's sailing the ocean blue in 1492 and 30 days has September. . . . Mnemonic devices such as these surely make it easier to prepare for closed book quizzes. Below, we consider mnemonics to encode certain information for easy retrieval.

All of the above will also help us take better notes in classes, in seminars, and during presentations. Speakers should make their topic and subtopics clear; regretfully these guides are not always given. We will, however, usually know the general subject area, for example, calculus or foreign policy. In such cases, we can always build on this by searching for principles and generalizations the speaker is trying to present. Then, we can use the spare time between thinking and speaking to build supporting evidence for these generalizations. In this way, we only need to take notes of the key ideas to have near total recall of the speaker's main thrust: the essence of the presentation.

Example

Consider applying this approach to a book by the noted psychiatrist Kubler-Ross [2]. In it she reviews the stages through which terminally ill patients progress; we can remember these stages by thinking of the acronyin, *DABDA*. Incidentally, these stages are applicable to the way we deal with crises.

D—Denying the forthcoming death
A—Anxiety about possible cures
B—Bargaining with God—as to permit living until a child's graduation or marriage
D—Depression
A—Acceptance, rational realization of the situation

In the Appendix, we will see how *SOMS* gives us the cues to a successful life as engineers. Similarly, by using these memory aids in note taking, you do not need to "record everything."

Let us review a work situation. Consider that you are trying to explain to an associate or perhaps your boss how your idea will solve his problem. Obviously, you must have listened properly to comprehend it. Furthermore, you must still give full attention to understand how the associate *perceives* the difficulty and what *feelings* might be involved. Search for meaning and clues that reveal other aspects. In this portion of the conversation, we must restrain ourselves from being overeager to offer our ideas as a solution. This eagerness could prevent our hearing and understanding important aspects of the problem. We must avoid the trap of hearing only a few sentences, assuming that all is already known, and simply waiting politely until the speaker is finished. How often have you listened to people talking and recognized that nobody is truly listening? They are politely taking turns presenting their own ideas.

One of our students offered an interesting observation: Good listeners are like good friends; they suppress their egos and give up trying to impress others, at least long enough to learn about the other fellow. A good listener learns much and is liked by all.

Let us now return to the work situation. Only after we are sure we have a mutual understanding are we in a position to fit our ideas to the problem being presented. In turn, recognizing our full understanding, our colleagues will more readily accept our ideas as possible solutions.

Concerning personal interchange, we recognize bonuses from being a good listener: we learn and pick up new ideas from others. Through effective listening, mutual understanding and a genuine rapport are developed. In turn, we are in a better position to fit our ideas to our colleague's problems, and they will more readily accept them. Also individuals sometimes reveal their secrets to attentive listeners, and thus they feel stronger ties of friendship. William James, a famous American psychologist, noted that appreciation is the greatest craving of humans. I offer that good listeners are appreciated most. Being liked by others enhances our self-esteem. Furthermore, it is generally recognized that superiors tend to promote those they like.

To summarize, in personal, academic, and professional life, success and happiness are highly dependent on our being effective listeners. This valuable capability can be developed by all of us. With practice, we can become good listeners by not yielding to internal and external distractions, by dismissing stray thoughts, and by using diligent restraint not to interrupt the speaker with our ideas built on incomplete data. Rather, we must use the spare thinking time to encode and maintain a running summary. The essential principle is that the better information is encoded (i.e., correlated in a familiar pattern), the easier and more accurately it will be retrieved. The better we listen, the better we understand and the better we can encode.

1-2 MATCHING ACTIVITIES TO CAPABILITIES THROUGH TIME MANAGEMENT

Throughout history, humans have sought to get the best results from their efforts. We engineers know that an energy source will deliver maximum energy when the load (receiver of the energy) is matched to the source. Similarly, people seem most productive and gain the most satisfaction when their activities are matched to their interest and capabilities. Our productivity increases with stress, up to a certain level. Appreciably above that level, it drops rapidly. Overloading is frustrating and causes our productivity to be significantly diminished for two reasons: the emotional state blocks our cognitive (awareness and judgment) processes, and correcting the attendant errors requires extra effort. This situation is highly stress producing. Behavior observed by Howard [3] is shown in Figure 1-1. You recognize that the critical point is our "matched condition" and that we can approach this ideal condition by close attention to two factors: (1) judgment in choosing our activities and (2) time management in scheduling these activities [4]. The following facilitate our doing this:

1 Clarify objectives, both long (i.e., a year or longer) and short range. Set daily goals, but only a few long-term objectives. Consider only the most important first!

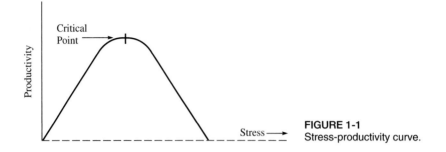

FIGURE 1-1
Stress-productivity curve.

Proper planning to achieve objectives must consider the whole day, week, year, yes, even your whole lifetime. Without objectives and plans we will not know what to do or when to do it. We drift from task to task without accomplishing any of the important items. As a milestone for today, implement only the first four of these suggestions. Assimilation is best, and all achievements easiest, when accomplished little by little!

2 Daily crises demand our attention and consume our time, leaving none for the important activities. *We tend to do the most pressing over the most important!*

3 The criteria for your long-term objectives must be
   a Measurable
   b Attainable
   c Challenging
   d Consistent with your value system and that of the people with whom you are most involved (i.e., family, organization)
   e Flexible when conditions change
The objectives should be written to help you keep them in mind; outline details and set deadlines for milestones in your long-term goals.

In planning, include the people on whom you have to depend to achieve your goals. By their sharing they will usually be committed to yours and their success.

4 Time comes in little pieces. Our plan, with its goals and objectives, permits your recognizing what can be accomplished in any particular time duration that becomes available. List everything you need to do and/or want to do, set deadlines and priorities for each task. In that way, any time available can be used on the most important activity that is suitable for that duration.

5 A written plan reinforces our memory; our plans are easily foiled by the numerous interruptions of the day. Keep the plan visible on your desk; the real reward comes at the end of the day when you see what you have accomplished. By establishing a favorable mental set, this reward enhances your competence: "The more you do, the easier it is to do more." Putting in "long hours" can indicate low productivity possibly caused by poor planning. *Efforts toward interruption control are extremely well rewarded.* If possible, ask that phone calls be held (or use a telephone answering system) for certain periods each day. If necessary, put your work in a folder or briefcase and go to a secluded part of your building or to another location. Libraries are great.

6 With anticipatory planning we can take advantage of favorable contingencies and avoid crisis-oriented reactionary actions. Chance favors those prepared! The key to a quick mind (action) is anticipation. This anticipation gives us the will to act—a requirement for success.

7 Make time to plan; a proper plan saves many hours in achieving all major goals, and many crises and errors are avoided. Correct action the first time saves the time of repeated correction. Develop the self-discipline of setting aside 15 min a day for planning; decide you will try it for 1 month. Enlist the aid of others; the success will quickly motivate you to further planning.

8 Deadlines are useful to establish positive stress. Unrealistic deadlines quickly put us beyond the critical point and set up frustration. A proper schedule maximizes the value we get for our time; it must include planned leisure. As Mackenzie has advised [4], "Get top value for your time!"

9 In setting priorities, use a principle attributed to Pareto. [5] Concentrate on the few critical tasks; i.e., 20 percent of our efforts produce 80 percent of the results and satisfaction. Concentrate on the important; do not let the urgent dominate your day! Realistic deadlines minimize crises and limit their effect. In crises we act in haste, without the benefit of deliberation or the opportunity to explore alternatives.

10 Rational decisions are based on models, but even adequate models are approximations. We tend to underestimate the time required for a task. We forget the nitty-gritty (essential, small details), which require only a small amount of time but together can account for 10 to 15 percent of the total. Try setting artificial deadlines earlier than the true ones. The cushion permits task achievement without crises.

11 In longer range projects, monitoring progress will signal the possibility of a critical milestone not being met. Early signals give time to act in a considered manner!

12 Maybe we attempt too much by not saying no. Of course the resulting frustration further diminishes what can be accomplished.

13 Properly organized work is easier to accomplish. It can be "left at the office" and does not consume our attention when at home with our family or friends.

14 Watch for the tendency to put off important but unpleasant or difficult tasks.

15 Ignore transitory problems; if they will go away, let them.

16 Limit your time at meetings. Come late and leave early, *if* it is possible to do so politely.

17 Don't surrender your time or your life to chance. Plan it!

You desire to realize your full potential; it is possible only by matching your activities to your capabilities, and easiest through time management! To quote MacKenzie: "Your time is your life; you waste your time—you waste your life!" [4]

Matching activities to your capabilities and time management reduce stress, which we study next.

1-3 STRESS: WOW, GOOD? WOW, BAD?

Anxiety is our response to a perceived threat, real or imagined. Frustration is our response to demands which we feel are beyond our capabilities. Both responses cause an emotional arousal (secretion of epinephrine and norepinephrine by the adrenal glands); the arousal affects one's effectiveness, as depicted in Figure 1-2. The term *stress* was borrowed from engineering by Hans Selye, M.D., [6] to define the effects of this arousal. It causes short-term physiological changes such as increased heart rate and blood pressure, rapid breathing, and other changes which prepare us for emergencies. Continual stress has long-term undesirable effects, which are discussed below.

Moderate stress enhances all of our capabilities, especially creativity and judgment. As shown in Figure 1-2, beyond a certain amount, depending on one's coping ability, more stress reduces these abilities; and too much (panic) can paralyze an individual, physically and mentally! Of course, as engineers we wish to optimize all systems, particularly ourselves.

Human systems, like other physical ones, have transient behaviors between equilibriums, i.e., between routine or steady-state operations. Also, as in other physical systems, the human system is frequently overstressed during times of transient behavior. Thus, stress can be defined as a system's response to any stimulus which disturbs equilibrium. In humans the stimulus or stressor may be pleasant or unpleasant, e.g., marriage or divorce. Generally, any significant change causes stress. It can also be a subconscious reaction to a mishap or any disappointment, even small. Although we are unaware of it, our capabilities are affected by stress.

Lord Kelvin said, "Little can be known about a subject unless it can be measured."[7] In an effort toward measurement, "stresstimation" is presented in Table 1-1. Other causes of stress are:

High noise level

Time pressure

Increase or decrease in the work required, especially if there is confusion or uncertainty about one's role or responsibilities, or both

Inconsistent demands from a superior or coworker, particularly if they conflict with one's values or morals

All of these causes are aggravated for the engineering student and the engineer by competitiveness and the necessity to make quick and accurate decisions. For the engineer or student with a family, if this anxiety is brought home, it starts a cycle. The added demands of the family can quickly make family members targets for work's (or school's) displaced anger and frustration. Naturally the situation disrupts the marital

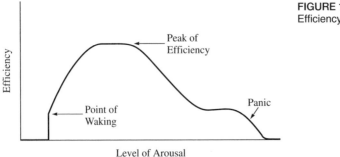

FIGURE 1-2
Efficiency-arousal relation.

TABLE 1-1
STRESSTIMATION

Check all of the items that have applied to you in the past year, and add up your LCU's, or life change unit points. This will show how improper responses to stressful changes in your life could affect your health. If you score 0 to 150 points, you have a 30% chance of a major health problem in the next 2 years; 151 to 299 points, a 50% chance; 300+ points, an 80% chance.

Event	LCU points
Death of a spouse	100
Divorce	73
Marital separation	65
Jail term	63
Death of a close family member	63
Personal injury or illness	53
Marriage	50
Fired at work or retirement	45
Change in health of a family member	44
Pregnancy	40
Sex difficulties	39
Gain of a new family member	39
Business readjustment	39
Change in financial state	38
Death of a close friend	37
Change in type of work	36
Change in number of arguments with spouse	35
Mortgage over $15,000 (1992 dollars)	31
Foreclosure of mortgage or loan	30
Change in responsibilities at work	29
Child leaving home, trouble with in-laws	29
Outstanding personal achievement	28
Spouse begins or stops work	26
Begin or end school	26
Change in living conditions	25
Revision of personal habits	24
Trouble with your boss	23
Change in work hours or conditions	20
Change in residence or school	20
Change in recreation or church activities	19
Change in social activities	18
Mortgage or loan less than $15,000	17
Change in sleeping habits	16
Change in number of family get-togethers	15
Change in eating habits	15
Vacations or special holiday	13
Minor violations of the law	11

This "social readjustment scale" was developed by Dr. Thomas H. Holmes, Richard H. Rahe, and their associates at the University of Washington's School of Medicine. They interviewed nearly 400 people, who were asked to rank the relative amount of readjustment required to meet various stressful events.

relationship and can produce problems with the children. The resulting discontent is then fed back to the workplace for another round of stress, causing reduced performance.

Other "aggravators" are smoking cigarettes and drinking too much coffee. Like fear and excitement, both of these reduce blood flow to the limbs and move one's system further along the stress curve.

Continual stress greatly increases the probability of heart attacks, strokes, and hypertension (see the "test" at the end of this section). Long-term effects also lead to increased cholesterol levels, headaches, skin rashes, ulcers, and colitis. More significant is the reduction of one's immune system, leaving the body highly susceptible to diseases ranging from colds to cancer.

Symptoms of too much stress are anxiety, irritability, depression, stomach and intestinal disorders, chest pains, shortness of breath, headaches, and decreased sex drive. These signal the need for stress control. Let us review two ways of coping:

1. Removal. Stress is primarily the result of an anxious, unrelaxed attitude. Toward its reduction, the first step, if possible, is to remove the cause of the anxiety. A common cause could be tense TV programs, dramas, or sports. Another might be a misunderstanding with a family member, a friend, your boss, or a colleague. In an honest and open way, talk with her about it, showing a willingness to admit error as proper.

2. Response reduction. Modern life has many unavoidable stresses. For them, the following methods are effective.

a. Deep muscle relaxation. Tense and then relax, sequentially, your forehead, jaw, neck, and shoulders. Then breathe deeply and hold it for a few seconds while also holding in your stomach.

Now while in a more relaxed feeling from technique (a), make a concerted effort to relax mentally; reflect on a pleasant event, past or anticipated. (Or perhaps you might visualize yourself sitting in the middle of a quiet grassy meadow with your back against a huge oak tree. A light breeze is caressing your face. You are totally relaxed, away from all activities.) These relaxation techniques seem helpful in going to sleep, especially the latter: reflecting on a pleasant event or situation.

b. Balanced life. Tolstoy (1828–1910), the Russian novelist and philosopher, noted that for a healthy mind one must have a balance between work and play, love and worship. Chapter 2, "Decision Analysis," offers thoughts on love; the same concepts are applicable to religion and worship. The remainder of this section deals with "work and play." Behavioral changes can be the first step in achieving their balance. Of special significance is the importance of scheduling breaks during your workday (naturally, you do not have to stop exactly on the minute of break time, but soon thereafter). And, of course, during the break do something pleasant. Another pattern change is to reduce the hassles in driving; alternate driving to work with one or more colleagues. If your schedules are slightly mismatched, use the waiting time for enjoyable reading, a little exercise, or just relaxing, i.e., "loafing a little." If possible, take a train or bus and use the travel time for relaxing. Similarly, when traveling (by auto) with family or friends, alternate drivers.

For another behavioral change, eat less fast foods and eat slowly, at least for most meals. *A scheduled, routine lifestyle is not thrilling, but it surely makes a strong mental base from which to launch into a challenge when it comes.* Also, most of the time, have a schedule for sleep—7 to 8 h a night. My experience indicates that fatigue is a nonlinear (i.e., nonproportional) response to extended activity. Staying up past the regular time requires a disproportionately greater period to recover. To determine the best amount of sleep for you, try varying the sleep periods by 15-min increments, and observe how you feel half an hour after getting up. For each increment, make the observations over at least 4 days; this interval allows the transients to die and the body to establish a steady state.

Diet and rest are paramount for a healthy mind and body. So often university medical clinics are crowded during the examination periods. They are filled with students already stressed by anxiety, often due to lack of earlier preparation. In a usually vain effort to compensate, the students study late hours but are further stressed by being generally "run down" and suffer from the resulting colds. It seems that common cold viruses are present everywhere and take over in any human whose immune system is weakened by poor diet or inadequate rest, both exacerbated by stress. Of course, inadequate rest and diet can also be problems for engineers who are pressed for time and work long hours on a project.

c. Friends. The techniques above are mostly preventive and are surely to be used first; however, stress is unavoidable because of ever-present randomness. When you feel anxious, restless, or depressed, seek a sympathetic friend.* Talking with a friend reduces the complexity of a problem. In the process you organize your thoughts. The organization will frequently solve the problem, or a new insight will come from the friend's comments and questions.

If for some reason you are isolated and a "true friend" is not available, or even before seeking a friend, talk with yourself by listing the facts about the situation. The list helps you to focus on the most troubling aspects and acts as a stress reducer.

Learn to accept that which cannot be changed, or cannot be changed with effort justified by a solution. This situation is sometimes frustrating when the methods and values of friends or colleagues are incompatible with yours. It is very helpful, however, to look for the best in others, realizing that none of us is faultless. We are all different, but "that's ok."

While you are following these methods it is also helpful to get away from an unpleasant situation for a period, even as short as 5 to 10 min: go for a walk, take an early lunch, or simply loaf a little. The situation will not necessarily have changed, but you will be more relaxed and recharged.

Sovereign among all methods for coping with stress is exercise! It is best to exercise regularly, in a routine way. First, one can increase exercise opportunities by running up steps or, if the stairway is crowded, walking up two steps at a time. Also, you might park at the edges of the parking lot at work or stores. A friend of mine parks about a quarter of a mile from his office and runs in.

Regular exercise is a great way to stay physically fit and gain the reward of a new, positive attitude. Home programs like the one offered by the Royal Canadian Air Force (RCAF) are excellent and convenient. One starts at an easy pace and builds to a level proper for his or her age. As with other programs, after building to the proper

---

*True friends will always be sympathetic, and developing them is an important part of preparing for adulthood.

level, the routine need only be done three times per week. Active sports are also good exercise, and many firms are providing fitness activities for their employees during lunch or after hours. The significant factor about exercise is not the burning of calories. It is that vigorous exercise elevates the breathing and pulse rate so important to the lungs and cardiovascular system. Exercise should be done for 20-min periods, three times per week. It is especially important for those over 40 years of age, after an "OK" from their physician. Quoting Charles B. Wilkinson, former consultant to the White House on youth fitness, "Physical fitness makes us work better, look better, and feel better. All of us should have enough self-discipline to spend a few minutes each day exercising. The RCAF program is an excellent method of attaining and maintaining fitness"[8].

Anxiety and uncertainty are surely major causes of stress; they cannot be eliminated, but the causes for anxiety can be minimized by using the techniques in Section 1-2 (Matching Activities to Capabilities through Time Management) and in the Appendix (How to Succeed in Life through and in Engineering). The effects of uncertainty can be minimized by using the methods presented in Chapter 2 ("Decision Analysis").

No one can or should completely avoid stress; below the critical amount, its increase enhances our productivity and creativity. Control of stress permits us to operate close to the optimum point. By taking the test below, you can determine your present stress level; the score will give you the measurements (feedback) needed to optimize your personal system through use of the techniques in this section. Wow! Stress can be wonderful.

ARE YOU DEALING WITH STRESS PROPERLY?

How well do you cope with stress in your life? Gauge your ability with the following quiz developed by George S. Everly, Jr., for the U.S. Department of Health and Human Services.

1 Do you believe that you have a supportive family? If so, score 10 points.

2 Give yourself 10 points if you actively pursue a hobby.

3 Do you belong to some social activity group that meets at least once a month (other than your family)? If so, score 10 points.

4 Are you within 5 lb of your "ideal" body weight, considering your health, age, and bone structure? If so, give yourself 15 points.

5 Do you practice some form of "deep relaxation" at least three times a week? This includes meditation, imagery, yoga, etc. If so, score 15 points.

6 Give yourself 5 points for each time you exercise 30 min during the course of an average week.

7 Give yourself 5 points for each nutritionally balanced and wholesome meal you consume during an average day.

8 If during the week you do something that you really enjoy and is "just for you," give yourself 5 points.

9 Do you have some place in your home where you can go to relax or be by yourself? If so, score 10 points.

10 Give yourself 10 points if you practice time management techniques in your daily life.

11 Subtract 10 points for each pack of cigarettes you smoke in an average day.

12 Do you use any drugs or alcohol to help you sleep? If so, subtract 5 points for each evening during an average week that you do this to help get to sleep.

13 During the day, do you take any drugs or alcohol to reduce anxiety or calm you down? If so, subtract 10 points for each time you do this during the course of an average week.

14 Do you ever bring work home in the evening? Subtract 5 points for each evening during an average week that you bring office work home.

Calculate your total score. A "perfect" score would be 115 points. The higher the score is, the greater will be your ability to cope with stress. A score of 50 to 60 points indicates an adequate ability to cope with most common stressors. Experts advise against using drugs or alcohol to deal with stress and instead advocate exercising, eating a balanced diet, and using relaxation techniques to minimize the effects of stress.

1-4 INTERPERSONAL RELATIONS

It is suggested that all rational decisions are based on models (i.e., representations of the situation). Thus, disagreement between rational people indicates different models or objectives, or both (e.g., disputes between labor and management). Perhaps the new cooperation in GM's Saturn plant will establish a "new order." In any case, unemotional relations between people are improved by an adequate model.

Models of People

People react in two ways: from emotions and from rational analyses. Emotions cause natural responses, imbued in early life; they are uninhibited and transitory. This natural behavior of people is probably governed by the part of the brain more nearly analogous to the ROM's (read-only memories) in a microcomputer. Lidz [9] suggests that the human mind is a programmed brain; according to Berne's behavioral model, behaviors are guided by the Child, Adult, or Parent ego states [10]. He designates them as P-A-C for Parent, Adult, and Child. Again drawing on the analogy to the microcomputer, the Child's behavior is a response to the part of the brain which can be represented by a PROM (programmable read-only memory); the Adult's response is analogous to the part of the brain which can be represented by the ALU (arithmetic logic unit). Berne presents the Parent as a pure memory, much like a computer disk, accepting data as given, with no logical analysis. To continue the analogy to the microcomputer, the Parent responds to the part of the brain which is similar to the readily programmable RAM (random access memory) and a PC's coprocessor. It does seem, however, that the memory part migrates to the ROM to become the person's cultural script [11]. For instance, it has been observed that in emergencies, a person in a foreign country will immediately revert to using his or her native tongue. In the emotional state, responses are from the Parent or Child. It is important to emphasize that emotions block judgment and the cognitive processes! For this reason, the second (rational) mode of behavior is almost always desired. Make every effort to avoid emotional power struggles!

The second way people respond, is from rational analyses; these responses are guided by the rational Adult or, to "beg the analogy," by their ALU. The significance of this development is that it seems a person can be modeled as a digital control system, with a soul (Figure 1-3.) Significant is that *the desired response is highly dependent on proper programming.* The programming can be guided by the axiom that in proportion to their intelligence, people do what and only what they find rewarding.* Maslow has formed a hierarchy of human needs [12, cf. Motivation, chapter 3]. First are the physiological needs, such as food, shelter, and clothing. Next, for some, is the affiliative need, a need to belong and be accepted by others. Then, for all, there are the growth needs, such as achievement and self-respect. These last needs involve the development of a self-concept. At birth no one has a self-concept; rather, it is developed over years primarily from the reflections of one's self as perceived from others. *There is no characteristic you have which is more important than your self-concept!* It is a psychological truism that our attitudes toward others are conditioned by our fundamental attitudes toward ourselves. Those who feel negative about themselves tend to be negative about others. Academics provide a very rewarding way for students to improve their self-concept. Doing well academically can be achieved by giving it top priority and by using the techniques presented in Section 1-2. Over many years, I have observed that a strong self-concept tends to give a person the confidence to be rational in an emotional situation.

Even in the rational mode, people under anxiety can act as differentiators, giving too much importance to small changes or even to noise (the randomness always present in nature). Recall our resonance phenomenon presented in the section on listening; because of it (responses controlled by characteristics developed in early life), we very nearly ignore certain information and give excessive importance to other, including noise. Conversely, as integrators, we respond more slowly, properly weighing and assimilating all the inputs, smoothing out the noise. Such is the optimum filter (in the

---

*Economists call this axiom "The Self-Interest Hypothesis."

FIGURE 1-3
A simplified human control system.

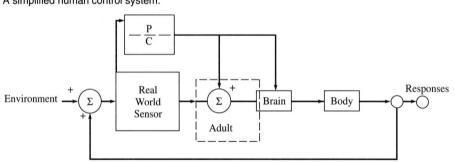

least-squared error sense) developed by Norbert Wiener for radar gun control systems in World War II. It is acclaimed to have won the Battle of Britain (against bombing by German aircraft) and has since been configured by R. E. Kalman for digital filters used for space navigation by NASA. The filter's principles are presented in Appendix A-2 ("Dead Reckoning") of Chapter 7. But a submarine captain once said, "if you take too long to decide [e.g., in assimilating the information], the target moves out of range." In physical control systems, a balance is sought; by accepting maximum allowable error, the response time is reduced by approximately one order of magnitude. We also need to develop judgment in how long to analyze before acting.

For further insight into personal interactions, let us expand Berne's P-A-C model. The Parent (memory) stores information from parental figures, largely programmed early in life. The Parent can be an asset, making many responses automatic, conserving considerable time and energy. Also, it acts as a PC's coprocessor freeing the Adult from making innumerable trivial decisions, leaving more time and energy for important issues. The Parent can be considered to have two aspects: the Critical Parent and the Nurturing Parent. As their names imply, after an accident (such as your spilling orange juice) the Nurturing Parent of a friend might say: "Those tall juice glasses are unstable and so easy to knock over." On the other hand, the friend's Critical Parent could have said: "Be more careful next time; this mess is so unnecessary! Now, I've got to clean it up." Such a criticism might well incite your Child.

The Child is in many ways the most valuable part of the personality; it also has two aspects:

1 The Natural Child is spontaneous, with charm, pleasure, and (most significant) creativity, but it can also be demandingly selfish.

2 The Adapted Child causes a person's behavior to be modified by parental influence, tending to be a little adult or perhaps retrogressing by withdrawing or whining. It is interesting to observe how alcohol intoxication usually first decommissions the Parent. Freed of the parental influence, the Adapted Child is transformed into the Natural Child, with poor judgment! Intoxication is also produced by fatigue, for example, from an active sport like boating.

The Adult is autonomous, directed toward objective appraisal of reality. It processes data in a nonprejudicial manner. It is necessary for survival; from data it computes probabilities which are essential for dealing effectively with the outside world. For example, when a driver comes to a busy intersection, the Adult processes the series of velocity data. Action is withheld until the computations indicate a high degree of probability of entering the intersection without colliding with other cars or people. It is important to note that the Adult regulates the activities of the Parent and Child, mediating objectively between them. In a mature person, the Adult is in control most of the time!

Transactional Analysis (TA)

Berne uses TA to describe the fundamental interactions between persons (Agent and Respondent). He defines the *stroke* as the basic unit of social interchange; a stroke is any act of a person implying recognition of another. A significant function of the

Adult is receiving strokes. It is recognized that strokes can be positive (ego-boosting) or negative. In positive ones, sincerity is important. By using positive strokes, you have a head start on engaging a person's Adult! Communication is enhanced by complementary transactions, as shown by the parallel stimulus-response vectors of Figures 1-4a and 1-4b.

Figure 1-4a represents complementary transactions, such as the one above concerning the orange juice spill. In it the Nurturing Parent of a friend soothed the Child of the friend who had spilled the juice. This is frequently the response of friends when they observe or you tell them of a misfortune. The simplest and most effective transactions are those in which both stimulus and response arise from the Adults of the people involved (Figure 1-4b). In a strained situation, however, the interchange should start from the Nurturing Parent to the Child (Figure 1-4a). This reduces the negative influence of emotions. Then with the Child of the respondent placated (i.e., with emotions diminished), the rational Adult to Adult transaction can be easily established. Accordingly, in our "orange juice" example, the friend would switch from the Nurturing Parent to an Adult by saying, "May I help you clean up?" This smoothly focuses on the realism (addressing one of the many unpleasantries in life) that someone does have to clean up. The Critical Parent of an Agent quickly activates the natural (perhaps unwieldy) Child; a subsequent Adult to Adult transaction (Figure 1-4c) would be crossed, and the resulting flare-up would require delaying further progress until emotions subside. Recall that, fortunately, emotions are transitory; thus, "changing subjects" seems best until the Child "settles down."

By being attuned to signals, we can recognize the ego states active in interpersonal relations. Accordingly, by utilizing the Nurturing Parent or changing the subject (to temporarily delay action), or both, we can move to establish an effective Adult to Adult interaction.

For any specific situation, achieving Adult to Adult interaction is highly dependent

FIGURE 1-4
Interpersonal transactions. *(From* Games People Play *by Eric Berne, M. D. Copyrighted 1964 by Eric Berne. Reprinted by permission of Random House, Inc.)*

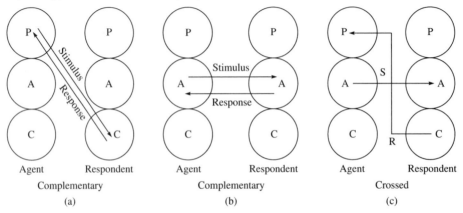

on the personality of the individuals. Furthermore, it appears that our personalities have been influenced by the order of our birth.

The Birth Order Phenomenon [13]

It seems generally recognized that parents, other significant adults, and siblings have a dominant influence on an individual's personality. The time a person spends within their sphere is so great. One's personality and its environmental state (i.e., its mood) control the individual's response to a given stimulus. The family environment is different for each child because of sibling interaction and their interactions with parents.

All of us seem to fall into one of three "birth order" categories:

a The firstborn (an only child tends to be a super firstborn)
b The middle child (being of the same sex as the next older sibling seems to strengthen characteristics reviewed below)
c The last born (being 5 or more years younger than the next older sibling seems to result in firstborn characteristics for a person of any birth order)

A psychological theory states that humans have three natural motivations:

1 To obtain rewards and recognition
2 To avoid pain and danger
3 To get even

Interestingly, these pattern Berne's ego states: Parent, Adult, and Child. Motivation (3, above) tends to be irrational, like "cutting off your nose to spite your face." All three motivations are operant in every birth order but, as reviewed below, to different degrees.

In order, we now review the general characteristics of each "birth order" category.

Perfectionist   Reliable, conscientious, driven to succeed, a list maker, critical, serious, scholarly, self-reliant, self-sacrificing, a good student, and a pleasant* and competent worker. Also, well-organized, conservative, strives to be in control, compliant, a very strong need for approval, believes in traditional authority, a people pleaser (but some first borns can be aggressive and inflexible).

Mediator   Negotiator, avoids conflict, independent but extremely loyal to the peer group, has many friends but feeling slighted at home, goes outside the family (to create another family) for values and attention. Although usually capable, the mediator often tries negotiating before working hard to achieve a goal.

Manipulator   Charming, blames others, a show-off, a people person, a good salesperson, socially precocious and engaging. Perhaps they did not have much pressure from parents for achievement and discipline. In turn, they are more oriented toward self-interest.

---

*As long as he is in control.

These characteristics have not been "confirmed" by research, but I have found being sensitive to their possibilities helpful in personal interactions and management, especially when interacting with firstborn peers. As feasible, let them "run it."

Lastly, Kevin Lehman cautions about marriage between two firstborns. From his counseling experience he observed, "Their relationship tends to be the opposite of the true concept of marriage: pulling together, sharing, melting into one." Furthermore, he adds, "For the better odds on a happier marriage, marry out of your birth order."[13]

1-5 CREATIVITY—THE DIAMOND OF OUR BEING*

Of his magnificent statue of David, Michelangelo said that he merely cut away the obscuring stone. All of us have creativity; we only need to bring it forth. Special conditions enhance and facilitate the process; this section reviews those conditions and how to establish them.

Creativity is the ability to bring forth many ideas new to the person making the effort, though not necessarily new to the world. A colleague trying to sell television sets in the 1950s designed a double paraboloid antenna to get a particularly weak UHF station (the only source of a major network). Later working with NASA, he learned that the antenna was a variation of a telescope developed by Cassegrain in 1672. The cassegrainian concept is used extensively for radar tracking of deep-space vehicles. Nevertheless, my colleague was creative.

As recent as the 1930s, it was thought that creativity was a magical gift or an inborn talent. At about that time, Alex Osborn of Batten, Barton, Durstine & Osborn, an advertising agency, came up with the technique now commonly known as "brainstorming." It was an extremely innovative approach. The creative person sees what everybody sees, but thinks what others have not. Osborn's methods involve two aspects: (1) brainstorming and idea gathering and (2) the analytic phases.

1. The brainstorming phase consists of gathering as many ideas as possible relevant to the problem being considered. Ideas tend to come easiest in a relaxed, happy, and unstressed atmosphere. Furthermore, new ideas beget others. While brainstorming, members of the group should avoid being critical of their own ideas or those of others, no matter how "far out" they may be. Often in life, a Critical Parent judges without knowing all the facts, i.e., the whole plan. The noncritical phase is very difficult, for an engineer characteristically thinks analytically, or in a judgmental mode. This is analogous to the Critical Parent of Berne's P-A-C personality model presented in Section 1-4. Recall that the Critical Parent activates the emotions of the Adapted Child and kills creativity—that valuable asset of the Natural Child. The significant point is that the engineer's judgmental mode must be shut down. When thinking in this mode, we progress logically, going from (1) to (2) to (3) etc., and are incapable of taking the creative leap to (7) or to (b), that is, to something that does not normally follow.

Research has shown that when 10 or more engineers are independently considering the same problems, they have a 90 percent overlap in the first 20 of 100 ideas, the best solutions will be between numbers 60 and 100, and the really top ideas will be among

*This section is based on an article by R. Donald Gamache, president of Innotech, Trumbull, CT.

those beyond the first 80. Again, a very significant element is that these truly top ideas come only in a climate in which members have no hesitancy to make "off the wall" suggestions. The members should be peers, so as to encourage spontaneity and confidence.

2. The idea-gathering phase is similar to fishermen who have cast a large seine, catching many and all species of fish in the huge net. The "good fish" must be chosen. After the large catch of phase one, the engineers use their normal judgmental mode to select the best ideas, analyzing each for technological and economic practicality. Here, they realize there is no one best solution; alternatives have to be weighed in trade-offs between advantages and costs.

Creative people are generalists with wide interests; thus, they do not reject new but different approaches. They recognize that techniques in one area can be applied to an entirely different area of technology.

Considering the following factors will help us chip away the mental blocks:

1 Often on quizzes we need to get *the* right answer; usually in life, especially in technology, there is no *one* solution to a problem or need. Looking for other solutions produces new ideas.

2 If we immediately think or "recognize" that the idea is illogical, we are being judgmental too soon. Such an attitude is death to generating fresh thoughts; new ideas fertilize best in the "loose soil" of a relaxed environment. Metaphors and analogies help in creating new concepts and techniques much as the mathematical models used extensively by engineers. In the last judgmental phase, we recall that models are always only representations.

3 Rules can make the illogical seem logical and limit our perception of the true situation, again killing our creativity. Ignore rules in the exploratory phase! It could be that a law of human origin is no longer applicable.

4 Being practical restricts the realm to the "possible," which itself can be too restrictive to develop new ideas.

5 Let us reemphasize that humor is conducive to creativity. Risk being foolish; it relaxes constraints and is a form of play. If need is the mother of invention, play is its father. Both are important.

6 Success and failure are parts of the same process: progress. Failures are rungs of its ladder. Thomas J. Watson, founder of IBM, said that "to succeed, double your failure rate." [14] According to the English poet John Donne, "Not failure, but low aim is crime."[15]

7 Fresh ideas in one field often come from people in another (not constrained by "deadening experiences"). We can get new ideas from others active in similar areas when just discussing our problems. Communication (the discussion) requires structuring our thoughts, and in a relaxed climate the process frequently leads to a new perspective and solution.

8 Persons who feel uncreative set in action a self-fulfilling prophecy. Those who feel they are not creative will not try a creative solution to an important problem. Studies show that creative people think they are! The less creative do not. Self-esteem is essential; a new idea puts you out there alone, taking risks of failure and ridicule. No one, even when moderately comfortable, likes changes; but the modern world imposes them, and catastrophic changes can best be avoided by creative people.

In youth, our minds ran free; with growth, rules narrowed our thinking. The rules are usually to our advantage; stored in our Parent, they lead to intuitive decisions which free our Adult of many trivial analyses. A significant limitation here, however, is that our Parent bases present decisions on past data (generals fight the last war); the modern world requires Adult decisions, aided by the creative Child while weighing information from the Parent.

1-6 SELF-DISCIPLINE AND ASSERTIVENESS

The techniques of personal growth presented in this chapter do us no good unless we act on them. Acting on prudent information requires two characteristics: self-discipline (control of ourselves) and assertiveness (control of our interaction with others).

Self-Discipline

Self-discipline is required to control our Child. The Child, and sometimes misinformation from our Parent, can override reasoned guidance from our Adult. Everybody struggles with self-discipline. We all rationalize why we "do not have to do what we don't want to do when we don't want to do it." Our Adult, however, can sometimes show that no action is best. For example, analysis can correctly indicate that continued study on an unyielding problem is unwise. Dismissing it by taking a break or having a good night's sleep is effective. It seems that our subconscious mind often works in its unstressed environment. On returning to the task, we often succeed quickly. More frequently, however, we *do* need to do what we would rather not. Regarding outside activities, we might say, "Wait until it is warmer (or cooler)"; or for those inside tasks, "When I am not so tired." The key to success here is that we can all make ourselves do a little of what we do not want to do, and subsequently do a little more the next time. In this way we can develop a stronger discipline. Stressing an earlier observation, the easiest way to accomplish an unpleasant or difficult task, or one that is both, is "little by little." Also, "little by little" affords the most pleasure from a pleasant activity, such as eating a delicious meal or dessert.

Our strong passion for personal worth is a factor that can facilitate* building self-discipline. Personal worth can be achieved through self-approbation and the approbations of others. Usually, however, the approbation, especially from others, must be perceived as merited. Thus, we can achieve small goals. For example, a student might study for 20 to 30 min before taking a break; a home owner might tackle a small cleaning task before doing something more pleasant. The key is use of reinforcers as self-approbations. The accumulation of small goals can build to be adequately significant to merit sincere compliments from others.

---

*Note again that many sophisticated terms are just simple words in another language; recall that *facile* is French for "easy"; interestingly *sophos* is Greek for "wise," as in *sophomore,* a wise fool. Roots of words are part of the fundamentals required for effective communications.

In summary, self-discipline and other behavior modifications can be developed easiest in small steps. The self-satisfaction from each reinforces and motivates further progress. This progress and the accomplishments give us the self-confidence which is a characteristic of an assertive individual.

Assertiveness

Essentially, assertiveness is a defense mechanism. Assertive persons have a proper balance between submissiveness and aggressiveness. They pleasantly exercise their rights in achieving a desired situation or avoiding an undesirable one. The rights of others, however, are never infringed.

Our ability to get along with others is by all measures a very important factor for our personal happiness and professional effectiveness. This ability depends upon whether we can love* and be loved, more than we hate. All people "hate," to some degree, the person or thing that frustrates their wishes.

Assertiveness is a skill essential to successful day-to-day personal management. It means following your plan for the day without often being coerced from it by others. Nonetheless, concern for others and their rights is a first requirement. Although, as noted above, this does not mean diminishing our own rights.

Assertive persons have self-confidence, are honest, act in a consistent manner, and take responsibility for their actions. They have the poise to accept compliments and admit errors. Their poise is supported by body language: neat, appropriate dress; erect posture; and only natural, undistracting hand movements. By trying to anticipate events, when they do occur, the assertive individual can unemotionally speak slowly, clearly, and with self-assurance. Generally, such persons are at ease with others and put them at ease. This mood is enhanced by always maintaining eye contact.

Self-confidence is a natural result of attaining goals, achievable through self-discipline. As alluded to above, confidence is strengthened by anticipatory planning.

Interacting honestly with others is recognized by mature people as essential. Honest people are characterized as dependable and trustworthy, virtues of great value in interpersonal and group relations.

We have reviewed the characteristics of an assertive individual; now let us review the techniques, which involve our responses to difficult or unpleasant situations. For assertive individuals, decisions must first be analyzed in a rational, unemotional manner, carefully weighing the short- and especially the long-term effects.

Primarily, there are three techniques:

1 Make the responses short, but pleasant.
2 Politely repeat them as often as necessary.
3 Always be friendly and firm. Empathize by first acting as a Nurturing Parent before your Adult presents your decision.

---

*Please recognize that true love has nothing to do with sex; sex is primarily mutual (perhaps unilateral) gratification. Genuine love, however, is concern for others, as much as or more than for ourselves. According to Peck, there is great probability that given a chance, love could cure all evil [16].

These techniques are easy when assertiveness is in order, that is, when even a considerate person has the right to say no, and maybe should, so that the other person will develop independence and responsibility. Furthermore, occasionally a responsible and thoughtful friend, in honest ignorance, might make an improper request. For instance, an individual, with his telephone on "call forwarding," was called while at a party by an acquaintance. The call involved about a 30-min discussion. Since the call had been forwarded, the acquaintance never knew he was imposing on the individual and would not have "consumed" that much time had he known the individual was not at home in his workshop. An assertive individual would have informed the caller in a pleasant manner, and requested a return call the next day.

Responsible persons are considerate and helpful. They earn a reputation that helps them be assertive, for when they decline a request, others know that they truly wish they could be helpful.

Assertive interchanges are quite varied: they involve your professors, your peers and superiors at work, friends, family members, and strangers. In all cases, the attitude "You're OK, I'm OK" gives you a head start; it immediately establishes mutual respect, and the other persons are more open to accept your viewpoints. Conversely, aggressive people put others on the defensive; consequently, they are resisted and resented, and are left frustrated and alone.

Consider two cases in which a quiz paper had been returned to you. On one, the points had been incorrectly added—to your disadvantage. On the other, you felt that too many points were deducted on an essay question. Be considerate. Do not interrupt the instructor while she is heading for another class or appointment. Ask for a convenient time to talk. After commenting that you know grading 30 (or perhaps more) papers is trying, state that your addition indicates a different quiz grade (a short, pleasant response). Probably repetition would not be necessary; but if the instructor wishes to review it, and possibly forgets to return it soon, you would need to offer a friendly but clear reminder.

In the case regarding the essay question, seek information from successful students about the significant aspects. Most instructors decide which major points the answer should include and grade according to how well they were grammatically presented. Using this information, analyze your answer; if the major points are covered well, unemotionally present your analysis to the instructor, asking that the essay be reevaluated.

Suppose a peer or an immediate superior requests that you attend a special meeting or perform an extra task. If your schedule permits, and you believe you can make a contribution at least equal to the time and effort the meeting or task will require, then you will grow personally and professionally in meeting the request. Conversely, be careful that your desire "not to refuse" does not bias your analysis. Most of us have a full schedule, which includes healthy breaks and relaxations. Thus, any added activity takes time and effort away from one's total resources, specifically from those not already committed. If you have a full schedule and the activity cannot be ranked higher than a present task requiring comparable time and effort, then it should be declined.

Particularly relevant today is the situation in which a male engineer asks a female secretary or a fellow female engineer to help choose a present for his wife or girlfriend. It is recognized that females have an advantage in shopping for gifts, especially those for other females. Should the secretary or female engineer try to help? Well, she should make an objective analysis. Will it frustrate her workload, including after-work household and social commitments?

Fundamentally, you need to determine whether the requested activity will require less effort or stress from you than from the person making the request. That is the essence of assertiveness interaction. To this extent, an assertive individual tries to help and support others. Since most of us are reluctant to offend, to balance this reluctance, at least initially, we should try to be slightly selfish in analyzing the specific situation. Also an important aspect in the evaluation is that small, often overlooked details always cause the total activity to require more time than first thought.

In the case above, the female's analysis could show that she should not actually choose and purchase the gift. She could, however, briefly tell the engineer that other commitments do not permit her helping as much as she would like but that she would enjoy helping decide on the gift. I feel that, generally, decisions require more effort than the action.

Occasionally, a stranger or acquaintance, such as a waitress or clerk, will err in determining your bill. As alluded to above, balance your "cost" against their "cost." If the error is small, ignore it to save their pride and enjoy your altruism. On the other hand, when the difference is significant, pleasantly request that they review the calculations with you. Again, showing anger or putting them in a defensive position can only aggravate the situation and undermine your effectiveness. Use humor where proper.

In summary, being assertive enhances your self-worth and strengthens your personal interactions. Through reasoned responses to emotional situations, assertive behavior shows respect for the rights of others. Openness, honesty, and pleasantness are its pillars.

1-7 SUMMARY

As has been noted before, we are primarily a product of our inheritance and culture. As also stressed, the single most important influence on our economic success and personal satisfaction is how we get along with people. No person is intelligent who does not master interpersonal relations! Its hallmark is being friendly but firm: as friendly as possible, no firmer than necessary. Engineers know about compromises, i.e., being flexible without being indecisive.

The engineering approach is a plan for action. This chapter presents information by which we can evaluate our present characteristics and, as needed, modify them to become who we want to be. Dominant among positive characteristics is love: forgoing one's pleasure or well-being for that of another. And as Peck [16] indicates, listening well is love in action.

DISCUSSION QUESTIONS AND PROBLEMS

1-1 THE DYNAMICS OF LISTENING

1 List an *external* distraction to listening, and discuss how its effect might be minimized.
2 List an *internal* distraction to listening, and discuss how its effect might be minimized.

3 In listening, what is significant about the difference in people's speed of thinking and speaking?

4 Review ways that speakers can develop their presentation to reduce the effects of internal and external distractions.

5 Name three advantages of listening well.

6 Give a mnemonic which encodes the clock setting made in the fall and spring to change from daylight savings to standard time and then back again.

1-2 MATCHING ACTIVITIES TO CAPABILITIES THROUGH TIME MANAGEMENT

7 Approach plates give information for instrument landings (in low clouds, etc.) at specific airports. For each airport the plates indicate *MARTHA: MA* is the procedure for a missed approach (required if the runway is not visible and clear at minimum altitude), *R* is the frequencies for radio communication with the controller and then the tower, *T* is the proper time duration (in which the aircraft at the proper speed is expected to arrive at the airport) for nonprecision approaches, *H* is the compass heading for the approach, and *A* is the minimum altitude at which a safe landing is secure. To minimize the stress of cockpit workload, pilots regularly simulate instrument landings. Discuss how regular practice (through simulation, mentally or on a simulator) influences the pilot's operating point on the stress-productivity curve. How would the occurrence of an unrehearsed emergency affect a pilot's "operation point" on the stress-productivity curve?

8 Discuss the criteria characteristics for long-term objectives.

9 How does anticipatory planning relate to (7) above?

10 We know that every joy (benefit) has a cost. Our Parent (Section 1-4) has been told that in any endeavor "planning" is *the* first step, yet we could spend so much time planning that we never "get anything done." Offer your thoughts on the trade-offs between time spent planning and the importance of the endeavor's outcome.

11 How might we reduce the pressures (or stresses) that are caused by the usual underestimating of the time required for activities, especially the small ones?

1-3 STRESS: WOW, GOOD? WOW, BAD?

12 a What are the physiological effects of stress?
  b How does stress affect us mentally?

13 Define stress in engineering terms.

14 Review how too much stress at work or home interacts with positive feedback, but in a detrimental manner.

15 What are the symptoms of too much stress?

16 Review two ways of coping with stress.

17 How did Tolstoy define a healthy mind?

18 Give your observations on the effects of diet and rest schedules.

19 Have you observed a correlation between exercise and your general attitude?

20 Determine your stress score from the quiz "Are You Dealing with Stress Properly?"

1-4 INTERPERSONAL RELATIONS

21 Which planet is farthermost from the sun? In 1979, a newspaper article stated that Pluto had come inside the orbit of Neptune and that Neptune will be our most distant presently known planet until 1999. Review what ego states are involved in accepting and storing this information.

22 What does Berne mean by a basic unit of social interchange?

23 Reflect on your recent social interchanges. Give a situation in which its TA (transactional analysis) could be described by Figure 1-4 a, b, or c, or by all three.

1-5 CREATIVITY—THE DIAMOND OF OUR BEING

24 In your own words, define *creativity.*

25 Review the process in the evolution of a creative idea.

26 Fear of failure is a barrier to creativity. Thomas Edison said that he never had a failure— only a learning experience. Discuss how building on his attitude could reduce this barrier to creativity.

1-6 SELF-DISCIPLINE AND ASSERTIVENESS

27 For a few days try building self-discipline "little by little." Then, in writing, discuss your results.

28 Contrast the characteristics of an assertive person with those of an aggressive person.

29 Present a scenario for being assertive. Identify each of the three techniques.

REFERENCES

1 McGinnis, A.L. *The friendship factor.* Minneapolis, MN: Augsburg, 1979, p. 15.

2 Kubler-Ross, E. *On death and dying.* New York: Macmillan, 1969.

3 Howard, J. H. Management productivity: Rusting out or burning out. *Business Quarterly,* Third quarter in 1977.

4 Mackenzie, R. A. Tape: Mackenzie on time. Allentown, PA: Day-Timers, 1979.

5 Humphreys, Kenneth K. *Jelen's cost and optimization engineering* (3rd ed.). New York: McGraw Hill, Inc., 1991, p. 512.

6 Selye, H. *Stress without distress.* Philadelphia: Lippincott, 1974, p. 102.

7 Plaque, Riggs Hall, Engineering Building, Clemson University, SC.

8 *Royal canadian air force exercise plan for physical fitness.,* New York: Pocket Book, Inc., 1972, Back Cover.

9 Lidz, T. *The person.* New York: Basic Books, 1968.

10 Berne, E. *Games people play.* New York: Grove, 1964.

11 James, M. & Jongeward, D. *Born to win.* Reading, MA: Addison-Wesley, 1976, pp. 78–79.

12 Maslow, A. H. *Toward a psychology of being* (2nd ed.). New York: Van Nostrand, 1968.

13 Lehman, K. *The birth order book.* New York: Bantam Doubleday Dell, 1985.

14 Kulisch, W. K. and W. L. Miranker. *New approach to scientific computation.* New York: Academic Press, 1983.

15 Rowe, F. A. *I launch a paradise: A consideration of John Donne, poet and preacher.* London: Epworth Press, 1964.

16 Peck, M. S. *The road less traveled.* New York: Simon & Schuster, 1978.

17 Boone, L. E. and D. L. Kurtz. *Management* (4th ed.). New York: McGraw-Hill, Inc., 1992.

APPENDIX: How to Succeed in Life through and in Engineering*

The engineering approach to a project or problem has five steps:

1 Define it.

2 Use fundamentals to analyze and understand it.**

---

*This section is largely based on an almost classic paper: "How to Succeed in Engineering" by Cochran, August 1977.

**Always strive for understanding; it is of primary importance in any endeavor.

3 Determine and evaluate possible solutions.
4 Implement the best.
5 Repeat these steps if probable gain seems worth the cost in time and effort.

This same approach can be applied to succeeding in life. Also, all five steps can be applied to refine each step.

A dictionary defines *happiness* as a state of well-being and contentment; it seems that this state requires a measure of success. The definition of success will differ with the individual, but a few thoughts from recognized sages give us guidance.

Success is achieving one's potential.

Dr. Max Lennon, president, Clemson University
Tau Beta Pi Initiation Banquet, 1987

Life consists not in holding good cards, but
in playing well those we do hold.

Josh Billings

The mind is its own place, and in itself can
make a Heav'n of Hell, a Hell of Heav'n.

John Milton

Human beings can alter
their lives by altering
their attitudes of mind.

William James

There is very little difference in people,
but that little difference makes a big difference.
The little difference is attitude. The big difference
is whether it is positive or negative.

Clement Stone

In these thoughts there seems to be a consensus that paramount to our feeling of success is a positive attitude. As stated earlier in Section 1-4, self-concept is our most important characteristic. It is a reflection of our attitude toward others and toward ourselves. Charles Schwab observed that people perform best under a spirit of approval and, however exalted, are diminished under a spirit of criticism [17]. The following rules help develop the positive attitude very important to success:

RULE ONE: ASSOCIATE WITH SUCCESS

Our associates are the most important element in achieving our potential and, thus, our success. Further, for most of us, our closest and most influential associate is our spouse. Techniques of Chapter 2 can be used to choose well.

Of course, most of our associations are with people, although the term *associate* also includes projects and companies. Persons involved with success are swept along with it. High performers are rewarded by increased pay and responsibilities, and in most cases so are the mediocre. Regretfully, top-notch people are undermined by failure, even when it is totally unrelated to their activities.

Of course, you may not be able to choose or discern whether a project (company, etc.) will be successful, but it should be your top priority to make every effort to improve that probability!

It is a truism that success breeds success, and failure induces more failure. This can be unfair, but it seems to be true. Thus, the first rule for success is to associate with successful people, projects, and companies.

RULE TWO: INTERCHANGE WITH OTHERS

Interchange with others at work, and seek their ideas and criticisms. This is really an extension of the first rule. It seems amazing how discussing your project, its problems, etc., with others clarifies your thinking and promotes mutual creativity. I believe that we learn mostly from experience: that of others, the experiences derived from simulation (i.e., exercising a model), and our own. Of course, the last is the most costly and often painful. As a junior engineer, you can avoid many mistakes and otherwise greatly benefit from the experiences of associates, especially those who have been with the firm 2 or more years. It will be complimentary for you to seek their advice and criticism. Seeking criticism is difficult, but your colleagues will respect you if your responses are positive. Be a good listener and learn; you will greatly benefit, and they will feel appreciated.

> The deepest principle in human nature is the craving to be appreciated.
>
> William James

> Appreciation is always appreciated.
>
> Arnold Glasow

An excellent way of showing appreciation is to give credit to your colleagues when you use their ideas. I offer what might be called the Praise Principle. Anytime a person is especially helpful to you, even a clerk in a store, tell this person's immediate superior, preferably in writing. Not only is this proper but it also increases her willingness to help you in the future. Recognizing competence and cooperation and giving sincere compliments enhance your life and those of your associates. Recognition is a very special form of appreciation! When proper, be especially appreciative to secretaries, technicians, and maintenance workers. They are a very important support group.

RULE THREE: LEARN TO COPE WITH TROUBLE

Learn to manage trouble, since it is inevitable for those who try to progress. There are three phases: (a) avoiding, (b) preparing, and (c) correcting. Recognizing the signals of trouble, discussed below, gives one an advantage in all three phases.

Avoiding Trouble

In the technical world, the most thorough, careful, and adequately accurate engineering analysis and design are the best investments. A poor engineering job will haunt the engineers and the project through repeated failures. Of course, it is highly improbable that all aspects of the project will be near perfect. For that reason, when problems arise, study them thoroughly to determine causes. Then make every effort to correct the problems. Do not treat the symptoms.

Generally, to avoid trouble, think adequately before acting. Carefully plan your work.

Preparing for Trouble

Early in the workday, your mind is fresh. Relax and try to anticipate possible difficulties, and plan what actions would be best. For many of us, the key to a quick mind is anticipation. Furthermore, reflect on your project, section by section; try to identify possible problems. In Chapter 5 we review CPM and PERT, two techniques for project planning. They both use networks to understand, control, and expedite the project.* The network facilitates identifying possible bottlenecks (critical resources of time and materials); use these networks to recognize special dependence on other people. Furthermore, unless past experience shows otherwise, do not blindly assume that others will deliver parts or complete their task perfectly on time. Even the weather, such as snowstorms, could be a factor; or a key person could be in an accident or get sick. Think how you would be affected. Consider what you could do to improve the possibility of their completing the task as planned. What alternatives are possible for you in case they do not?

Probably the most direct way of coping with trouble is to involve your supervisor. Then your troubles will be your boss's troubles as well. Being at the next higher management level, she will have more resources available.

This involvement will not be an imposition or improper; for in your efforts to avoid trouble by using recommendations from your boss and other colleagues, you will have already done an excellent engineering analysis and design. Also, in preparing for possible troubles, tell your boss, your fellow engineers, technicians, and (as appropriate) the secretary about them, giving a "worst case" analysis. Then if any problems occur, your colleagues will be more inclined to help; furthermore, it is highly probable that the problem will not be the worst case, and everyone will be thankful. This guide also works well in family life and with other friends.

Lastly, try to arrange for the resources you might need, such as spare parts, repair kits, and other equipment.

Correcting, or Getting Out of Trouble

Here the important aspects are honesty and openness. Seek help, hopefully from those already advised of the possibility that trouble might occur. Do not try to hide the problem from others or from yourself. Partial information will leak and probably have a much worse effect. People tend to be understanding of others in trouble and respect those who are open about it. They readily empathize with the difficulties we all have, but scorn those who act superior. This is especially true when the situation has been withheld and is later discovered. Let others benefit from your mistakes; they will appreciate your openness.

Do not expect troubles to go away; sometimes they will, but not often. Thus, the next step is to take undelayed action, based on your engineering analysis and preplanning. These actions will be unemotional and will reflect your earlier consideration. Also, remember that early, thorough, and complete correction requires less effort, time, and material resources. Of course, it is unnecessary and wasteful to do better than needed, but extreme care is required to keep laziness from influencing our judgment.

Look for *simple* causes first; then consider the more complex. Realize that two or more simple problems can occur simultaneously, complicating the analysis. If you recognize a simple problem, but realize that the troubles are more complex, still correct the easy one. The progress is encouraging, and the cause of other difficulties could come into focus.

---

*Above all else, remember that understanding is most important in solving problems and controlling projects.

RULE FOUR: BE SENSITIVE TO SIGNALS

Troubles seldom come abruptly with a "bang." I believe there are always signals, a sequence of events or an assemblage of facts, so often ignored by all of us. Usually trouble comes quietly—a trickle that grows like the hole in Hans's dike. If not recognized, faced, and corrected, it could become overwhelming. Without being obsessive, we must keep a high sensitivity to signs that indicate things are not going right. Naturally, in their early stages, troubles are easiest to correct; a thorough analysis gives us the courage to act quickly and sufficiently to correct the true cause.

As suggested above, when all is going well, take time to relax and try to anticipate problems and plan corrections. Relaxation is significant, for being overanxious can paralyze our ability to act. Remember that emotions block our thinking processes. Again, anticipation is the key to a quick, clear-thinking mind.

Information retrieval is often a significant factor in daily successes. In an earlier section, it was noted that the ease and accuracy of information retrieval are strongly correlated with how well the information is encoded in our memory. Furthermore, it seems that encoding is made easier by using acronyms.

Thus, in summary, *SOMS* is used to encode the salient aspects of this section on success. The letters in the acronym are decoded accordingly:

S is for success; remember the importance of our making a strong effort to seek people, projects, and firms which have high probability of success.

O is for others; those who can give us valuable information, creative ideas, and constructive criticism, all for only the small cost of listening well and showing appreciation.

M is for managing trouble; first avoid it by using thorough engineering practices, particularly by making adequate analyses to determine the true cause. Watch for the tendency to correct only the symptoms. In addition, prepare for troubles: preplan so that when the inevitable comes, you can act in a considered, unemotional manner. Next, act as quickly and openly as your preplanning allows.

The last factor related to managing trouble is so important that I have placed it separately; it concerns signals.

S is for signals; have a high sensitivity to the unusual events which signal the onset of problems. Usually the first perceptions of trouble come in your relaxed review of the project or even when engaged in other activities. Fortunately, the habit of maintaining this sensitivity gives the self-confidence which promotes the desired, relaxed approach to the project.

DISCUSSION QUESTIONS

1 Personally define success.
2 In the engineering approach, step (4) is "implement the best solution." Give your thoughts on how this step might consider the value of the solution versus the difficulty of its implementation, i.e., a sort of benefit/cost analysis.
3 Write a brief summary of "How to Succeed in Life through and in Engineering." For ideas on how to improve your writing, skim through Chapter 4.

Courtesy C. R. S. Serrine Engineers, Inc.

CHAPTER 2

DECISION ANALYSIS

CHAPTER OUTLINE

2-1 DECISION-MAKING TECHNIQUES

Nonanalytic Techniques

Rational Decision Making

Nature of the Decision-Making Environment

2-2 ANALYTICAL DECISION MAKING

Expected Value (EV)

Home Fire Insurance

Utility Function

Determining Utility

Bayes' Theorem

Subjective Probabilities

Group Decisions

Decision Trees

Analytic versus Judgmental

2-3 APPLICATIONS TO PERSONAL DECISIONS: CAREER CHOICE OR
CHANGE

A Fundamental Approach

Self-Concept and Career Choices

Choosing a Matching Career

2-4 SUMMARY

DISCUSSION QUESTIONS AND PROBLEMS

REFERENCES

BIBLIOGRAPHY

Successful engineers are satisfied with their values and personal lives. The satisfaction is due largely to their control of both their daily and professional endeavors. Furthermore, this control is highly dependent on their usually making good, if not the best, decisions. Developing that capability is the main objective of this book. All sections act as foundations for that competence. This chapter presents the specific fundamentals used for managerial decisions. These fundamentals are equally applicable to personal decisions.

As noted later in Chapter 3, a significant role of management is decision making. Using the methods of decision making, we engineers can manage our personal lives and careers. Mature individuals realize that we cannot have or do "everything," since every individual and organization has limited resources. We must make choices about allocating our time and other resources. Living requires choices, made consciously or unconsciously. Over a lifetime a conscious, rational approach will keep us in control of our affairs.

2-1 DECISION-MAKING TECHNIQUES [1]

A decision is a choice between alternatives. Choices can be made by nonanalytic or analytic techniques. The nonanalytic methods are adequate for less important decisions. Decisions involving an irreversible allocation of significant resources, on the other hand, justify the greater time, effort, and expense of the rational and logical analytic techniques. After reviewing the essential fundamentals of both nonanalytic and analytic methods, we will consider an important decision: choosing or changing a career. Although this text is directed to those committed to the engineering profession, we recognize that "career choice" needs lifelong attention. A carefully chosen career can lead to a mutually rewarding long-term commitment with a firm, a commitment which promotes personal and professional growth and happiness. These important aspects of a successful life are also strongly influenced by the choice of a spouse. This choice justifies decision analysis before you date someone many times.

Nonanalytic Techniques

These are based on either intuition or judgment. We first review intuition. It is an ingrained belief about a situation. Intuitive decisions are made without conscious consideration; they are based on what "feels right." We recognize that this type of decision is guided solely by our Parent ego state. It uses data from past experiences which involved similar situations. Transactional analysis is presented in Section 1-4. In it we note that the Parent ego state saves the Adult's energy for the "important decisions." But there is danger here: new conditions could well be different. For example, when we wish to slow an automobile, it feels right to continuously apply the brakes, and our Parent so directs. But if the road is icy, such an action could be catastrophic. Therefore, except in matters of small consequence, intuitive decisions should be used with extreme care. The other nonanalytic technique, judgmental decisions, is appropriate only for recurring situations. The judgmental decision is a first step toward a rational approach. In developing judgment a person builds on past experience and general knowledge. General William Westmoreland is credited with the observation that [2] "judgment comes from experience, and [bad] experiences come from poor judgment." In judgmental decisions the decision maker (DM) uses judgment to *consciously* reason the probable outcomes of the possible alternatives. In turn the DM chooses the alternative which is expected to give the most desirable outcome. Since organizational situations tend to recur frequently, judgmental decisions are very useful. In similar situations, the DM is confident that past solutions will again give the desired results. This similarity of problem situations permits the effective use of programmed decisions.

Nobel prize winner Herbert Simon borrowed the term *programmed* from computer technology. In programmed decisions the process for choice of alternatives is highly structured. Like a computer algorithm, the process consists of a sequence of clearly stated steps. Such programmed decisions can lead to the broad guidelines of corporate policies. Nonetheless, over time, the "program" must be modified to consider situational changes. Judgment is the basis of the programmed decision, but judgment alone is inadequate when the situation is unique or very complicated. On a long-term basis, only a nonprogrammed rational approach will minimize poor results.

Nonprogrammed rational decisions are required for new, perhaps nonrecurring problems. Usually they are more complicated, involving a large number of factors. Thus, they are unstructured and can involve situations in an unknown environment. When the state of the environment is unknown, the decisions are categorized as "decisions under uncertainty"; these are reviewed below.

Only correct actions produce the desired results, and correct actions require correct decisions. The probability of the right choice is greatly improved by a rational approach. And the foundation of a rational or logical approach is a model.

Rational Decision Making

This analytic technique is a specific application of the engineering approach.

Recall that there are five steps:

Recognize and Define the Decision Situation   The necessity of a decision must be recognized. The stimulus could be positive or negative: For example, surplus earnings need prudent investing, or malfunctioning equipment needs repair. For a failed piece of equipment, the objective is obvious: get it operating. Even in this case, and particularly for the example of the surplus funds, the occurrences of the problem could offer opportunities. An alternative machine or method could improve the preproblem situation. A careful consideration of the decision's objectives and a search for opportunities are needed for the best decision.

Basic in problem recognition is the need to correctly define the problem. For example, the falling rate of acceptance of a company's new product might be misinterpreted. It could be thought to be caused by little public interest or need. As noted in Chapter 5, the real cause of poor acceptance could well be lack of public awareness of the product and its capabilities.

Identify Alternatives   Of course the purpose of choosing an alternative is to produce the desired results. Experience will suggest the routine methods of achieving these results. Innovative alternatives should also be considered, and gathering them can be enhanced by creative processes, such as brainstorming (see Section 1-5). To correct the product acceptance problem reviewed above, a member of the review group suggested an innovative advertising plan. It would build on the adage that the best advertisement is provided by satisfied customers. In this plan, by offering an appropriate gift or other incentive, the firm would make contacts with potential users through present users.

Evaluate and Select   After brainstorming, each alternative needs to be evaluated. Based on the alternative's fundamental cause-effect relations, the evaluation needs to assess the impact of its implementation and constraints. For example, the choice of new equipment could disrupt present production or require greater capacity of the electrical or computer system, or both. Economic, legal, and ethical factors may also be involved. These are reviewed in Chapters 6, 8, and 9. The decision objectives and the underlying constraints, then, set the criteria for selection of the best alternative. As stressed in Section 1-5 on creativity, research has shown that both quantity and quality of alternative ideas increase with greater time between their conception and evaluation.

Again, considering the problem of the falling rate of product acceptance discussed above, many alternatives were considered. They included cutting prices, purchasing more magazine and newspaper advertising, and trying to further improve the product. Convinced of the present quality and usefulness of the product, the group chose the idea in which potential users were contacted by "satisfied customers."

Implement Selected Alternative   This step stresses an obvious but important aspect in decision making. The opportunity made possible by selecting the best alternative can be realized only if the decision is fully implemented. An effective implementation requires acceptance by the people affected. As alluded to in Chapter 3, people resent being "railroaded," or having another's plan forced on them. Their natural reaction is to undermine the plan. In the example of the product acceptance problem, the group felt no coercion. They fully endorsed the plan. Present users were quickly contacted. For informing three colleagues about the product, they were offered a large discount on a second product.

There is another aspect which affects alternative implementation: people's natural resistance to change because of insecurity, inconvenience, and fear of the unknown, particularly anxiety about their present status. It is highly probable that implementation of the most carefully analyzed plan will always have unexpected difficulties such as greater cost or some mismatch between the chosen alternative and the present systems. These conflicts can be minimized by following the suggestions in the Appendix of Chapter 1: "How to Succeed in Life through and in Engineering."

Evaluate Decision Results, Continue Improvement   Naturally, this step is made after the initial decision. Recall what Lord Kelvin said, "We know little about something unless we can measure it." [3] So, as far as possible, we need to make quantitative observations of the results of the decision. Often the results cannot be quantified. In such cases we have to use judgment. In both situations we compare the actual results with those desired and make corrections. The results could be totally unsatisfactory. Even in such a case, we learn from experience. Let us review three possible responses:

1 A previously considered alternative might be recognized as the proper action.

2 Our insight could now show us that the problem was incorrectly identified. Thus, the decision-making process would be restarted.

3 Alternatively, a review could show that more time is required for the chosen alternative to effect the expected results. This is frequently the case.

It is interesting how we continue to learn as we get involved with a project. It is truly satisfying, and gives continuous guidance for improving the original and subsequent alternatives. Of course, this is the action of the feedback control system, which is used to model humans (Figure 1-3) in Section 1-4. Feedback, over time, gives important information. This information reduces the uncertainty prevalent in all contemporary organizations and their environments.

Nature of the Decision-Making Environment

Decisions can be categorized by the environmental conditions under which they are made. There are three categories: certainty, uncertainty, and risk. The last two categories are not synonymous but have special meanings in decision analysis.

Certainty This state never truly exists. Nonetheless, managers often feel "certain" of the outcome of each alternative being considered. As an example, for an industrial manager of a firm's financial assets (or for us personally), a U.S. Treasury note is "certain" to pay the note's face value on the due date. In engineering, we refer to this situation as "deterministic." Although the note's interest (rate of return on the investment) might be 10 to 12 percent, a well-planned project might give a return on investment (ROI) of 30 to 40 percent. The success of the project, however, is uncertain.

Uncertainty This is the state of most significant decisions. The DM does not know all the alternatives, possible results, or probabilities of the results of the known alternatives. Uncertainty is often induced by the novelty and complexity of the rapidly changing technology associated with modern projects. This is especially evident in solid-state technology and the defense-industrial complex in which this technology is widely used. Complete uncertainty, however, is usually not the case.

Here the DM has two options. One is to use intuition and judgment based on present knowledge and that of colleagues. This approach might be called for under time pressure, if the needed information is unavailable, or if the cost of the information could be too great. The second and usually preferred option is to reduce the uncertainty by searching for additional relevant information. From this and from experience, subjective or inferred probabilities can be assessed. Use of these probabilities can change the decision from the category of uncertainty to that of risk.

Risk   In this state the outcomes of a decision are random, but the probabilities of the outcomes are known. This category is well suited to an analytic approach. We study this decision in depth. Outcomes depend on what is termed the *state of nature*. A state could be the toss of a die being greater than "3" or the occurrence of a fire. The probability of an *i*th state, $s(i)$, is between 0 and 1. Recall that probability indicates the degree of likelihood of the outcome. A value of 0 indicates an impossible event; a value of 1 indicates certainty. We know that objective probabilities are generally determined two ways: (1) by classical theory, that is, by dividing the number of possible "successes" by the number of equally likely, possible outcomes (for example, the probability of drawing a spade from a well-shuffled deck is $\frac{13}{52} = .25$); or (2) by statistical analysis. In the latter method the number of successes is divided by the total number of trials. This is the "relative-frequency" approach, and its accuracy increases asymptotically with the number of trials. Accordingly, since insurance companies use the nation's entire population as an actuarial base, they have considerable accuracy in decisions for setting premiums. Unlike insurance companies, most firms, especially small ones, will have inadequate information for objectively determining probability. Thus, the experience of management, with judgment, is used to assign subjective values of probability to the outcomes of the alternatives. These values can be used in a rational, quantitative approach for decision making. The techniques are generally described as analytical.

2-2 ANALYTICAL DECISION MAKING

Naturally, a decision is guided by the criteria for the desired results. Intangibles are considered later. A quantitative analysis requires quantifying the results or outcomes. Any numerical scale is suitable. Costs or monetary gains are often proper and will be used first. In dealing with risk, there are uncertain states with known (or estimated) probabilities; only the expected value of gains or losses resulting from those states is representative.

Expected Value (EV)

This value has a specific meaning; it is the average or mean value of the rewards (or losses) of a probabilistic endeavor. This value is determined by summing the products of the values of the rewards and their respective probabilities. In a lottery, let $r(i)$ be the reward which results from the *i*th state in an event that has n possible states. The probability of each state is $p(i)$. The expected value of the reward r for n different possible states is

$$E[r] = \sum_{i=1}^{n} p(i)\, r(i)$$

Example 2-1

Consider a game of chance. In it a die is rolled, and you will win $10 times any number "turned up" over 3. You lose $20 times any number "turned up" less than or equal to 3. Would you like to play this game? How much would it be prudent to pay for the opportunity to play?

Solution: By the classical approach, we recognize that the probability of each number is $\frac{1}{6}$. Let us evaluate the expected value of the game.

$$E\,[r] = [-20 \times 1 \times \tfrac{1}{6} - 20 \times 2 \times \tfrac{1}{6} - 20 \times 3 \times \tfrac{1}{6}$$
$$+\, 10 \times 4 \times \tfrac{1}{6} + 10 \times 5 \times \tfrac{1}{6} + 10 \times 6 \times \tfrac{1}{6}] = \$5$$

The frequency approach to probability (reviewed above) shows that its accuracy for predicting average outcomes is heavily dependent on the number of trials. Obviously, in the game just considered, on the first try you could well lose $20 by "rolling" a 1. By using the expected value as a pattern in life for making decisions, however, you can average a gain. The criterion is to only pay less than the expected value for any "game." We might think, "I will just not play games of chance." In reality, we have no other option. Living requires choices; the expected value approach is a rational way to make decisions. Below we will review how the approach needs to be modified for individual characteristics about risk.

Insurance is a technique of exchanging the cost of a misfortune at a particular time for a greater cost spread over a long period. We will show that, if feasible considering one's total wealth, on a long-term basis, it is always more economical to self-insure.

Home Fire Insurance

A home owner wishes to analyze the cost of fire protection. For a total insurance of $200,000, covering the dwelling and its contents, the annual premium is $340. From the local fire commissioner, the home owner learns that for houses in the neighborhood, the probability of a fire causing total loss is .001. The data for this unsought-for "game" are given in Table 2-1. Such a table with probabilities of the outcomes is often called a *pay-off matrix*. The home owner's total wealth, i.e., savings, stocks, and equity in the house, equals an amount which we represented by W. For simplicity, we will consider only two states: s(1), a total loss by fire of $200,000; and s(2), no fire (the

TABLE 2-1
INFORMATION FOR INSURANCE ANALYSIS

	Results		
Actions	s(1):fire $p(1) = .001$	s(2): no fire $p(2) = .999$	EMV, expected monetary value
Self-ins.a(1)	$W - \$200{,}000$	W	$W - \$200$
Buy ins. a(2)	$W - \$340$	$W - \$340$	$W - \$340$

only loss is cost of the insurance). The probability of each is given in the table. Consider two actions: a(1), self-insure; or a(2), buy insurance from a company for $340. For a(1), we determine the EMV, expected monetary value: EMV[a(1)] = .001 × (W − $200,000) + .999 × W = W − $200. For a(2), the net wealth for both states is the same (neglecting small unavoidable costs in case of fire). Thus, its EMV is unaffected by a fire.

Frequently, fires do not cause total loss of the dwelling. A similar analysis could be made by getting typical amounts of partial losses and their probabilities.

From the analysis above, the "expected monetary value" decision rule indicates that self-insuring is the better action. Nonetheless, aside from mortgage requirements, most home owners feel better buying insurance. If the value of the house is a very small portion of the home owner's total wealth, she might self-insure. This implies that the worth of money depends on its magnitude compared with a person's total wealth and that person's risk-taking characteristics. Bernoulli (1730) suggested that one endeavors to maximize utility rather than the amount of dollars. A person's utility function, $u(x)$, gives that person's relation between utility and the magnitude of any good for which it is developed. Typical goods might be commodities or money. The most widely accepted approach to studying risk is based on expected utility theory. It concerns an individual making decisions under risk.

Utility Function [4, 5]

This function is often developed and presented graphically. As implied above, the function gives a person's worth or "utility" for goods such as money versus the magnitude of the goods or money. Von Neumann and Morganstern (1944) stimulated a renewed interest in Bernoulli's concept of utility. They assumed that the DM (a rational person) possessed preferences for the outcomes of alternatives. Also, the DM had preferences for gambles involving the outcomes. They further formulated a set of axioms [5, p. 63; 6, p. 23] from which two tools can be developed. These tools are essential to explicit quantitative decision methodology: (a) a cardinal utility scale (i.e., a numerical value versus ordinal relations) for measuring the DM's preferences, and (b) a decision rule, maximizing expected *utility*. By the axioms, this rule ensures that the DM will be consistent with his preferences. The expected utility value (EUV) is calculated in the same way as above for the expected monetary value (EMV). The value of the utility of the outcomes of each uncertain event is multiplied by the probability of that outcome. The expected utility is the sum of those products for all possible outcomes. We make this evaluation in an example below.

Determining Utility

There are various methods for obtaining utility functions. The function can be for an individual, or for an organization or group. The Basic Risk Model is defined by four elements: certainty amount, gain amount, loss amount, and probability. The gain must be equal to or greater than the amount of the actual risk. Also, the loss must be equal to or greater than any amount in the actual situation. When an evaluator specifies any three of the four elements for a series of risks, the DM, on giving the fourth, establishes her utility function. Usually the bounds for utility are set as 0 and 1 (or 100). The properties of the function would be unchanged under a linear transformation. In such a transformation, a constant could be added to all utility magnitudes or the magnitudes could be multiplied by a constant, or both.

Two methods are generally used to assess points on utility curves. Both involve the concept of a "certainty monetary equivalent" (CME). This equivalent is the sum of money the DM would pay or accept to avoid a specific risk situation. One of its specifications is its expected monetary value (EMV). For most of us, the CME is usually less than the EMV of the particular risk situation. We define the excess as the risk premium (RP):

$$RP = CME - EMV$$

For individuals of usual wealth, the RP will be negative (an outflow) for EMV positive, and vice versa. After developing a utility function for the home insurer, we will show how the function can be used to explain the RP.

People are generally categorized as risk-averse, risk-neutral, or risk-seekers. This characteristic is made evident by a person's utility curve. As noted above, the curve gives the person's utility for the goods or money, as a function of its magnitude. As would be expected, it seems that one's feelings toward risk change with its magnitude. By definition, for a risk-averse individual, CME < EMV. For the risk-neutral, CME = EMV; and for the risk-seeker, CME > EMV. The first situation requires a concave utility curve (it curves downward, between the middle and the ends). For the risk-neutral, the curve is linear; and for the risk-seeker, it is convex (curves up). In serious business matters, most individuals are risk-averse, Thus, a little familiarity with the DM will help in developing his utility curve. Although the development is somewhat an art, two tenets are helpful. (1) For the risk-averse, CME < EMV; and (2) if a DM prefers A to B and B to C, then for that DM and some value of probability p, $u(B) = p \times u(A) + (1-p) \times u(C)$. This second tenet is established by a relatively simple axiom set presented by Luce and Raiffa [6, chapter 2; 5, p. 63]. Let us now consider the methods for determining utility curves.

One of the methods for determining the utility curve is to present a series of hypothetical 50-50 two-way gambles to the DM. The DM might first be asked for the CME (certainty monetary equivalent) he would accept (or pay) to avoid a risk: for example, lose $10 or win $20, each with a 50 percent probability. Note that the EMV = .5 × $20 − .5 × $10 = +$5. Thus, for the risk-averse, we know that consistency requires CME < $5. In using an actual risk situation, the highest utility value would be set equal to the maximum possible gain. The lowest utility value would be set to the smallest gain or maximum loss. As noted earlier, the extremes of an individual's utility function, $u(x)$, are arbitrary. For convenience, the lower extreme is usually set to 0, and the upper is set to 1.0, or 100 if you prefer. Let us use an example to develop this method.

Example 2-2

Consider a small construction business; two contracts, A and B, have been offered to its manager. Skilled personnel and other resource limitations prevent accepting both contracts. Data for the choice are presented in Table 2-2. In this simple example, there are only three possible outcomes for each contract. Henceforth, we refer to the contracts, respectively, as A or B.

Solution: As a reference, we determine the EMV of each. In thousands

$$\text{EMV(A)} = \$90 \times .50 + \$20 \times .30 - \$35 \times .20 = \$44.00$$
$$\text{EMV(B)} = \$60 \times .55 + \$35 \times .25 - \$5 \times .20 = \$40.75$$

The DM could always decline both contracts; that action would give an outcome of 0. The "maximum EMV" rule indicates that A should be chosen. However, since the firm is small, the DM is anxious to avoid the possible loss of $35,000 from A. Hence, the DM accepts the smaller possible EMV of B. Consider how the DM's utility function confirms this action. The DM's function is developed next. Please see the decision trees below.

Solution: With the 0 outcome for accepting neither contract, there are seven possible outcomes. The outcomes are expressed in thousands of dollars. We set the utility of the greatest possible gain (90) at a value of 100, and for the greatest loss (−35) at 0. The first risk we offer the DM is a 50-50 gamble between a gain in net assets of 90 and a loss of −35. The DM specifies a CME = 18. As we will note later, it is proper to assign a utility of 50 to this CME. Now we use this CME = 18 as the lower and 90 as the upper amount of a 50-50 gamble to get the utility of 75 utils,* halfway between 50 and 100. For it, the DM offers 35. Next, for 25 utils, halfway between 50 and 0 we use the CME = 18 from the first offer as the upper amount and the −35 as the lower. For it, the CME = 5. To closer approximate the DM's feelings for large amounts, for 90 utils, we offer a 50-50 gamble between +90 and +35. Its CME is 58. These are shown below with their respective gambles. The decimal values near the decision arms are their respective probabilities (all .50).

*The term utils is used to indicate units of utility.

TABLE 2-2
SUMMARY OF DATA FOR CONTRACT OPTIONS

Outcomes	Contract A		Contract B	
	Pay-off	Probability	Pay-off	Probability
y(1)	$90,000	.50	$60,000	.55
y(2)	20,000	.30	35,000	.25
y(3)	−35,000	.20	−5,000	.20

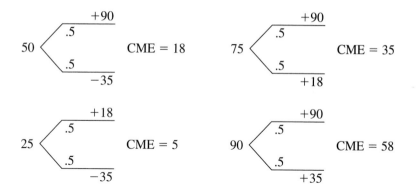

Now we will plot these points, with $u(-35) = 0$ and $u(90) = 100$. The curve is shown in Figure 2-1. The six data points are connected by straight segments, based on the assumption of small changes in the DM's risk characteristics between the data points. Recall that all continuous functions are quasi-linear for small perturbations. This property is made evident by a Taylor series. All continuous functions can be represented by a Taylor series. When the change in the independent variable is small, all the higher-order terms of the representation are negligible, and only the linear terms remain. Quasi-linear analysis is the foundation of the perturbation methods used widely in quantum mechanics and the small signal analyses of electronic and mechanical systems.

Corollary

All systems have a linear response to adequately small excitations and a nonlinear (saturated) response to adequately large excitations. For our DM, the assumption was checked by proposing other gambles for changes in net worth of $+70$, -2, and -15. The corresponding utilities of 95, 15, and 7 given by the DM's are also plotted on Figure 2-1.

We now use this utility curve to determine the EUV's for contracts A and B. For each of the seven possible outcomes we use the curve to determine the DM's utility. The utilities were determined by linear interpolation between the plotted points. They are presented in Table 2-3.

The EUV's for contracts A and B are determined in the same manner as the EMV's:

$$\text{EUV}(A) = 100 \times .50 + 53 \times .3 + 0 \times .20 = 66$$
$$\text{EUV}(B) = 92 \times .55 + 75 \times .25 + 17 \times .20 = 73$$

For this DM, contract B with a 7.5 percent smaller expected monetary value has a 11 percent greater expected utility value. This analysis shows the advantage of develop-

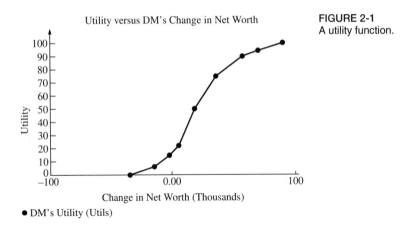

Utility versus DM's Change in Net Worth

Change in Net Worth (Thousands)

● DM's Utility (Utils)

FIGURE 2-1
A utility function.

ing utility curves. Also, their use provides a systematic and consistent method for considering the DM's personal risk characteristics.

A second method for determining a DM's utility is not to specify a 50-50 probability but rather to use the CME as the given or independent variable. The DM is asked to give the probability she would accept. For example, she might be asked at what probability would a CME of $5 for a hypothetical gamble be accepted: lose $10 or win $20. Obviously, for a probability of 1, the "gamble" would be chosen. For a small probability, the CME would be chosen. From this viewpoint, it is recognized that there is some value at which the DM would be indifferent between the gamble and the CME. In subsequent offers the CME and the possible losses and gains would be increased in magnitude. It is the practice to use the probability values specified by the DM as her utility in utils [3, p. 66]. The curve is sketched-in between these points. For both methods of determining the utility function, the wins and losses would change the DM's net worth *W*. Recall that net worth is the individual's or the organization's total assets minus all debts.

Using the second method, I developed the utility function for a friend. She is professionally established and has a net worth of approximately $300,000. Her utility curve is depicted in Figure 2-2. We will use it to analyze her "DM's behavior" in insuring her home. The straight line between $0.00 and $300 (thousands) shows the

TABLE 2-3
UTILITIES AND PROBABILITIES FOR POSSIBLE CONTRACT OUTCOMES

Outcomes	Contract A			Contract B		
	Pay-off	Prob.	Utils	Pay-off	Prob.	Utils
y(1)	$90,000	.50	100	$60,000	.55	92
y(2)	20,000	.30	53	35,000	.25	75
y(3)	− 35,000	.20	0	−5,000	.20	17

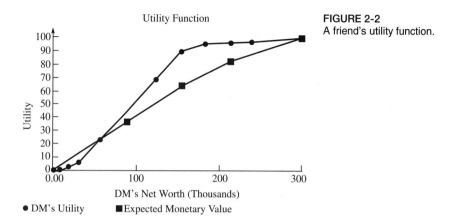

FIGURE 2-2
A friend's utility function.

utility of someone who is risk-neutral, often referred to as an EMVer, that is, someone whose decisions depend only on the EMV of the gamble. The sigmoid* shape of her curve shows that for small amounts she is slightly risk-seeking. The near straight portion for the middle range indicates her being risk-neutral, and the curving down for large magnitudes shows her being risk-averse for great possible losses. For example, my friend "loves" flying in small airplanes. Since adulthood, however, she will not fly at night in a single-engine plane. Although the probability of engine failure for airplanes is extremely small, she feels that the possible loss is too great.

For major material losses, insurance can be used to reduce the changes the losses will make in our net worth. The reduction can be used to keep the changes to a percentage of our net worth which is within the linear portion of our utility function. Buying an insurance policy with "deductible" amounts reduces possible losses to the amount of the deductible. For instance, with a deductible (amount) of $100, in the case of a greater loss such as $1000, we pay $100 and the insurance company pays the rest. We insure ourselves for an amount equal to the deductible. As noted earlier, on a long-term basis, self-insurance is always more economical—when we can afford (and emotionally tolerate) the loss. The linear portion of our utility curve identifies the amount which we are willing to "tolerate." Self-insurance is more economical for us because we do not have the insurance company's many overhead expenses (such as rent, maintenance, and personnel) and do not have to make a profit. Accordingly, our cost to self-insure is about one-half the premium required for the insurance company to be profitable.

An affordable way to self-insure is to buy insurance with deductibles. Since small claims (for losses) require almost as much paperwork and other overhead as larger ones, insurance companies give considerable discounts for increasing deductible amounts. This is shown below and in Problem 7b. Example 2-3 shows that regular insurance (with minimal deductible amounts) is often advantageous; it is also considered in Problem 7a.

---

*The suffix *oid* is Greek for "like"; thus, *sigmoid* means "sigma" or "s-like" (curved in two directions).

We know that curves and equations, such as $y = 5x$, transform the magnitude of one variable to the corresponding value of the other. So it is for an individual's (or for a corporation's) utility curve. The greatest value in using this curve is that it permits determining, for a given net worth, the ratio of the DM's utility to that of an EMVer. The ratio reflects the risk characteristic of the DM. We will see that this characteristic usually depends on the magnitude of the possible loss as a fraction of the DM's net worth.

Example 2-3

For this example, return to my friend's insurance decision presented above. She made a choice about whether or not to self-insure for the possible $200,000 loss in case of fire. She chose to insure. Does her utility curve show the prudence of that action? The cost of fire insurance is $340.

Solution: As noted above, her net worth is approximately $300,000. Other information is given in Table 2-1, repeated here. Let us closely examine the upper portion of her utility. It is shown in Figure 2-3. Note that her curve can be considered linear for small decreases from $300 (thousands) and for small increases from $90 (thousands). The EMV utility (the lower straight curve) is for a person who is risk-neutral, that is, an EMVer. The EMVer's curve is always linear, but it is of greater slope than hers on the high end and of lesser slope on the low end. Here, we need to consider only the upper end. Specifically, for a decrease in net worth of $10,000 to $290,000, the utility of the EMVer is 97. For this value of utility, her curve indicates a net worth of $250,000, a decrease of $50,000. We recognize that in this range her CME is five times that of the EMV for this risk. From Table 2-1, the expected value of the loss is .001 × $200,000 = $200. Five times that amount is $1000. By the equation above, her RP (risk premium) is $1000 − $200 = $800. Thus, the insurance cost of $340 with a premium of only $140 (above the EMV of the possible loss) is pleasing to her. It is not higher for two reasons: (1) The insurance company's very large net worth (compared with that of the home owner) makes its utility function nearly linear, even for losses of $200,000. (2) Competition prohibits companies from charging home owners the full worth. The lower insurance cost is available to the home owner if she is prudent in checking with a number of companies. However, my friend must be assured of their financial

TABLE 2-1 (Repeated)
INFORMATION FOR INSURANCE ANALYSIS

	Results		
Actions	s(1):fire p(1) = .001	s(2): no fire p(2) = .999	EMV, expected monetary value
Self-ins. a(1)	W − $200,000	W	W − $200
Buy ins. a(2)	W − $340	W − $340	W − $340

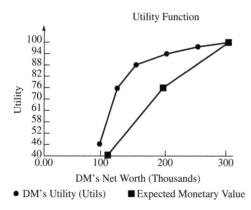

FIGURE 2-3
The upper portion of a friend's utility curve.

strength. This can be done by reviewing their financial report (see Chapter 7). The A. M. Best ratings also give the needed information.

To consider the advantages of using deductions for self-insurance, let us now examine the lower segment of my friend's utility curve. It is shown in Figure 2-4. For these amounts, she is risk-seeking. The reductions in annual premiums for the deductions on $200,000 fire insurance (house and contents) are given in Table 2-4.

Using this data and the curve (Figure 2-4), we will consider the $500 deductible. To make this example more interesting, we will use a different approach than the method used above. Enter the utility curve for $500. The value of its utility is 1.4. On drawing a horizontal line from this point to the EMVer's straight-line "curve," we observe that her CME (certainty monetary equivalent) associated with the utility of 1.4 is only $250. This indicates that to her the "thrill of the game" is worth the difference of $250. Using .07, the probability of a $500 loss, we multiply it by her CME of $250 to get the EMV of this loss (for her). It is $17.50. This is the loss she will effectively have per

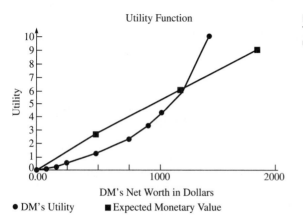

FIGURE 2-4
The low end of a friend's utility curve.

TABLE 2-4*

Deduct. amount	Incr. in deduction	Annual premium	Annual savings	Prob. of loss deduct.	Exp. loss of incr. in deduct.
Required: $ 100		$340			
Optional: 250	$150	271	$ 69	.100	$15
Optional: 500	400	244	96	.070	28
Optional: 1000	900	204	136	.050	45

* *Source:* Ms. Pat Hix, Dean Davis Agent, Allstate Insurance Co., Seneca, S.C. 29678.

year on a long-term basis, the $250 worth of "thrill" having been subtracted from the actual loss when it occurs. Thus, the $17.50 must be deducted from the annual $96 saving from the reduced premium. The resulting $78.50 gives an additional after-tax interest of 15.7 percent on the $500 held in a savings account established for the possible loss. The effective rate could be increased by putting the $500 in an interest-bearing account although that interest is taxable (see Chapter 7). For simplicity, the effects of the required $100 deductible have been neglected. The net savings possible for the other available deductions can be determined in a similar manner. They are listed in Table 2-5 with the value just determined.

Note that her after-tax interest rate decreases as the deductible amount is increased. This reflects and results from her being less risk-seeking for increased possible losses. Nonetheless, a nontaxable interest rate over 9 percent is excellent and, of course, as with the other deductibles, is in addition to any taxable interest earned on the "reserve" account.

Note in Figure 2-4 that the lower segment of her utility curve is convex; i.e., it is below the straight-line curve of an EMVer. As mentioned earlier, for small amounts (or possible damages) she and most of us are not risk-averse. It is for possible casualty losses of these magnitudes that self-insurance is appropriate and financially advantageous.

As noted in the previous examples, the probabilities of alternative outcomes are significant in analytic decision methods. Often objective determination of their values is not possible or worth the cost. Managers depend heavily on inferred or subjective probabilities. They can be greatly improved in a bayesian manner.

TABLE 2-5

Actual loss (the deduct.)	Her equiv. (from curve)	Her EMV (ave. cost)	Annual savings	Annual net gain	% of deduct.
$ 250	$100	$10.00	$ 69	$59.00	23.6
500	250	17.50	96	78.50	15.7
1000	900	45.00	136	91.00	9.1

Bayes' Theorem

This theorem or formula was developed in 1763 by Thomas Bayes, a Presbyterian minister. The same formula was independently developed by Laplace in 1774. Bayes suggests that probabilities based on earlier judgments should be combined with probabilities formed by later observations or relative frequencies [7, p. xx]. This theorem is very important in decision analysis. It can quantitatively improve subjective probabilities by using conditional probability, that is, the probability of one result conditioned on another result. As reviewed in the next paragraph, our subconscious is very competent in using additional information, in a bayesian manner, to update prior subjective probabilities.

Subjective Probabilities

There is a degree of uncertainty for all decisions. The probabilities of the possible outcomes and their pay-offs (results) are needed to change decisions of uncertainty to ones of risk, i.e., those amenable to quantitative analysis. As noted above, decisions under risk are those in which random outcomes and their pay-offs can be associated with their respective probabilities. Difficulties in their objective determination lead managers to use subjective probabilities. These are also called *inferred* probabilities, for they are inferred by managers from experience. Typical inferred probabilities are the "percentages" given by friends or experts concerning the odds on a football game. In a bayesian manner the odds change as one gains "experience" from the team's performance during the season. Recall that the odds $= P/(100 - P)$. For example, a 25 percent probability gives odds of 1 to 3; even odds, 1 to 1, are given by $P = 50$ percent.

In the previous paragraph, we noted that using judgment, managers infer probabilities from experience. The experiences of colleagues and others improve the manager's judgment. Using the experiences of a group can greatly improve the subjective evaluations.

Group Decisions [1, p. 40]

Today more and more organizations use groups in making important decisions. The most common types of groups are interacting, Delphi, and nominal.

An interacting group is the most common in decision making. The group is either an existing or a newly designated one. The latter is often analogous to ad hoc committees, created for relatively narrow purposes and existing for only a short term. Naturally, here the purpose is to make a decision. As noted in Chapter 3, the members start with pleasant interchange and then go through cycles of arguing and agreeing Finally, they reach a consensus and make a decision. Also, as noted in Chapter 3, constructive interaction is fertile in sparking new ideas and promoting understanding of the problem and of each other.

The Delphi method was developed by the Rand Corporation. In this format, a panel of experts is solicited individually and anonymously for information and ideas. The method is mostly used for determining the best estimate of a forecast, particularly one involving a technological breakthrough. Nonetheless, the technique is applicable to all types of decisions or for evaluating probabilities. The contributions of the experts are combined by the leader to form an accord. In a second request, those whose opinions varied widely are asked to defend their ideas. The explanations are passed on to all those in the panel. When the results stabilize, they are accepted. Obviously, the time and expense required for this method make its use unwise except for significant decisions.

The nominal group is similar to the interacting group; it is used to generate many creative ideas. Unlike the Delphi method, the members interchange freely, although this group also consists of experts. After the manager or leader reviews the problem, each group member is asked individually to write as many alternative solutions as possible. In the next step, members are requested to present their ideas, which are recorded on a chalkboard or flip chart. After an extensive and open discussion, the alternatives are ranked by votes. The alternative with the highest rank is considered to be the consensus of the group. Often the manager reserves final acceptance.

Group decision making takes time and is costly. The significance of the decision must be adequate to justify these greater requirements. Another disadvantage is that "Sherman tank"–type people can appreciably delay the group's progress, and some members cower to others to avoid conflicts. This tendency can lead to "groupthink," which occurs when the group's desire to reach a consensus and reestablish harmony detracts from its reaching the best decision.

An advantage of group decision making is that greater information is available. The free interchange of ideas has a synergistic effect and encourages the exchange of information. Also, in explaining their ideas to the group, the members clarify concepts in their own mind. Furthermore, there is greater commitment of the members to the decision. By understanding the alternatives, the members are better able to communicate to those outside the group who must act on the decision. The enhanced communication with nongroup members improves the decision's acceptance. Research indicates that decisions made by groups are better than those made by individuals [1, p. 142].

Decision Trees

This approach to decision making is especially helpful in decisions which are serial; that is, future actions depend on the outcomes of previous ones which have not yet occurred. As an example, let us return to the manager in Example 2-2.

Example 2-4

The manager in Example 2-2 restudies the specifications of contracts A and B in the hope of being able to accept both (instead of only B). Engineers with special design competence make up a group designated as group 1. Not having other engineers with their competence restricted the firm's accepting both contracts. By using the techniques in Chapter 5, the manager determines that it is very probable that

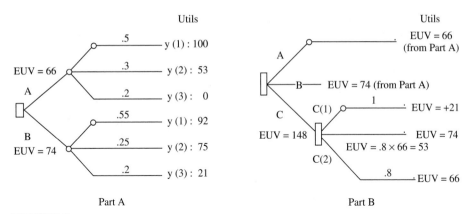

FIGURE 2-5
A decision tree for comparing action C with A and B. Rectangles represent the manager's decisions; the dots are those decided by chance.

contract A could be completed in 9 months, 3 months less than the 12 allowed in the contract. Also, group 1 can complete their work on B in the first 3 months. Thus, they would then be free to start on A. In order for group 1 to immediately start on A, however, in those first 3 months other engineers would have to prepare detailed plans for A. The plans would cost $5000, money that would be totally lost if group 1 was unable to finish their work on B in the scheduled time. The alternative for trying to carry out both contracts is defined as choice C. It has parts C(1) and C(2). C(1) represents the activities by other engineers planning for A. C(2) is group 1's completion in 3 months of the work on B and their being ready to start A. Giving special weight to the estimates of group 1, the consensus of the entire staff was .8 for the probability of their success if choice C is chosen.

Solution: The situation can best be analyzed using a decision tree, shown in Figure 2-5, Part B. Part A, for review, represents only the information in Table 2-3. Since Part A does not involve a sequence of decisions, a table would be equally suitable. In the new option, however, later chance outcomes need to be considered in the initial decision. Accordingly, the decision tree of Part B is appropriate. In both Parts A and B, the utils of this DM are used. From the expected utility values, Figure 2-5 indicates that for the possible alternatives and risk characteristics of this DM, alternative C is best: EUV(C) = 148 > EUV(B) = 74 > EUV(A)= 66.

Analytic versus Judgmental

The manager's insight is needed to weigh trade-offs, that is, the greater expected profits or savings of the analytic method versus the fast, inexpensive judgmental method. Both use models; in the judgmental method, the model is subconscious, evolved from past experiences. It is similar to Griffin's administrative model, not always logical and rational [1, p. 129]. Naturally, the effectiveness of both the analytic and the judgmental methods depends on the accuracy of their models.

We realize that all models are approximations. Their development has to be a compromise, with adequate detail to represent the relevant aspects of the situation but not too complex for feasibility and full understanding by its user.

There are other limitations to rational decisions: managers sometimes filter information (remember the "resonant filter" in Chapter 1?), ignoring misunderstood but relevant facts or overreacting to others. The lack of knowledge or even lack of familiarity with the situation biases the manager's judgment. We humans tend to reject what we do not understand. Finally, political forces bias manager's decisions.

Research identifies four forms of political behavior [1, p. 496] in organizations. The first is inducement. For example, a manager promises to support another's plan if she will endorse his proposal for a product modification. Second is persuasion. It is based on emotions and logic; often, however, the logic is illusory, based on subjective and personal reasoning. Next is the creation of obligation. An individual tries to accumulate "Brownie points" by supporting the plan of an influential manager in the hope that he will be able to collect on the "debt" when he needs the support, especially from one in a stronger position. Finally, coercion is used to get one's way, often in the form of threats.

We have considered the fundamentals of both nonanalytic and analytic decision-making techniques. The emphasis has been on the professional needs of managers. All of the techniques, however, are suitable and helpful in our personal lives.

2-3 APPLICATIONS TO PERSONAL DECISIONS: CAREER CHOICE OR CHANGE

Emotions diminish our rational processes. This section examines the use of unemotional analytical techniques for personal decisions. As noted above, their use requires models, which could be mathematical or the individual's understanding and considered assessment of results under possible conditions. Furthermore, over time, probability assessments used in the model can be improved in a bayesian manner. An initial requirement in analytical methods is a quantitative criterion. For personal decisions, I suggest that the duration of the time affected by the decision is a suitable index by which to judge a decision's importance. For decisions in the "uncertainty" category which are of lasting importance, a variation of the engineering approach is applicable [7, p. ix–x]:

1 Determine your preferred results (goals).

2 Determine which alternative actions are possible to achieve those goals.

3 Gather information relevant to the possible outcomes of each action.

4 Assess the probabilities (probably subjective) of each action's giving the desired results.

5 Choose and make a commitment to the action which has the greatest expected utility.

Only step (3), gathering relevant information, is not primarily based on your personal values. Accordingly, only step (3) is considered in the following paragraphs. The relevant fundamentals for a very important personal decision are reviewed. The decision involves choosing a career and keeping that choice current during your professional life. As suggested above, the ordering of a decision's importance is based on the duration of the time it affects. Accordingly, marriage probably has greater importance. After studying the presentation on career choice or change, you should be able to make a rational decision in choosing a spouse.

Since this book addresses university juniors or seniors in engineering programs and graduate engineers, it might seem inappropriate to have a section on choosing a career. However, there are two reasons: first, the specific aspects of a career change throughout one's working life span; and second, over that period it is important that your career *match* your changing needs, abilities, interests, and values. This match also greatly determines your life's happiness and productivity, both largely dependent on your self-esteem. This and other *self* terms are clarified below.

Engineering education and experiences are significant assets in many other careers or professions. Thus, in early adulthood, an engineering student or graduate engineer has high career mobility. Nonetheless, every change "casts a long shadow"; the sooner a prudent change is made, the less the cost, economically and emotionally.

It is essential, then, that as soon as possible, one should determine a profile of her or his needs, abilities, interests, and values, and subsequently gather the corresponding information about possible careers. This information should include projections of supply and demand in that occupation. Such data are available in reports from the U.S. Department of Labor, which can be found in a university's career planning office, a university library, or a public library. Specific Department of Labor reports are discussed later.

According to Maslow, our needs fall into two categories, security and growth, both of which involve recognition for accomplishments [8]. Medicine and engineering are excellent for meeting these needs, and I believe that for a given level of satisfaction, engineering requires much less training. Law should be considered also, but at present the field seems overcrowded [9].

A Fundamental Approach

Approach your career choice as you would an engineering project; keep a journal, and date the entries to help you assimilate the information. Do start now, surely within the next week. *Preparing ahead of need is a pillar of success.* Self-assessment, which we will discuss later in this section, is the first step; it leads us to goals. Furthermore, goals give meaning to and motivation for our present activities. The real advantage of long-term projects is the learning and deeper understanding that come with the endeavor. Keeping a journal enhances this. Naturally, use the engineering approach discussed in the Chapter 1 Appendix. Chapter 5 gives techniques of project management that will allow you to enjoy and gain satisfaction from definite progress toward your life goals. As noted above, for this project, we need to study fundamentals and gather relevant information. We start by defining the word *career*. It is unlike a job. A job is an endeavor pursued primarily for economic rewards and for set periods of time such as 8 a.m. to 5 p.m. A person's career, however, is the totality of the work-related experiences, behaviors, and attitudes over one's adult life. A career does not stop with retirement. Furthermore, the individual's psychological involvement is an important aspect of the endeavor. A career is a long-term commitment, with goals that can be achieved only in a sequence of steps guided by a lifetime plan. Monetary reward is secondary.

These characteristics of a career are the hallmark of professionalism: the professional practices a learned profession essentially under one's own direction, and in an ethical manner, for the benefit of others. *Techno,* from the Greek word *techne,* is equivalent to the Spanish word *hacer* and the French word *faire:* to make or to do. Thus, *technology* is the study of how to do something; a *technique* is a method, usually employed without understanding. Through the use of scientific principles, *engineering gives understanding to technology.* The understanding is based on mathematical representations or models, such as Newton's law of gravity. *Engineering* might be defined as "the use of models to design new systems or to improve existing ones, all for the benefit of mankind."

In engineering, and seemingly in most of life, interactions are most effective under matched conditions. This situation is familiar in electrical and mechanical systems and also applies to human beings: in Section 1-2 we suggest matching one's activities to one's capabilities. The matching is particularly important for an individual's career, the totality of a person's activities over one's adult life. The importance of one's self-concept, as stressed in Section 1-4, is particularly significant in making the best match between the individual and a fulfilling career. We next study the components of one's self [10, p. 12].

One's selfhood, or real self, is who the individual truly is. We all distort reality to some extent. Thus, although the real self exists, it is difficult to know.

The self-concept is the self "as seen by the self." Although it is in part developed from interactions, e.g., successes and failures within the environment, the self-concept primarily develops from the reflections from others. If a person is taught to be of great worth, the individual will see a self of great value. The reverse is also true. This is important in human development, for as one matures, the self-concept is self-perpetuating. We tend to see what we have been conditioned to see. It is our self-perception.

Self-esteem is the individual's evaluation of the self. This requires a standard for comparison: the ideal self. The ideal self reflects what one would like to be. It commences in childhood, identifying with a parental figure. It then moves through stages of romanticism and fantasy during late childhood and early adolescence, and culminates in late adolescence as a composite of desirable characteristics. These may be symbolized by an attractive, competent young adult, or simply an imaginary person. The ideal self integrates goals and aspirations and these provide direction to one's life. It is a healthy state when the ideal self is within one's capability. Too many differences between the self-concept and the ideal self cause conflict in the person and could lead to unusual behavior. Studies [10, p. 14] show that agreement between the self-concept and ideal self leads to positive adjustments in school, career, and social life. A close agreement gives the individual a strong advantage in any endeavor.

As noted above, the individual's self-concept is self-perpetuating. Similar to the "resonant filters" presented in Section 1-1 our self-concept causes us to have selective perception. Using it, the person maintains order in the world and consistency in self-evaluation. Thus, the self-concept is fairly stable from late adolescence until late maturity.

Self-Concept and Career Choices

The premise of this section is that individuals need to implement their self-concept by choosing a career that will allow self-expression. A career is a major process in maintaining and enhancing the self-concept. This process consists of a series of life stages: growth, exploration, establishment, maintenance, and decline. The exploratory stage can be divided into three phases: fantasy, tentative, and realistic. The establishment stage can also be divided into three phases: trial, modification, and stability. Development through the stages can be guided and facilitated by continual testing and enhancement of one's self-concept.

Career and, in turn, life satisfactions depend on the extent to which individuals find adequate outlets for their abilities, personality traits, and values. In other words, satisfaction depends on how well the individual's personal characteristics are *matched* to the requirements of one's career! Holland offers a theory: To the extent that one's career and personality are congruent (matched), one will experience increased satisfaction, stability, and achievement [11]. Holland's personality types and their "suitable" careers are reviewed next.

As a result of the early and continuing influences of genetics (one's innate ability) and its interaction with one's culture, a hierarchy develops of habitual and preferred methods of dealing with social and environmental tasks. Accordingly, Holland [11] presents six personality types: realistic, investigative, artistic, social, enterprising, and conventional. It is not probable that any one personality type will describe you. Most likely, however, you can form a weighted average of your preferences, which reflect your values, capabilities, and attitudes.

The realistic type prefers explicit, well-ordered, and systematic activities, those with minimum uncertainty and requiring minimum judgments from abstract models. Realistic personality types include mechanics, technicians, service personnel, and lower-level managers.

The investigative type has a preference for activities which involve gathering facts, observing results, and creatively developing a theory or model of physical, biological, or cultural phenomena. This type seeks understanding and control of the phenomena, and has an aversion to persuasive, social, and repetitive activities. These kinds of people are scientists (social or physical) and engineers.

The artistic type prefers activities that are ambiguous and unconstrained, with little initial order. These activities give them freedom to create art forms or products. Such characteristics are common in artists of all types: writers, composers, engineers, and inventors.

The social type prefers interacting with others to inform, train, develop, cure, or guide them to attain a goal. These individuals can meet their needs through the practice of medicine (nurses, dentists, and physicians), teaching, and counseling.

The enterprising personality type seeks activities that require persuading and even (perhaps) manipulating others to attain organizational goals or economic gain. People of this type are often salespersons, politicians, labor leaders, and executives.

The conventional type likes activities which entail explicit and ordered details, such as the systematic manipulation of data. Examples of these activities include record keeping, filing, and reproducing material.

Naturally most people are a combination of several personality types. Let us now review the desired "blend" for the various responsibilities.of engineers.

With a bachelor's degree one could work as a plant engineer, manufacturing (also known as production) engineer, sales engineer, or application engineer. Some may even advance to the level of advanced design engineer.

The plant engineer is responsible for the physical facilities, such as lighting, heating and air-conditioning, and plumbing. For a large plant there are many challenging problems. As with all problems, creative solutions (artistic type) are needed; however, the greater needs are for tested solutions (realistic type) and spare parts inventory control (conventional type).

The manufacturing or production line engineer needs to be well-ordered and systematic (realistic type), but the position also requires a high level of ingenuity (artistic type). This inventiveness is needed to interface the capabilities of new machines with those of older ones, always to achieve the highest performance at the lowest possible costs.

Application engineers work closely with production line engineers; they improve present equipment, introduce machines of greater capability, and interface them with those already in place. Their activities require a moderately high level of creativity (artistic type) as well as an orderly and systematic personality (realistic type).

Sales engineers have responsibilities and requirements similar to those of the application engineer. Obviously, the former works for the seller of machines and the other for the buyer.

Finally, the engineer in advanced design works closely with engineers holding master's degrees and doctorates. They are all primarily involved in research and development. This work requires those traits associated with the investigative and artistic personality types.

The kinds of engineers described above work in private industries and for governments. Their interdependence in private enterprise is presented in a sample technical article in Chapter 4 ("The Economic Roles of the Engineer").

Choosing a Matching Career

Most of us must work to fill the basic needs of food, shelter, and clothing. The higher-level needs must also be met for an individual to be fulfilled and to achieve a truly satisfying life. Properly chosen careers meet the needs at both levels. These careers match your values, interests, aptitudes, and capabilities. Naturally, our choosing an optimum career requires self-knowledge and accurate career data. The term *optimum* is used here, for even a properly chosen career will involve compromises and imperfect data. These are discussed later.

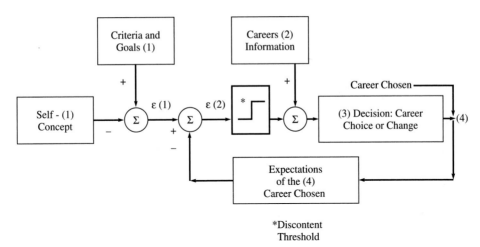

Type of errors, ε's:
ε(1) = difference between goals and self-concept.
ε(2) = discontent between goals and career expectations.
Note, if ε(1) = 0 or ε(2) < threshold, there is no action.

FIGURE 2-6
A block diagram of the engineering approach (applied to career choice or change).

The matching process has five major aspects:

1 Self-assessment
2 Career information
3 Decision making, based on (1) and (2)
4 Action (implementing the plan)
5 Review and improvement (as needed)

The process is depicted in Figure 2-6. The numbers in parentheses correspond, respectively, to the first four aspects listed above.

Self-Assessment   This phase has two parts: (1) reviewing general personality characteristics and (2) determining how your personal profile reflects them. The review is primarily based on values, interests, aptitudes, and personal qualities. Use *VIA-PQ* to remember these. Values are the principles which guide important decisions. They are feelings which relate to facts, things, people, and many broad concepts which are important in your life. The importance of each will vary. Criteria for decisions are based on one's values. Murray proposed that our behavioral tendencies are correlated to a seven-component value vector [12, p. 186]: (1) body or physical well-being; (2) property and wealth; (3) authority, such as decision-making power; (4) affiliation, such as love and acceptance; (5) knowledge and understanding; (6) aesthetics, such as beauty; and (7) ideology, such as philosophy and religion. Below is a list of personal values which could be components in your value vector; you might add others. In your self-analysis, quantify them to help clarify their importance. Assign values such as 4 for very important, 3 for important, 2 for moderately important, and 1 for not important. You might prefer using a range of 1 to 10.

Personal Values

Achievement	Marriage
Advancements	Morality
Aesthetics	Physical appearance
Autonomy	Recognition
Commitment	Religion
Creativity	Travel
Esteem	Stability
Ethics	Status
Honesty	Understanding
Independence	Wealth
Leadership	Work ethics
Location	

Interests are persons, activities, places, or things which we like and for which we have positive feelings. Their importance to us is related to our values. Note, however, that we can value something but still may not like it. For me, parts of this book, physical exercise, and certain foods are in this class. The list below presents your probable personal interests; add others and quantify each on a scale of 1 to 4. The time you spend in certain activities and with certain persons could indicate your interest preferences.

Personal Interests

Athletics	Planning
Being alone	Reading
Challenges	Relaxing
Competing	Solving problems
Controlling	Speaking
Coordinating	Sports
Creating concepts and things	Traveling
Deciding	Working with your hands
Helping others	Working with your mind
Managing	Writing
Outdoor activities	

Aptitudes are closely linked with intelligence and muscular coordination. Mathematical analysis and dancing are examples of each, respectively. Theses aptitudes suggest the ease with which one can become competent in a given activity or in a certain area. The Department of Labor job analysts have defined 11 different aptitudes [13, p. 39]. Most aptitudes are for physical activities, just as IQ is related to mental ones. For the engineer, however, the dominant aptitudes are those evaluated by the Scholastic Aptitude Test (SAT): verbal ability and mathematics. You have probably taken this test and know your scores.

Personal qualities are developed in late adolescence and are strongly influenced by our values, interests, and aptitudes. We like (have interest in) that which we do well (have an aptitude for) or that which a significant person (whom we value) has helped us learn. A significant adult also influences our values such as honesty, dependability, and courtesy. These and other personal qualities are listed below. Personal qualities reflect our personality.

Personal Qualities

Achiever	Enthusiastic	Persuasive
Adaptable	Extroverted	Pessimistic
Aggressive	Fair	Pleasant
Alert	Flexible	Practical
Ambitious	Follower	Quiet
Analytical	Follows through	Rational
Articulate	Friendly	Realistic
Calm	Goal-oriented	Relaxed
Carefree	Gregarious	Reliable
Caring	Hard working	Respectful
Cheerful	Honest	Responsible
Competent	Imaginative	Self-confident
Competitive	Independent	Self-starter
Confident	Innovative	Selfish
Conforming	Intellectual	Sincere
Conscientious	Introverted	Sociable
Controlled	Leader	Soft talker
Courteous	Logical	Stable
Creative	Mature	Supportive
Dependable	Methodical	Tactful
Disciplined	Optimistic	Tenacious
Discreet	Organized	Tolerant
Domineering	Patient	Trustful
Emotional	Perceptive	Trusting
Enterprising	Personable	Workaholic

Allport defines personality as the dynamic organization of the psychophysical* systems that determine one's characteristic behavior and thought [12, p. 263]. The self is the basis of reference around which the individual makes the "dynamic [changing] organization." From this frame of reference, the present situation is perceived. One's

---

*Again, word etymologies (roots) help our understanding. *Psycho* is Greek for "mind," "soul," or "spirit"; *neuro* is Greek for "nerves."

personality is similar to the transfer function of automatic control systems. That function gives the response of a system as a function of both its stimulus (input) and its state. We have all taken advantage of this phenomenon when we waited for an individual to be in a good mood (state) before making an important request.

Dispositions and traits are the prominent factors of one's personality. Allport defines a trait as a neuropsychic structure which causes many stimuli to produce the same forms of adaptive and expressive behavior [12, p. 265]. In the mature person, traits are both conscious and rational. As we see later in considering the traits predominant in certain careers, a particular trait can apply to many people. On the other hand, a disposition, as defined by Allport, is a trait unique to an individual. Both traits and dispositions have to be inferred from observed behavior. Attitudes are predispositions which result from genetic factors and learning. They are linked to specific objects, persons, and concepts; traits and dispositions are always general. Attitudes imply acceptance or rejection of the specific object, person, or concept. One's personal qualities make up an array of the individual's traits, dispositions, and attitudes. The array of personal qualities above should be helpful in identifying the qualities which you portray.

It is of great importance that this image of yours be correctly evaluated by others. As alluded to in the next paragraph, your special care to this aspect will be well rewarded. Judgments about people influence employment decisions and the degree to which colleagues seek professional and personal interchanges with them. Résumés give the education, experiences, and capabilities of an individual; but interviews are almost always used to assess one's personal qualities. Some experts suggest that the assessment of these qualities influences 90 percent of the decision to offer employment. I suggest that the intelligent person recognizes the importance of being personable and puts forth the effort to learn how to "get along" with people, at least 95 percent of the time. A complete self-assessment integrates one's education, experience, and *VIA-PQ:* values, interests, aptitudes, and personal qualities. These constitute one's personal profile. Its development is a discovery process.

Self-Discovery   The needed information can come from you, your parents, or close friends. Parents and friends are considered later. Your "selfhood" is the basis of your personal profile, and much of the selfhood lies in the subconscious. It can be accessed in two ways: by brainstorming and by writing. These really are "two sides of the same coin:" Brainstorm and write about yourself. What you record will bring forth fresh ideas for further reflection.

As discussed in Section 1-5 on creativity, brainstorming is a process by which you can access your subconscious. Recall that the general technique is similar to daydreaming; let your mind explore any and all ideas, but permit *no* evaluations or any other type of criticism. Relax. Hopefully you have started your career project early in college. Nonetheless, whenever you start, your progress will be best in an unhurried, easygoing frame of mind.

At the beginning, over several days or a week, for 20 to 30 min a day, write about yourself in your journal; let it become your autobiography. It's fun. Relax and let your mind run free. Develop the narrative chronologically; give explanations and probable rationale for various life transitions, such as a change of geographical location or a new direction in your major interests and skills development. While writing, note your values, attitudes, interests, and personal qualities. Do not try to categorize them; we will consider grouping them later. For now, avoid structured thinking. After completing your life story, write about courses you enjoyed and those you did not like, as well as about activities and people you did or did not like. Among the activities include part-time work, projects, and community service, such as tutoring or cleaning up the environment. Remember, you are the only one who needs to see your writings; put your inhibitions aside. Be completely open and truthful with yourself. Below we discuss how friends can be helpful; hopefully, discussions with your friends will be mutually beneficial. Writing is *work,* but here the rewards are very great. As is characteristic of long-term projects, there is no pressure if one uses time management (see Section 1-2). Just enjoy learning about your closest friend: yourself.

As noted above, for an insight about your interests, keep a time log for 3 or 4 days. It has been suggested that we tend to spend the most time on activities we find most interesting.

In another session, dream a little; in three or four (handwritten) pages describe *success.* Record what you think an ideal life and career would be. Project your dream 20 to 30 years into the future. The average engineering student or graduate engineer can achieve almost any goals by planning and starting early. Recall that an IQ is the ratio of one's mental age to one's chronological age. Norbert Wiener at age 19 received a doctorate from Harvard. If his IQ was 150, he could achieve mentally at age 10 what the average person could not accomplish until age 15. On the other hand, we average people can accomplish the same feats as those who are brilliant by starting early or sticking with the task longer, or perhaps both. The old adage seems to hold true: "Whether you think you can or you can't, you're probably right."

After 2 or 3 days, review your writing and pick out and group separately your values, interests, aptitudes, and personal qualities. A review of the items listed above will help identify your *VIA-PQ*. Finally, estimate their importance by assigning values: 4 for very important, 3 for important, 2 for moderately important, and 1 for not important.

You will now have completed a search through your most knowledgeable source: you. By being honest you will minimize the bias always present to some extent in a self-analysis.

Other significant sources are close friends, especially fellow students or graduate engineers. If still in college, seek information from professors, those under whom you did well and those whose courses you found difficult. The efficacy (the power to achieve the desired results) of your interchange could probably be improved in several ways. First, a few days before your meeting, tell your friends or professors its purpose. This gives them time to formulate the information latent in their memory. Then, be prepared to guide discussions to their evaluations of your values, interests, aptitudes, and personal qualities. Stress that you will appreciate their honesty. Often in appraisal meetings such as these, you will catch yourself being defensive, thus not fully benefiting from the discussion.

The activities just presented will give you the raw data for an in-depth self-assessment. This assessment naturally leads to your setting goals. It is likely that ideas about goals have come to mind as you made your self-analysis. In any case, before proceeding, over the next several days, schedule relaxed sessions to formulate your ideas. For later quantitative evaluations, give a decimal weight to each goal such that the total of all the weights equals 1. Dreams can be a first step to practical realization of a fulfilling career. "Don't stomp on your dreams."

Your self-assessment and goals make up your personal profile. By using the numerical evaluations, you have the basic data for matching your personal profile to the corresponding characteristics needed for success in possible careers. This we will consider next. As you review these career characteristics, it might be advantageous to go back, reconsider, and perhaps reassess your profile. Recycling is an important advantage of the engineering approach. Each iteration will probably give you a deeper understanding of both the career options and yourself.

Career Exploration   This phase consists of gathering information about careers of interest to you. There are several pertinent sources. They are publications (including computer data bases), people, and personal experience.

Publications   These are usually better than personal sources because the information has a broader scope. The publications are objectively prepared after observations of a large sample of people and occupations. Most of the professional engineering societies (ASCE, ASME, IEEE, and others) publish books, booklets, and various materials about careers in their fields. These are often available in university libraries or can be obtained directly from the societies. University career counseling centers also provide career publications, along with other relevant information. Another excellent source is the U.S. Government Printing Office.

Career choice is a systematic process of exploration. From the data available, you can narrow the many career possibilities to the few best for you. Let your search progress from the general to the specific, i.e., those careers compatible with your personal profile. Accordingly, sequentially review three reports prepared by the Department of Labor: *Dictionary of Occupational Titles (DOT)*, the *Guide for Occupational Exploration (GOE)*, and the *Occupational Outlook Handbook (OOH) for College Graduates*. These are available at most large city and university libraries. In addition, the JIST Corporation has published the *Enhanced Guide for Occupational Exploration* (1991), which includes job descriptions. The Consumer Information Center (Department 89, Pueblo, CO 81009) also offers a catalog which has a section on careers.

The *DOT* lists over 20,000 occupations, along with descriptions. Each has been assigned a nine-digit code. The descriptions indicate what the job is and how "the employee does it." These descriptions are organized into a structure called the occupational group arrangement (OGA). The nine-digit code consists of three sets of three digits each. A period is used to separate the first and second set; a hyphen, to separate the second and third.

The first set indicates the OGA. In it, the first digit identifies an occupational category. There are nine categories, 0/1 through 9. The first and second digit together place that occupation in a division in the category denoted by the first digit. All three digits in the first set place the occupation in a group.

For example, the *DOT* code for aeronautical engineering is 002.061-014. The first digit (0) places this occupation in the professional, technical, and managerial category.

The second digit (0) places this occupation in the division (of that category) of engineering and architecture. The third digit (2) places this occupation in the group for engineers. Under each breakdown is a description for the category, division, and group, respectively. Similarly, under further breakdowns the corresponding descriptions are given (in this case, for aeronautical engineering). Each description, condensed to a single paragraph, offers valuable information on the processes used or the aptitudes required, or both.

To locate a particular occupational description in the *DOT,* use the Alphabetical Index of Occupational Titles. It is in the back section and gives the nine-digit code for each occupation. Use that code to locate the specific description.

For the careers of interest in the *DOT,* continue your sorting by reviewing the *Guide for Occupational Exploration (GOE).* It classifies occupations by personal characteristics. The occupations (of the *DOT*) are organized in the *GOE* into 12 areas and 66 groups. A six-digit code is used (three sets of two). The first set indicates the area, the second set the group, and the third the subgroup. The alphabetical list at the back of the *GOE* gives the six-digit codes for the occupations. As in the *DOT,* there is a description for each breakdown.

For the next step in narrowing your search for a suitable career, consider the *OOH.* The *Occupational Outlook Handbook for College Graduates* is revised annually. Its up-to-date information, which uses the same codes as the *DOT,* makes it valuable for both career planning and changing. There are four sections. "Nature of the Work" gives the duties, assignments, and day-to-day activities of those in specific career fields. "Qualifications" gives the aptitudes and educational degrees necessary in that field. Often the personal qualities are also specified. This section is similar to the requirements in an employer's job description. "Employment Outlook" includes information on supply and demand. It also discusses growth and replacements needed in the near future. Finally, "Earnings and Working Conditions" gives entry-level, mid-point, and high-end salaries.

As noted above, the *OOH* is revised annually. There is also a quarterly report: the *Occupational Outlook Quarterly.* Together, they are the most useful reports to study.

Computer Data Bases   These are programs which organize a large amount of stored data about careers. Many have interactions which facilitate your self-assessment. These data bases are often available in a university library and at college counseling or career centers. Four well-known bases are DISCOVER, the Guidance Information System (GIS), the System of Interactive Guidance and Information (SIGI PLUS), and CHOICES. There could be a charge for their use, but it would probably be an excellent investment.

People   In the 1940s an auto company used the slogan, "Ask the person who owns one." It was so effective it put the company out of business. I believe generally that in making decisions, such as the choice of a career, it is very worthwhile to seek information from those who have done or are doing the kind of work you are considering. A professional learns early that time is a very valuable asset. To save your time and especially that of the interviewee, plan your questions before the meeting. This is proper and advantageous even when seeking information informally. Naturally, you want a career which offers *success* and *satisfaction.* Success indicates the degree to which the person's capabilities meet the needs of the employer. Satisfaction indicates the degree to which the person's profile (*VIA-PQ*) is met by the career. Your questions need to determine, first, whether the interviewee's personal profile is close to yours; then determine whether the person finds the career fulfilling.

Experience   Unquestionably, experience is the most effective teacher. Often, though, it is not the best because of the cost and trauma involved. In the previous paragraph we reviewed how to learn from the experiences of others; personal experience, however, is far superior. It enlarges one's self- and one's career awareness. I believe that work experience in any area is beneficial because it usually involves interacting with people. Reflecting on and learning from one's mistakes enhance the ability to get along with others, probably the most valuable of all skills. Part-time work in the career area under consideration is usually possible but can often require special effort. However, the significant benefit of the technical experience, in addition to the "people" experience, usually justifies the extra effort. Cooperative education programs, available at many universities, offer internships in career areas. They are excellent supplements to the on-campus theory courses.

Decisions   The materials you have gathered give the information required for a rational decision. Details for these decisions are presented in the first part of this section; only particular ideas are reviewed here.

One's personal profile is used to make the first selection of possible matching careers. Naturally, your capabilities and either completed or expected-to-be completed academic preparation will also be a factor. These careers could be selected from the *DOT*. Your goals act as criteria by which to quantitatively compare career options. The comparison will require that you subjectively assign numerical values for each goal of the career options. The range could be 0 to 10. Choose the values so as to reflect how you perceive a particular career to meet each goal. These are not "Mickey Mouse" numbers used to play games. As you seriously consider these values and the weights given to your values earlier in this chapter, you are quantifying your self-knowledge. As in psychological tests, after reflection, you might wish to revise these numbers. Keep all evaluations in your journal for convenient reference. Finally, multiply each goal value by the decimal weight you gave that goal earlier. The total for each career option gives it an index between 0 and 10. With this number, you can compare the option with others. To illustrate, consider this oversimplified two-goal example. The comparison index is determined for a career in electrical engineering design (*DOT* code 003.061-026).

Career values and goals with their weightings follow:

Goal	Value	Weighting	Index component
Annual salary over $30,000	9	.4	3.6
Creative projects	8	.6	4.8
Index value for this career			8.4

This same procedure can be used to determine an index value for the other career possibilities. For those choices remaining after the first elimination, further information can be sought from the *OOH*. Subsequently, the above procedure can be used again for a new set of indices. In normal economic conditions, there are a great number of opportunities open to engineers. Thus, career choice is a process of repeated eliminations. Because economic times are not always "good," the personal investing concepts in Chapter 7 are important to achieve a stable career.

Expectations of Chosen Career   Here you simulate (mentally visualize) your expected life at work and at home. Over a period of several days, try to consider both in detail. Think about commuting to work, types of probable work projects, and the knowledge you need for and will gain from these projects. Naturally, some career aspects, especially the intangible ones, will not be quantifiable. Judgment could be used to give them values or you could qualitatively assess the numerical data and other aspects to arrive at a decision. The human mind is quite capable of integrating disparate information.

Iterations and Improvements   All human decisions are compromises. If the difference between the "dreamed" expectations and visualized reality exceeds your threshold, it is advantageous to go back to the beginning (self-assessment). The beauty of simulation is that even when expensive computers are required, we can acquire considerable knowledge of reality at a fraction of its normal cost. In many careers an advanced degree is very advantageous.

Graduate Study   The B.S. degree opens career opportunities not otherwise possible. Similarly, advanced degrees offer levels of opportunity usually not realized with only a B.S. Generally, education for the bachelor's degree imparts an understanding of the "language of engineering" and an ability to "think like an engineer." Industry uses these capabilities and guides the recent graduate into an environment in which engineering skills are developed. The master's degree student will more rapidly acquire most of those skills and will gain high competence in a particular area of interest.

By pursuing the master's degree, the student gains a greater understanding of the concepts used in undergraduate work and acquires the ability to do advanced engineering. With this deeper understanding, the student can also communicate more effectively, both orally and in writing. With these advantages, the engineer with a master's degree will start at a salary about $3000 a year higher than that for a bachelor's degree [14]. On a long-term basis, industry pays for performance, and advanced degrees determine the rate (slope) of career progress. Regardless of the individual's performance, the "absolute" value of career progress will be greater by starting at a higher base. Here, the term *career progress* is used to indicate total satisfaction, not just higher salary.

In my personal experience, nothing has contributed more to feeling competent and, in turn, to having a higher level of satisfaction than "understanding the situation." The greatest understanding and the ability to "be your own teacher" come in a Ph.D. program. The Ph.D. is a broad-based degree. The candidate develops depth of understanding in at least four fundamental areas. Accordingly, the graduate acquires the competence to teach and do state-of-the-art research in those areas. The starting salary (about $10,000 [14] a year higher than that for a B.S.) is probably the least significant of all the advantages.

The decision plan presented in this section is equally applicable to changing a career. Remember, career planning is a lifelong endeavor and does not stop at retirement. As the saying goes, "Failing to plan is planning to fail."

In conclusion, the prudent choice of a career requires considerable effort. First, self-analysis is necessary; then information needs to be gathered about the requirements, challenges, and rewards of possible career choices. It is vital that you develop a match between your abilities and the career requirements, its challenges and rewards, and your attitude and values. Fortunately, a first mistake in this choice, while costly in money, time and emotions, is many orders less traumatic than mismatching with a spouse (a choice amenable to decision analysis).

Peck [15, p. 44] notes that one's adult life is a series of personal choices, i.e., decisions. If we avoid these decisions by letting chance or others set our course, we become unable to cope with life. To become free people, we must learn to make and accept responsibility. Using the techniques described in this chapter, you can consider your individual preferences and with confidence make logical, unemotional decisions.

2-4 SUMMARY

This chapter presents the elements of decision making. On a long-term basis, rational decisions are recognized as giving the best results. Significant decisions justify the greater cost of analytic techniques. These techniques include developing utility functions to model the risk profile of the individual or corporate decision maker. To illustrate the ideas discussed here, an analytical approach is applied to the significant decision of choosing or changing a career. The approach is applicable to choosing a spouse. For both choices, happiness and stability depend on your matching values, attitudes, interest, and personal qualities. In progress and change, errors are inevitable. Well-considered decisions minimize these errors.

DISCUSSION QUESTIONS AND PROBLEMS

2-1 DECISION-MAKING TECHNIQUES

1 a Give a definition for the term *significant decision*. How does it differ from other decisions?
   b Discuss the bases of nonanalytic decision-making techniques.
   c Under what conditions are judgmental decisions usually adequate?
2 What is the foundation of rational or logical decision-making methods?
3 a Apply the decision-making process to a significant past decision in your life. Examples might be the university you chose to attend or your major field of study.
   b Regarding part (a), review what relevant information "20-20 hindsight" has provided. With a little more effort on your part, could this information have been considered in a probabilistic manner before the decision was made?

4 a Define the three categories in the decision-making environment.

b How and by what method can a decision "under uncertainty" be changed to one "under risk"?

2-2 ANALYTICAL DECISION MAKING

5 a What is the expected value of the number of dots which "turn up" when dice are rolled?

b In Example 2-1, a game of chance is presented concerning the roll of a die. The expected value of the reward $E[r] = \$5$. Determine the expected value if all gains and losses are doubled. *Hint:* In linear systems, mathematical operations can be commuted; e.g., the sum of the products (each term multiplied by the same factor) equals the product of the factor and the sum (of the terms).

c Careful planning can improve the probabilities of a project's outcomes. Determine the expected value of the reward in the original game (Example 2-1) if the probabilities of all gains are increased by 50 percent (i.e., to $\frac{1}{4}$) and those of all losses are halved (i.e., to $\frac{1}{12}$). *Hint:* The linear rule can be applied separately to the gains and then to the losses.

6 The utility function (such as in Figure 2-1) shows the DM's risk characteristics. It also reflects the law of diminishing returns (recall your introductory economics). To continue in this vein, consider the expected utility value (EUV) of this DM for increasing returns.

a Determine the DM's EUV if all outcomes of contract A (Table 2-3) are doubled. By what percentage is the EUV increased over its original value of 66? To facilitate interpolation in Figure 2-1, consider the horizontal scale as $29,400/cm; the vertical, 28.6 utils/cm.

b Repeat part (a) with all outcomes tripled instead of doubled.

7 Shortly after graduation, by using the information in Chapter 7, you could well have a net worth exceeding $30,000. For this problem, however, assume that your net worth is that amount right now. Also assume that Figure 2-2 and its upper portion in Figure 2-3 represent the shape of your utility function. The numerical values for your utility in utils are read from the y-axis, with the x-axis reduced by a factor of 10; e.g., the $300,000 at the upper end of the curve becomes your $30,000. The x-axis scale is $6000/cm.

You have just purchased a late-model low-mileage used car for $10,000. Probably you paid cash to escape the high interest rates of installment purchasing. In any case, you wish to have it covered by collision insurance to avoid the possible loss of this significant portion of your net worth. As reviewed in this chapter, on a long-term basis, self-insurance is always the more economical. One way to self-insure for amounts we can afford is to get insurance with greater deductibles. Since small claims (for losses) require almost as much paperwork and other overhead costs as larger ones, the insurance companies give considerable discounts for increasing deductible amounts. Furthermore, most companies have a required $250 deductible. The premiums for $10,000 collision insurance in South Carolina (1991) are similar to these:

Deductible amount	Six months premium	Increase in deduction	Annual savings
Required: $ 250	$105		
Optional: 500	94	$250	$22.00
Optional: 1000	70	750	$70.00

Using risk ratios typical of those given for the casualty insurance industry by Value Line, an information source reviewed in Chapter 7, the following probabilities apply:

$$\text{Case A: Pr[\$10,000 > loss > \$250] = .105}$$
$$\text{B: Pr[\$10,000 > loss > \$500] = .094}$$
$$\text{C: Pr[\$10,000 > loss > \$1000] = .070}$$

a Using the probability given above for case A, determine the expected value of the loss of $9750 ($10,000 minus the deductible of $250). Using the slope, determine your utility expressed in dollars (the value that the slope of your utility shows you would be willing to pay to avoid the risk). Does this value show the insurance cost to be rational? *Hint:* Review Example 2-3 (the first part).

b The probabilities given in cases B and C for the insurance company's (large) losses can be considered approximately the probabilities of the smaller losses (the deductibles) by you, no *x*-axis reduction by 10. Also consider that Figure 2-4 represents your risk characteristics regarding small losses. Determine the expected value of each loss and use these values with your curve to get your corresponding (CME) utilities in dollars. Recognize that the "utility in dollars" is the value you would pay yourself for the greater risk of the corresponding deductible you choose. *Hint:* See the latter part of Example 2-3.

8 List and briefly review the common forms of groups presently used by organizations to make group decisions.

9 Present the pros and cons of analytical and judgmental decisions.

10 Discuss the political forces which influence industrial decisions.

2-3 APPLICATIONS TO PERSONAL DECISIONS: CAREER CHOICE OR CHANGE

11 Discuss the engineering approach to career choice or change.

12 Make a preliminary analysis of choosing your own career.

REFERENCES

1 Griffin, R. W. *Management.* Boston: Houghton Mifflin, 1990.
2 Jurgurson, E. B. *The inevitable general.* Boston: Little Brown & Co., 1968.
3 Plaque, Riggs Hall, Engineering Building, Clemson University, SC.
4 Moore, P. G. *The business of risk.* Cambridge: Cambridge University Press, 1983.
5 Lifson, M. W. *Decision and risk analysis for practicing engineers.* Boston: Cahness, 1972.
6 Luce, R. D., & Raiffa, H. *Games and decision.* New York: Wiley, 1958.
7 Raiffa, H. *Decision analysis.* Reading, MA: Addison-Wesley, 1968.
8 Maslow, A. H. *Toward a psychology of being* (2nd ed.). New York: Van Nostrand, 1968.
9 U.S. Department of Labor and Bureau of Labor Statistics. *Occupational outlook handbook* (revised annually). This report gives prospects for employment.
10 Pretrofesa, J. J., & Splete, H. *Career development: Theory and research.* New York: Grune & Stratton, 1975.
11 Herr, E. L., & Cramer, S. H. *Career guidance and counseling through the life span.* Boston: Little, Brown, 1984.
12 Hall, C. S.,& Lindzey, G. *Theories of personality* (2nd ed.). New York: Wiley, 1970.
13 Wineforder, D. W. *Career planning and decision-making for college.* Bloomington, IL: McKnight, 1980.

14 Collins, A. S. *Professional development through advanced degrees.* Career Day, Clemson University, 1988.

15 Peck, M. S. *The road less traveled.* New York: Simon & Schuster, 1978.

BIBLIOGRAPHY

Excellent supplements to material presented in this chapter.

Bolles, R. N. *The three boxes of life and how to get out of them.* Berkeley, CA: Ten Speed Press, 1981.

Bolles, R. N. *What color is your parachute: A practical manual for job-hunters and career-changers.* Berkeley, CA: Ten Speed Press, 1981.

Ellis, D., Lankowitz, S., Supka, E., & Toft, D. *Career planning.* Rapid City, SD: College Survival, 1990.

Powell, C. R. *Career planning today* (2nd ed.). Dubuque, IA: Kendall/Hunt, 1990.

U.S. Department of Labor. *Guide for occupational exploration (GOE)* (2nd ed.). 1984.

Courtesy Fluor Daniel, Inc., Greenville, South Carolina

CHAPTER 3

MANAGEMENT

CHAPTER OUTLINE

Management is important to the engineer for professional and personal reasons. There are two aspects: self-management and the management of others. Most of Chapter 1 concerns self-management. In this chapter we address the second type, management of others. The chapters are complementary; in both we study the needs and aspirations of the individual and groups.

On first entering an industrial or public enterprise, you will be working with a supervisor or first-line manager. It is highly probable that you will later become a project leader or first-line manager, and it is also likely that you might choose to move up in management. The following classic quote by Fletcher Byrom (1976), president of the Kopper Company, is a motivation for this text. It is in regard to his first transition to management [1, p. 549] :

> As an engineer, I was suddenly in a whole new world. In those days, an engineer got exposed to very little. I had a survey course in business law and economics, but all I remember about economics was there is decreasing utility in adding more fertilizer to land. I didn't know how to read a profit and loss, or a balance sheet,* and here I was 33 years old. I had never done any significant philosophizing.

The information in this chapter will give you an understanding of the thinking and responsibilities of first-, middle-, and top-level managers. The roles and skills required for management are also helpful in achieving goals and satisfaction in your personal life. Accordingly, these are presented in Chapter 1 and are referenced in this chapter. Effective communication is of great significance in both management and personal life. It is presented in Chapter 4. Engineers can move into management and advance, provided they can manage their "self," get along with others, and communicate well.

As noted in the Preface, this book is designed for those of us who cannot take as many nontechnical electives as we would like. The chapters here present the normally adequate fundamental essences of their respective subject areas. A full semester course in management, however, provides additional information in many areas important for personal growth. Accordingly, I suggest taking Management. It is an important nontechnical elective. If that is not possible, study this chapter's references; they are very readable and informative.

Perhaps management is best defined in terms of systems theory. The systems approach is a "way of thinking"; it is holistic,† considering the particular situation in its entirety. This approach helps engineers and managers better understand the total interactions of projects and organizations. In turn, we first review the concepts of systems.

A system is an entity composed of independent parts; each contributes to the unique characteristics of the whole. In most systems the independent components are coordinated to achieve a goal. Everything external to it is considered its environment. Thus, a system and its environment constitute the universe. There are two types of systems: open and closed. A closed system has definite boundaries and is independent of its environment. For example, a hermetically sealed quartz watch is nearly closed.

---

*Profit and loss statements and balance sheets are reviewed in Chapter 6.
†*Holos* is Greek for "whole," as *holocaust* is Greek for "wholly burnt."

Actually, there are few truly closed systems, but it is often helpful to consider one as closed in an initial analysis.

An open system is one which interacts with its environment. An organization is a group of two or more people working in a coordinated fashion to attain a set of goals. It is an open system, with inputs of information, materials, capital, and human resources. Outputs include products or services, monetary profits or losses, and social responsibility. Management is the set of activities directed to setting goals and coordinating the organization's independent parts toward these goals. Coordination by competent management achieves the organization's goals in an efficient and effective manner. An efficient manner uses resources wisely, with minimum waste. Obviously, an effective manner achieves the goals. As in career planning, goals give direction to the activities required to attain them.

3-1 LEVELS AND PHASES OF MANAGEMENT

There are four major phases in the management process: planning and decision making, organizing, leading, and controlling [2, pp. 7 & 13]. Managers are those whose primary activities support the management process. Although we will mostly be using the more descriptive system theory, the classical management model (Figure 3-1) is

FIGURE 3-1
A typical industrial organization.

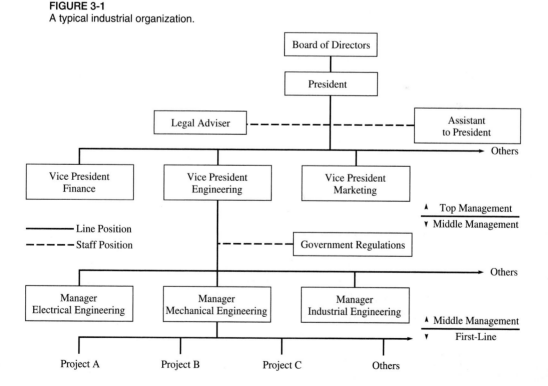

better for indicating who reports to whom in the organizational structure. Line managers are those in the direct chain of command. There are three levels in the chain: top, middle, and first-line managers. This chain is responsible for achieving the organization's goals.

Levels

Top Managers   They are responsible for the organization as a whole or for a large segment of it. They make the major decisions as Petersen did at Ford when he became president in 1980. With clear vision of the environment facing the U.S. auto industry, Petersen made hard decisions about reducing the work force, controlling costs, and improving product quality.

Middle Managers   They constitute the largest management group in most organizations and are primarily responsible for implementing the policies and plans made by top management. They are also responsible for supervising and coordinating the activities of first-line managers.

First-Line Managers   First line, supervisory, or operating managers supervise and coordinate activities of operating employees. As the name implies, this is the level immediately above nonmanagerial engineers or workers. Most who enter management for the first time are at this supervisory level.

Staff Positions   Upper (top and middle) managers often have assistants who provide expertise, advice, and support for line positions. Staff positions can usually be classified as advisory, service, or personal. The last is assigned to someone like the president to relieve the burden of mechanical-type work. Staff positions are shown by dashed lines in Figure 3-1.

A *board of directors* is required of all corporations. Many other organizations have comparable boards: universities have boards of regents; hospitals and charities have boards of trustees. For corporations, the board is elected by the stockholders and is responsible for overseeing the general management to ensure that the best interest of the stockholders is served. Inside directors are full-time employees of the firm. Outside directors are elected to the board for specific purposes, such as financial or legal.

Phases

As noted earlier, there are four phases of the management process:

1 Goals, planning, and decision making
2 Organizing
3 Leading
4 Controlling

Goals, Planning, and Decision Making   In *planning* there are two aspects: setting goals and developing methods (plans) to pursue them. As for an individual, goals for an organization are set in accordance with its mission: purpose, premises, values, and directions. Reflecting its mission, both an organization's goals and plans are considered strategic, tactical, or operational. We will review each type of goal and plan.

Strategic goals are set by and for top management. They focus on broad, long-term, and general issues. For example, Sears was Singer's major purchaser of hand power tools. Singer made a significant policy change to seek other purchasers so as not to be largely controlled by Sears' tool designs or its merchandising success.

A strategy is a detailed, comprehensive, and integrated plan to ensure that the organization's mission and general objectives are met. Note that as opposed to tactical goals (discussed below), a strategy (Greek for "lead an army") is concerned with general, long-term goals. They are set by the board of directors and top management (see Figure 3-1).

Strategic plans are long-term and are concerned with three types of decisions: resource allocations, setting priorities, and the sequence of activities necessary to achieve the organization's strategic goals.

Tactical goals are arranged by and for middle managers. These goals are the short-term achievements required to carry out the organization's strategic plan. For example, tactical goal for Singer (to reduce dependence on Sears) would be to obtain purchase contracts from 10 other buyers.

Tactical plans are those designed to achieve the tactical goals. At Singer, middle management, along with the first-line managers, could develop a plan to determine the type of power tools sold by other retailers such as K Mart. With these data, the plan could require design and production of tools with the desired specifications.

Operational goals are arranged by and for first-line or supervisory managers. The goals are set to implement the tactical plans. Accordingly, Singer's first-line managers would procure equipment and develop techniques to produce the new design of tools.

Operational plans are developed to carry out the tactical plans and achieve the operational goals.

Decision making is a very important part of planning. It is equally important in the other three phases. It involves choosing the course of actions to implement plans. Wise decisions are perhaps of greatest importance in the engineer's personal and professional life. For this reason, the techniques for making objective decisions are presented in Chapter 2. The next step in implementing plans is organizing personnel and resources.

Organizing   This is the coordinating of activities and resources as required to implement the organization's plans. For this aspect, managers must make decisions concerning six basic building blocks: designing tasks, grouping the tasks, deciding on reporting relationships between tasks and groups, distributing authority between groups, coordinating the groups, and differentiating between them.

Designing tasks involves specifying what is to be done, by whom, and the criteria by which the results are to be judged as successful.

Grouping of the tasks is often known as departmentalizing. The tasks are arranged in a logical manner. Such a grouping is shown in Figure 3-1 under the vice president of engineering. In large organizations, departmentalization is necessary to effectively and efficiently supervise others.

Reporting relations complement the chain of command. This chain is the clear line of authority for the positions in the organization. Someone in the organization must ultimately be responsible for every decision. The number of people reporting to each particular manager is defined as the "span of management." An organization with only a few reporting to the managers has a narrow span; if the organization is large, it will of necessity require a "tall" structure. Obviously, if the span is broad, the organization will be "flat." The best compromise must be determined by upper management. An early writer, A. V. Graicunas, quantified the problem [2, p. 285]. He noted that managers must deal with three kinds of interactions with and among subordinates: direct (manager to subordinate, one-to-one), cross (among the subordinates), and group (between groups of subordinates). The number of these interactions I is given by the expression

$$I = N(2^{N-1} + N - 1)$$

where N = number of subordinates

Distributing authority among positions is concerned with establishing power that can be conferred by the organization. The distribution primarily reflects a higher-level management decision to decentralize or maintain centralization. Decentralization is a process of systematically delegating power and authority throughout the organization to middle and lower-level managers. Centralization is the reverse. It is a systematic process of retaining power and authority in higher-level management. While there are no clear guidelines for determining the better of these two, the tendency today is toward decentralization. Power and authority at lower levels permit the faster decisions needed to meet new situations.

Task specialization of the various departments is required for activities to be most efficient. Coordination, however, is the required process for effectively guiding the departmental activities toward the organization's goals.

Finally, concerning differentiation between positions, a retired naval admiral once told me that "a tight ship is a happy ship." So it is in other organizations. Many critical decisions are delayed and possible achievements lost because one or more individuals feel that their "territory" (authority) is being encroached. One example of this involves line and staff positions. Line authority is the formal or legitimate authority established by the organization's chain of command. It is reviewed in this chapter. The staff members' function is typically "to advise"; the line manager may or may not seek that advice or follow it. Staff must often spend time selling line managers on the worth of their advice. Often, however, top management extends staff authority to include compulsory consultation. In that case, line must discuss relevant situations with staff before acting or bringing a proposal before top management.

Leading   This is a particularly important management function. It is of little value to have the best plans and organizational structures unless managers can influence people to perform the tasks needed to achieve the organization's goals. Leading is a set of activities by which management influences persons at all levels to work together toward furthering the interest of the organization. There are four areas of activities: motivation, interpersonal relations, communication, and leadership.

Leadership is a process of influencing the behavior and goals of persons and groups. A leader has the ability to implement the leadership process, which is described in detail below. An effective leader is also an effective manager, one who uses the other three areas (motivation, interpersonal relations, and communication) to support the process. After reviewing each of these areas, we will develop the idea of leadership as the coordination of all three.

Motivation   Motivation is an inward drive to act and is the result of a complex set of needs. The effective manager must determine and understand the general needs of people and the specific needs of her subordinates. Accordingly, she must offer activities (directed toward the organizational goals) and organize them so that the subordinates' needs are satisfied through their performance. We will review the theories about needs next.

Early in this century, Fredrick Taylor suggested the use of an incentive pay system to meet the workers' needs and thus achieve strong motivation [2, p. 442]. This has been defined as the *traditional approach;* its premise was that work is inherently unpleasant for most of us. Taylor's ideas, however, have been recognized as having too narrow a view of the motivational factors.

The *human relations approach* to motivation grew from studies at the Hawthorne plant of Western Electric (1927–1932). They were made by Elton Mayo and his associates from Harvard. In these experiments Mayo changed the level of illumination for one group of workers but not for another. Surprisingly, the productivity increased for both groups [2, p. 50]. With this and other studies, Mayo concluded that the human element (the need for attention) was more important than had been previously realized. Advocates of this approach advised managers to make workers feel important, to keep them informed, and to allow a modicum of self-direction and self-control. Abraham Maslow (1943) and Douglas McGregor (1960) advanced the human relations approach [2, p. 52]. Later, we will discuss the significance of team participation in decisions.

The motivation process begins with needs that reflect a deficiency within the individual. As alluded to in Chapter 1, Maslow presented the concept of a need hierarchy: physiological, safety and security, social or belongingness, esteem, and self-actualization.

Physiological needs are the basic ones, such as food, shelter, clothing, and air. We must meet these needs before turning our attention to others. In an organization, basic needs are generally met by a proper physical environment and adequate wages.

Safety and security needs include those for protection against physical and psychological threats in the environment. Also included here is confidence that the future physiological needs will be met.

Social or affiliation needs concern our wanting to belong to and be accepted by others, receiving affection and support from them.

Esteem relates to the need for a positive evaluation by one's self. It includes self-respect, achievement, and recognition.

Self-actualization is mainly dependent on fulfilling one's potential and continuing to grow. These are internal needs; but a manager can help promote a climate conducive to this growth by including subordinates in the decision-making process and promoting opportunities for them to learn more about their endeavors.

Maslow suggests that as lower-level needs are met, there is a tendency to "automatically" move up to the next level. Research indicates, however, that Maslow's five levels are not always present and that the order is not as postulated by him.

Clayton Alderfer has proposed a different hierarchy. In it, the levels are condensed to three: E, R, and G. Respectively, they stand for existence, relatedness, and growth. Alderfer's proposal is the ERG theory of motivation. It differs from Maslow's in two important ways: first, ERG theory suggests that more than one level can be active at the same time. For example, a person might simultaneously seek money (for existence), friendship, and the desire to learn a new skill [2, p. 441].

The second difference is a frustration-regression element. Recall that according to Maslow, one moves to a higher level only after the needs at the lower level are met. The ERG theory suggests that if a higher-level need remains unmet, the person can revert to a lower level and begin again. Research supports the ERG theory and indicates that it may be a better model for motivation than Maslow's.

Fredrick Hertzberg (1959) recognized the incompleteness of the earlier models and developed a "two-factor theory." From interviews with 200 accountants and engineers in Pittsburgh, he inferred that different sets of factors correlate with satisfaction and dissatisfaction. For example, low pay could cause dissatisfaction, but higher pay would not necessarily produce satisfaction. Building on the work of Maslow and Alderfer, Hertzberg suggested that satisfaction required factors such as recognition and accomplishment. Hertzberg's theory is now considered inadequate by present researchers. Nonetheless, it was instrumental in increasing managers' awareness of the importance of human relations in motivation. More recent developments in this area are based on three theories: attribution, equity, and expectancy theories.

Attribution theory is presented by Kelly [2, p. 466; #28]. He suggests that an individual observes the behavior of himself and that of others. Based on his self-perception and the observations, he attributes cause and effect relations to them. For example, if he perceives himself as being intrinsically motivated by the challenging nature of a task, he will seek those types of tasks. Conversely, if he feels that external factors such as salary motivate him, he will seek tasks with monetary rewards. Accordingly, the individual seeks tasks which enhance and reward his self-attributes. Managers aware of these effects can observe employees and decide what motivates them. Below, we see how these relations enter into the implementation of expectancy theory.

Equity theory [2, p. 465; #24] is based on the belief that individuals want to be treated fairly. Some may even demand it. Their frame of reference is what they know or think others are receiving in proportion to their efforts. We define the term *outcome* as the net (the good minus the bad) results of an effort. Typical positive outcomes might be the subjectively quantified values of pay, recognition, promotions, and social relationships. Negative outcomes could be the reverse of the positive ones, or factors such as routine or unchallenging requirements. Attribution theory acts as the basis for quantification. To make comparisons, employees determine the ratio of their outcomes to their inputs. The inputs could include factors such as time, experience, effort, education, and loyalty. The idea here is to compare one's ratio to that of a colleague. For example,

$$\frac{\text{Your net outcome}}{\text{Your total inputs}} \overset{?}{=} \frac{\text{the other's outcome}}{\text{the other's inputs}}$$

Perceptions of inequality are detrimental to the employee's motivation and satisfaction. In turn, these could lead to discontentment and poor performance.

Expectancy theory [2, p. 465; 3, p. 361] uses the concepts of attribution and equity theories to evaluate one's motivation. The expectancy model provides a way of analyzing this motivation and predicting the action of an individual. The quantified motivation, based on the individual's attribution and equity characteristics, is the expected value of a set of outcomes. The expected value is defined and used in Chapter 2. We review it below. The essence of the expectancy model is that the motivational "force" required for a certain behavior is a multiplicative function of two terms. One is the expectancy (probability) that a person assigns to the occurrence of the desired result from a certain behavior; the other is the person's valence (value) of that result. Expressed as an equation, it is

$$MF = E \times V$$

where MF = the motivational force
  E = the expectancy (probability) of the desired results
  V = the valence or value of the desired results

This equation indicates that MF is the expected value of the worth of the desired results. For most behaviors, however, there will be multiple outcomes, some with desirable results and some undesirable. For these more realistic situations, Vroom offered a second model; it is based on instrumentality-expectancy theory.

$$\text{Net MF} = \sum_{k=1}^{n} E(k) \times I(k) \times V(k)$$

where net MF = the net results (expected value) of both positive and negative $V(k)$ of possible outcomes
  $E(k)$ = the expectancy (subjective probability) that the efforts expended will achieve the kth outcome
  $I(k)$ = the instrumentality (subjective probability) that the kth outcome will produce $V(k)$
  $V(k)$ = the individual's valence or worth of the kth outcome (in satisfying a need)

Example 3-1

Consider three engineers carpooling to work. There would be many outcomes, each with different values. To demonstrate the principle as simply as possible, we consider only six. Each valence is assigned a value from 0 to 10. The motivational effect of multiple outcomes is determined by summing, respectively, the products of the expectancies, the corresponding instrumentalities, and the valences for each outcome. It seems logical to subjectively quantify the V's, I's, and E's in a reverse order of that in the equation. Thus, the V's are considered first.

Solution: For the positive valences, $+ V(k)$:

$V(1) = 8$. All three are already good friends, and their company for the 45-min ride would be pleasant.
$V(2) = 4$. The operating costs of getting to work would be reduced to approximately one-third.
$V(3) = 6$. The risk of damaging your auto or a breakdown would be reduced to about one-third.

For the negative valences, $- V(k)$:

$V(4) = 4$. Would have to leave a little earlier and get back later.
$V(5) = 6$. Would get to work late when a rider is detained by family complications.
$V(6) = 4$. Would get home later when a rider is tied up at quitting time.

The instrumentality $I(k)$ (estimated probability) that the kth outcome will produce the resultant worth $V(k)$ is considered next.

$I(1) = .9$. Although the three are good friends, they have never been together in a car for 45 min, 5 days a week.
$I(2) = .95$. The reduction in operating cost (by one-third) is contingent on all three riding together every day. Sickness or an occasional need for a rider to travel at a different time would reduce the savings.
$I(3) = .95$. This value is the same as number (2), for it is contingent on the same factors.
$I(4) = 1.00$. It is certain that it would require more time for three riders to be picked up and travel in one auto.
$I(5) = 1.00$. It is also certain that when a rider is detained, all will be late. Leaving a tardy rider behind, as agreed by the group, would reduce this instrumentality.
$I(6) = 1.00$. Of course, it is certain that a rider who "cannot" leave at the normal time will delay the return trip.

The expectancy or probability $E(k)$ that the kth outcome will be achieved by the effort expended is estimated as follows:

$E(1) = 90$ percent	$E(4) = 100$ percent
$E(2) = 95$ percent	$E(5) = 10$ percent
$E(3) = 95$ percent	$E(6) = 20$ percent

The net motivation for carpooling is the expected values of the valences, that is, the algebraic sum of the products of the valences of each outcome and its probability. Recall that the probability of independent events is the product of their respective probabilities. Here that probability is the product of the expectancy (the individual's subjective probability of that outcome) and its instrumentality (the subjective probability of the valence for that outcome). The net motivation follows:

$$
\begin{aligned}
\text{Net MF} &= \sum_{k=1}^{6} E(k) \times I(k) \times V(k) \\
&= .9 \times .9 \times 8 + .95 \times .95 \times 4 + .95 \times .95 \times 6 - 1 \times 1 \times 4 - .1 \times 1 \times 6 \\
&\quad - .2 \times 1 \times 4 \\
&= 6.48 + 3.61 + 5.415 - 4 - .6 - .8 = 10.11
\end{aligned}
$$

This positive value indicates the desirability of this activity for the person whose values were used. Naturally, different persons would probably have a different value, maybe even a negative one. If a rider declined to be in the pool, the factors would probably change, as would the net MF for the others. Also, a rider could possibly choose to take a bus or train. She could make an evaluation for that alternative, compare it to the one for carpooling, and use the comparison to make a decision.

Interpersonal Relations   These involve an interchange between persons, individually and within groups. The first type was discussed in Chapter 1 in the section on transactional analysis. Here, we consider interaction in groups [4].

Groups are the building blocks of society and organizations. The fundamentals of group dynamics are very important to the effective manager, for a group can powerfully influence a person's behavior. To the degree that one values membership, group norms, discussed below, can modify an individual's behavior in a given situation. There are three basic kinds of groups: functional, task, and informal. Functional groups are set up by an organization. They have an indefinite time horizon and are established for the accomplishment of numerous organizational purposes. Some examples are marketing and design groups. Task groups are also set up by organizations. Their purposes, however, are of a narrow range and for a definite time period. Most engineering projects are accomplished by task groups.

Informal groups can have their origins inside or outside an organization. These groups are of two types: spontaneous and interest groups. Spontaneous groups have no continued existence. They may or may not be relevant to organizational goals and could exist only as long as necessary to promote or demote a manager.

Interest groups, as their name implies, focus on a mutual interest, such as baseball, a common alma mater, or a desire to develop oratory skills as a toastmaster. Informal groups can be powerful, and a manager will benefit by understanding their behavior.

Another important characteristic of groups is their maturity. In a group which has functioned for a long period of time, the members know each other's strengths and weaknesses and are secure in their roles. Naturally, most groups are immature at first; each member is unfamiliar with most of the other members. But as the group matures, a "state trajectory" has been observed [2, p. 518]. It is analogous to the trajectory of a physical system moving from one energy state (level) to another. For newly formed groups, there are four levels in the "state" progression: forming, storming, norming, and performing [4, p. 477].

Forming is a period of hesitancy and testing. Members become acquainted and begin testing for proper interpersonal behavior. In this stage, ground rules and a tentative group structure evolve.

Storming characterizes the second state. From confidence gained in the first state, members make bolder statements and healthy disagreements might arise. For example, a member might try to establish himself as the leader or assume a major role in the group. Recall, however, that in "storming," emotions could block the cognitive (rational thinking) processes. The situation could cause a breakdown in the group or at least a considerable delay in progress. Since emotions are transitory, it would be advantageous for someone to be prepared to change the subject. Perhaps use humor; it is unmatched in allowing emotions to die. There is a danger, though, that one could become known as the group comedian or clown, a role of questionable value.

Norming is the state in which group cohesiveness is developed. Here, members recognize and accept their roles and those of others. There is a sense of unity. However, before gaining long-term stability, the group could go back to the storming stage. Roles could also change: an accepted leader could be rejected and another evolve, not necessarily in an unpleasant way.

Performing is the final state. In the roles established in the previous state, the members cooperate in focusing on the task for which the group was formed. After insecurity has been overcome in the earlier states, the group proceeds, with mutual acceptance and trust, to exchange ideas and information, to define goals, to clarify tasks, and to make the decisions needed for the tasks to be accomplished.

Group Commitment and Involvement   Although the above gives the group's intrinsic "state trajectory," the probability of good group experiences can be greatly increased by considering the following conditions; they encourage commitment and involvement by the participants:

• Sharing is essential. Members' thoughts, feelings, and experiences are the raw materials for the group. Feelings are closely related to emotions. One's uninhibited but discreet expressions encourage openness in other group members. Senior participants with considerable industrial experience seek innovation from the younger members, who are not inhibited by the "realities" of such experience.

Paul B. MacCready designed the Gossamer Albatross, the first human-powered airplane to cross the English Channel (June 1979) [5]. Later, he noted that the main factor in their success was the vehicle's structure. Many of their competitors were significantly more competent in structures; however, regarding his group's advantage, he stated: "It is useful to approach a problem with knowledge of fundamentals but without the deadening influence of prior, detailed expertise and prejudice."

• Expressing negative feelings can, on occasion, also be helpful. Unexpressed, they set up blocks and cause internal distractions.

• Respecting and supporting each person in the group is vital. The more confidence each feels, the more anxiety diminishes and the deeper the group can explore the topic.

- Support needs to be expressed. Do not presume that people somehow know you will be supportive. They will not know unless you show it.
- Putting people down closes them up and is counterproductive for the group. No repartees!
- Positive confrontation is constructive and needed. To confront means to present someone with a new or opposing idea. This promotes agreement, contradiction, or clarification, and shows where people stand and what they feel is important.
- Confront others publicly with their (perhaps unused) *strengths;* confront them privately with their weaknesses. Both will help them develop and, in turn, contribute more to the group's goals. Also, for everyday life, I offer a Praise Principle: When someone does well, let their immediate superior know, preferably by letter or note.
- Accusations and ridicule only engender *hostility* and emotions which block the individual's and the group's progress. Avoid them.
- Avoid forcing your viewpoint by an overbearing attitude or barrage of arguments. Sometimes needed clarifications can lead to this extreme attempt at persuasion. As objectively as possible, present your thoughts and make clear that you want to accept the wisdom of the group. It seems that people tend to compete in greed or graciousness; a gracious approach will usually promote a similar response.
- Keep in mind that the goal is not for any individual to win but for the group to succeed. *It is amazing how much can be accomplished when members put the benefit of the group over individual recognition.* The resulting atmosphere can be improved by members freely, but justly, giving credit to the sources of ideas, even when those sources are outside the group. The NIH (*not invented here*) effect can narrow the horizon of possible solutions to problems.

Finally, creativity contributes greatly to all activities, especially engineering. Section 1-5 stresses that every engineer has some creative talent. The trick is to use two types of thinking, sequentially: first, the exploratory, relaxed, or wandering thought processes; second, the judgmental or analytic. The second type stifles the first, but must be used, finally, to winnow out the chaff.

Communication   This is very important to managers and engineers. The success of almost all human endeavors depends on effective communication. Because of its great importance, an entire chapter of this text is dedicated to it (Chapter 4).

Leadership   This is the process by which one person influences another to perform an act. In management, it is defined as leading with minimal conflict. Conflict produces stress, which below a certain magnitude gives positive results. The relationship between performance and conflict is depicted by an inverted U-shaped curve such as the one shown for stress in Section 1-3. On occasion, conflict might be encouraged, such as competition between groups. Finally, leadership involves the influence to shape organizational goals and encourage others to act toward promoting them.

Power is defined as a leader's ability to influence the behavior of others. In the organizational setting, in order of ascending effectiveness, there are five kinds of power: legitimate, coercive, reward, expert, and referent (charismatic) [6, p. 466].

Legitimate power is authority granted and defined by the organization's hierarchy. This kind of power is based on the follower's internalized values which convince him that the influencer has a legitimate right to influence. This power is the core of the traditional influence system. It appeals to the follower's safety, security, and affiliative needs. Two examples are the teacher/student and the policeman/driver relationship. There is a danger, though, that followers can use the leader's influence to relieve them of thinking for themselves, making their own decisions, and assuming responsibility for those decisions.

The processes of coercion and reward are familiar to us. Coercive power forces compliance by psychological, emotional, or physical threats. Reward power is the ability to give or withhold rewards, such as salary increases, promotion recommendations, praise, recognition, and desirable job assignments.

Expert power is derived from information or expertise. One example is a physician who prescribes medication for a certain ailment, or an engineer with unusual expertise in interfacing a new computer with a customer's older equipment.

Finally, and most effective, is charisma or referent power. It is held by one who* you feel has a warm and caring concern for you. Also, we are more easily influenced by those who have personalities, backgrounds, and attitudes similar to ours.

Frequently a leader will have all five kinds of influence. Their use has been illustrated by behavioral models, such as Douglas McGregor's X and Y theories of leadership, developed in the early 1960s. Extending McGregor's theories, Blake and Mouton presented a managerial grid model. Later, Likert developed a model based on four leadership systems. More recently (1976) McClelland and Burnham have developed a model which describes successful leaders in terms of their needs for power relative to other needs. [7] The newest models show a leader sharing the decision-making power. The first of this type was developed by Vroom and Yetton. That model was later improved by Vroom and Jago [8, p. 519]. We will review these models next.

Douglas McGregor [1, p. 169] hypothesizes theories X and Y as the extreme philosophies of leaders. For autocratic theory X, the leader has these beliefs:

- The average human inherently dislikes work and avoids it.
- Thus, people must be coerced, controlled, directed, or threatened to get them to put forth adequate effort toward the organization's objectives.

Theory X–type leaders tend to centralized authority; the work of subordinates is as structured as possible. These leaders maintain close supervision and use psychological pressure to ensure performance. For democratic theory Y, the leader has these beliefs:

- Expenditure of physical and mental effort in work is as natural as in play or rest.
- External control and threats are not the only methods of influencing people to work toward organizational objectives.
- The average person, in a proper environment, not only accepts but seeks challenges and responsibilities.
- The capacity to use imagination, ingenuity, and creativity in solutions of organizational problems is characteristic of most people.

Organizations characterized by the democratic style have a highly decentralized authority. Subordinates actively share in decision making and enjoy wide latitude in executing tasks.

*Note, whom is not used, for *feel* is used as as intransitive verb: please see "Polished English," Chapter 4.

The managerial grid of Blake and Mouton [9] considers the leadership perspective as having two dominant facets:

1 *Concern for production.* Setting up structure and pressure for performance, such as by assigning particular tasks, specifying procedures, making clear expectations, and scheduling work. This perspective can be associated with authoritarian managers or an authoritarian style. Obviously, this facet is close to McGregor's theory X–type leader.

2 *Concern for people.* Creating an environment of psychological support, and generally promoting subordinates' welfare by generating warmth, friendliness, and approachability. Again, this follows McGregor's model, of a theory Y–type leader.

Consider the graph (grid) in Figure 3-2, with facet 1 on the horizontal or *x*–axis and facet 2 on the vertical or *y*–axis. It has been observed that production $P(m)$ peaks under managers described approximately by $x = 9$, $y = 6$ [9, p. 197]; however, morale $M(v)$ is low. Morale seems to be high and changes only slightly for *x* between 4 and 7, while productivity increases proportionally with *x*.

Interestingly, Blake and Mouton found a high correlation between a manager's style and his or her parents. For good or bad, without extreme effort, we have a strong tendency to be like our parents (or early life parental figures).

Likert characterized organization in four management systems [6, p. 494]. The parental role somewhat parallels the sequence of these systems, as children develop from infancy through adolescence and into the maturity of adulthood.

- *Exploitive-authoritative*—gives commands, accepts no feedback, has no confidence in subordinates
- *Benevolent authoritative*—asks for feedback, either after the decision or (insincerely) before
- *Consultive-democratic*—has moderate interaction, asks for suggestions and ideas, but makes final decision

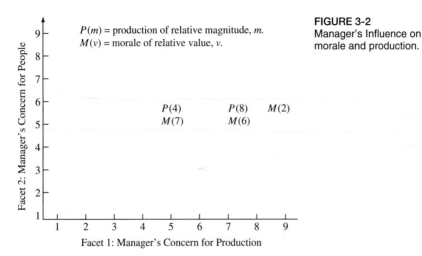

FIGURE 3-2
Manager's Influence on morale and production.

$P(m)$ = production of relative magnitude, m.
$M(v)$ = morale of relative value, v.

Facet 2: Manager's Concern for People

Facet 1: Manager's Concern for Production

• *Participative-democratic*—has complete confidence in letting subordinates make decisions, and then helps implement them

The effective leader uses the type of management system appropriate for the situation but is primarily people-oriented. This leader is not an "over the shoulder" observer, but rather one who defines the task and leaves as much as possible to the worker's initiative, always exercising the least power. These leaders make people feel special and indispensable by appealing to their self-image. They can, however, be tyrannical in emergencies; but normally, by planning, they get acceptance from their workers before such a need arises. Everything is open; nothing is done under cover. They realize that people need and deserve credit for what they do. Under leaders such as these, work is highly satisfying and enjoyable.

Needs Profile of an Effective Leader*   McClelland and Burnham (1976) developed a model that describes successful leaders. It is expressed in terms of their needs for power relative to other needs and according to how they go about exerting their power [7]. The model has four dimensions, each of which refers to a different aspect of the use of power.

The first dimension can be called *need orientation*. Effective leaders tend to have a high need for power and moderate needs for affiliation and achievement. Ineffective leaders have a moderate or low need for power and either high or low needs for affiliation and achievement. This combination of needs associated with effective leaders is not obvious. Many people remark that they can understand why leaders should have a high need for power because the job of a leader is to influence people to do things—to delegate tasks. They wonder, however, why someone working with people should not have a high need for affiliation (a desire to like and be liked by others) and a high need for achievement (a desire to get things done). The reason a high need for affiliation tends to be detrimental to effective leadership is that it is possible for a leader to like people too much. Such a leader finds it difficult to make the unpopular decisions that sometimes have to be made for the good of the group or the organization. Furthermore, highly affiliative leaders tend to be "taken in" too readily by people's tales of woe, true or not. For this reason, they are often seen as playing favorites or as being suckers for hard luck stories.

The problem with too high a need for achievement can also be explained. Need for achievement is defined as the desire to perform a task by oneself and to receive timely feedback on how well the task was performed. However, as mentioned above, the leader's job is to make sure that work is done by others, not by the leader. Leaders with a high need for achievement fail to delegate because they want to do the work themselves. The high need for achievement is responsible for many manifestations of the Peter Principle, which states that people tend to be promoted to their level of incompetence. Excellent engineers sometimes fall victim to this. They are often promoted to leadership positions as a reward for their high performance as engineers. Their excellence at engineering can frequently be traced to the fact that they have a

---

*John K. Butler, professor of management, Clemson University, contributed significantly to the ideas in this section.

high need for achievement. They like to design, develop, and repair devices and see the results of their efforts as soon as possible. When they become managers, they find that they have to let go of the hands-on work; and sometimes months or years pass before they receive any feedback about their performance. Consequently, their need for achievement becomes frustrated. When this happens, *a company loses a good engineer and gains a poor manager,* especially if the engineer never cared much for influencing other people. Thus, leaders with a high need for power and moderate needs for affiliation and achievement tend to be the most effective.

The second dimension of McClelland and Burnham's model relates to the leader's focus of power. Leaders can use power in a socialized manner, for the good of their groups or organizations, or they can use it for personal gain. Clearly, the former tends to be more effective than the latter in terms of accomplishing organizational goals. Self-gainers are too common in organizations—entering positions of power, shaking things up, and then going on to other jobs, leaving the mess for others to clean up.

The third dimension of the model considers leadership style. As with Likert's model, reviewed earlier, McClelland and Burnham's model proposes that leadership style can be defined along a continuum bounded by "democratic" on one end and "autocratic" on the other. This leadership style reflects the extent to which a leader shares power. It is a relatively simplistic notion, but it is consistent with other approaches, including the one we present later. McClelland and Burnham found that effective leaders can use a democratic style, while less effective ones tended to be autocratic.

The fourth dimension is represented by the maturity of the leader, ranging from "immature" to "mature." The model, not surprisingly, indicates that the mature end of this continuum is best for effective leadership. Not so obvious, however, is the definition of maturity: i.e., a four-stage life-cycle process. The first, or least mature, stage is characterized by dependence on others for guidance and strength. In this stage, similar to the situation of a small child, one has no power at all and is completely subject to the power of others. The second stage reflects autonomy, and frequently characterizes adolescents and young adults in their quest for independence from authoritative figures, for control of their own destiny, and for power over themselves. The third stage, influence, is marked by a desire to have power over the behaviors of others. People who have reached the fourth stage of maturity, altruism, have moved beyond the egocentric desire of controlling others. Instead, they feel a need to use their power for the purpose of helping others.

In summary, the profile of an effective leader consists of four dimensions that describe different aspects of power. Effective leaders tend to have a high need for power relative to their needs for achievement and affiliation. They focus on the use of their power for the good of their organizations rather than for themselves. They share their power, using the democratic leadership style. They have reached the stage of maturity in which they no longer desire to control others but, preferably, try to help them. Leaders with this profile tend to reflect the need to influence behavior for the good of their groups or organizations. They tend to be most effective because they get the most out of their people.

We have attempted to clarify the concept of power and stress its importance for effective leadership. Now we will turn to the ways in which leaders can share power with subordinates. We will also look at the situations that call for sharing versus withholding power.

A Model for Determining the Type and Amount of Subordinate Participation in Decision Making   This model is based on situational leadership theory. It concerns only the decision-making part of management and the motivational aspects of leadership. We consider how and to what extent a leader shares power with subordinates in making decisions, and how sharing affects their motivation.

Important decisions can provoke considerable anxiety because we often have to live with their consequences for a long time. Our most crucial life decisions, as reviewed in Chapter 2, involve choices such as which career to pursue and whom to marry. We realize the importance of our having a significant part in decisions that affect our lives. Also, we are likely to accept decisions in which we participate and resist those that are made for us by decree, regardless of their quality.

Some leaders believe that high quality is the only criterion for decisions. They strive for the best decisions possible, even if they know that their followers will vigorously resist putting those decisions into practice. Many of these leaders seem to be totally unaware of, or at least unconcerned with, the feelings of their followers. Other leaders disagree with that approach. They feel that decisions need not be the best possible as long as their followers are committed to the decisions, since the followers will have to implement them.

Who is correct? For example, is it better for a firm's engineering department to have the best laboratory equipment possible, even though the engineers do not like the equipment and resist working with it? Or is it better to have less capable lab equipment as long as the engineers like it and use it effectively? How do you vote? The answer is that "it depends on the situation." More generally, the situation dictates the decision-making style that a leader should choose.

"What about the situation?" you may ask. That is, how can we define the situation in terms relevant to choosing a decision-making style? Furthermore, what are the decision-making styles from which a leader can choose? A third question concerns matching style with the situation. If we have correctly defined the situation and we have a list of possible styles from which to choose, how do we go about choosing the appropriate decision-making style to fit the situation? Vroom and Yetton (1973) developed a model that helps to answer these three questions. This model was later improved by Vroom and Jago (1988) [8].

Vroom and Yetton identify three criteria that can be used to define a situation facing a leader: quality, commitment, and timeliness. The quality criterion refers to how good, or how rational, the decision has to be.

Example 3-2

Suppose a section head supervises 10 equally competent electrical engineers. One of them must be sent to Nome, Alaska, for 2 years on an undesirable assignment. In this case, any one of the 10 engineers could do the job equally well. Thus, in the terminology of the model, there is no "quality requirement." In contrast, if the engineers varied in terms of specialty or in degree of skill, and a specific type of engineer was needed in Nome, there would be a quality requirement.* Please keep this example in mind as we review the other criteria and eight problem attributes.

The second criterion, commitment, means that followers not only accept the decision but will exert a high level of effort in order to make it work and will feel strongly about maintaining rapport with the decision maker. Before commitment to a decision can occur, the decision has to be communicated accurately to the followers. Furthermore they have to understand fully both the decision and its consequences. In the case of the decision about who gets transferred to Nome, commitment would be critical because an uncommitted transferee might consciously or unconsciously sabotage the job, or leave the company for another position.

The third criterion, timeliness, refers to the urgency of the decision. How much time can pass before it is too late to do anything? In the case of the transfer to Nome, the deadline might be defined in terms of the number of days or weeks before someone is needed there (not particularly urgent). In contrast, an out-of-control automated machine, turning out dozens of expensive defective parts every minute, must be adjusted or shut down immediately!

Thus, the decision facing any leader can be defined in terms of its need for quality, commitment, and timeliness. Next, we will address the question about decision-making styles.

In the Vroom-Yetton model decision-making styles (Table 3-1) are defined in terms of how and to what extent leaders can involve their followers in making decisions. There are three fundamental styles: autocratic (A), consultative (C), and group (G). Further more there are two variations on each. They are designated with Roman numerals (e.g., AI and AII). Thus, Vroom and Yetton assume that you, as a leader, can choose a decision-making style from among those in Table 3-1.

The five decision-making (leadership) styles can be matched with the situations. Matching maximizes the probability of effective outcomes. That is, the quality, acceptance (commitment), and timeliness of decisions are likely to be greater when the appropriate style is used. The decision tree portrayed in Figure 3-3 (the later Vroom-Jago model) indicates how the various decision-making styles should be matched with situations. Thus, the model is normative in that it prescribes leadership styles for given situations.

The eight problem attributes are listed below Figure 3-3. These attributes define the situation facing a leader in terms of two of the three situational criteria previously mentioned: the required decision quality and the followers' commitment. Quality requirement (QR), leader's information (LI), problem structure (ST),* and subordinate information (SI) refer to quality. Commitment requirement (CR), commitment probability (CP), goal congruence (GC), and subordinate conflict (CO) refer to commitment.

---

*For a structured problem, the current state, the desired state, and the mechanisms for transforming from the current to the desired state are all known. The decision maker knows what information is needed and how to get it without help from the followers. In contrast, an unstructured problem requires diagnosis to understand the present state, consultation to clarify the goals of the desired state, and creativity for the mechanisms to change from the present state to the desired one.

TABLE 3-1
DECISION-MAKING STYLES FOR LEADERS

AI	(Autocratic I). You make the decision yourself, using the information that you have at the time and without obtaining any information from your followers.
AII	(Autocratic II). You make the decision yourself, after obtaining information from your followers. You do not share the problem with them.
CI	(Consultative I). You make the decision yourself, but only after sharing the problem with your followers and obtaining information from them individually. Your decision may or may not reflect their input.
CII	(Consultative II). You make the decision yourself, after sharing the problem with your followers and obtaining their ideas and suggestions in a group meeting, as opposed to meeting with them individually. Your decision may or may not reflect their input.
GII*	(Group II). You and your followers discuss the problem as a group. Then you become a member of the group, making a special effort not to influence the group to accept your ideas more than others. Your combined expertise forms a joint position which all believe to be best. You agree to accept and support the group's decision.

*"I" indicates no group meetings, not applicable for G.

These problem attributes define the nodes of the decision tree. The combination of answers to the questions at the nodes defines the situation in terms of quality and commitment.

The third criterion, timeliness, is not considered by the nodes. It is considered by the leadership styles (AI, GII, etc.) at the right of the decision tree. The style listed at the endpoint of the tree is the least time-consuming in the set which is feasible in the given situation.

Solution: Let us use the above situation of the transfer to Nome to show how the tree diagram (Figure 3-3) can be used to choose a decision-making style for the section head. Assume that the 10 engineers vary in competence, so that the first node is significant. Thus, the answer to QR is "high." The next question is CR, whose answer is also "high" because the transfer will have an impact on someone's life for 2 years. This "high" branch leads to LI, whose answer has to be "yes" since the leader must be familiar with the competence and specialties of all 10 engineers. This "yes" path goes to the top of the diagram to CP, whose answer is "no" because the transferee would not feel committed to such a decision made by the leader alone. The next node is GC, and the answer here is "yes" if the engineers agree with the goal that the effort in Nome be successful. The answer at the next node, SI, would depend on how well informed the engineers are with the job in Nome. Unless it is a secret mission, the answer to SI is probably "yes," and this leads to GII. These decisions are enclosed by squares at the nodes in Figure 3-3. Thus, our definition of the situation facing the leader can be summarized as "high, high, yes, no, yes, yes," with the appropriate style being GII. (If the answer to GC had been "no," then the appropriate style would have been CII.)

Under GII, the section head would join the group and share the problem with its members. Together, they would generate and evaluate ideas and attempt to reach a consensus, which the section head would be willing to accept and implement. A good

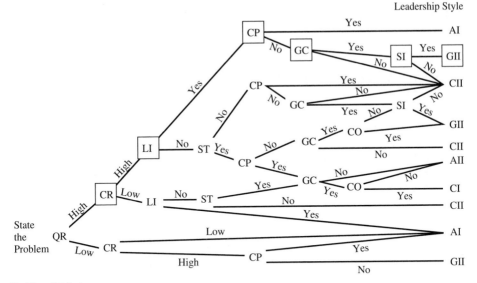

Problem Attributes

QR	Quality requirement:	How important is the technical quality of this decision?
CR	Commitment requirement:	How important is subordinate commitment to the decision?
LI	Leader's information:	Do you have sufficient information to make a high-quality decision?
ST	Problem structure:	Is the problem well structured?
CP	Commitment probability:	If you were to make the decision by yourself, is it reasonably certain that your subordinate(s) would be committed to the decision?
GC	Goal congruence:	Do subordinates share the organizational goals to be attained in solving this problem?
CO	Subordinate conflict:	Is conflict among subordinates over preferred solutions likely?
SI	Subordinate information:	Do subordinates have sufficient information to make a high-quality decision?

FIGURE 3-3
Vroom-Jago decision tree. (Reprinted from *Leadership and Decision-Making* by Victor H. Vroom and Philip W. Yetton, by permission of Leadership Software, Inc. Houston, TX and the University of Pittsburgh Press. © 1973 by University of Pittsburgh Press.)

start would be to share information about the new job and the location, and ask for volunteers from among those qualified.

At this point, the section head will be able to determine whether the engineers share the organizational goals (GC) and whether to use GII or CII. If commitment is important and choice of styles is somewhat uncertain, it is preferable to begin with a more autocratic style (CII in this case) and then move to a more participative one—rather than move in the opposite direction. Followers find it extremely frustrating to learn, after hours of GII decision making, that their leader intended to use CII (hold final decision to one's self) all along; or, perhaps, that the decision has already been made (with an AI and AII style).

Controlling   This process is instrumental in keeping an organization efficiently and effectively progressing toward its goals. It is a managerial application of the automatic control theory prevalent in engineering and biological systems. A familiar example of the latter is the thermal system of the human body. In it, heat is produced (as in exercise) or dissipated (as in perspiring). The human automatic control system keeps them balanced so as to maintain the body temperature at 37°C (98.6°F). A review of automatic control theory in engineering systems will give insight into its application to an organization (an open system). The block diagram (Figure 3-4) is the general scheme for a *closed-loop* engineering control system. There also are *open-loop* systems. These terms should not be confused with the open and closed systems discussed earlier. After defining open- and closed-*loop* systems, we review the definitions to clarify the possible confusion with open and closed systems.

In an open-loop control system no information is fed back for comparison between the value of the controlled output and its desired value. For example, early model washing machines used timers to control the water level (i.e., the level to which the tubs were filled). On the day of installation, the machine's timer was set for a duration so that the rate of water flow at that time and location filled the tub to the proper level. After installation, however, changes could cause improper filling. There might be a change in the rate of water flow, or excess water might have been added to the tub. The point is that in open-loop systems there is no *sensor* (Figure 3-4). Thus, it cannot adapt to changes, internally or externally. In later models a weight sensor is used to "close the loop" and cut off the water flow when a certain weight of water is obtained. As you know, 1 fl oz of (fresh) water weighs 1 oz. In Figure 3-4, the symbol ϵ represents the error or difference between the criteria ($+$) and the sensed outputs ($-$). For the washing machine, ϵ goes to 0 when the water weight rises to the criterion set in the machine's "brain."

To summarize, in an engineering control system, such as that represented in Figure 3-4, the "plant" can be any physical apparatus to be controlled. A transfer function characterizes the intrinsic relation between the plant's output(s) and its input(s). Using control theory fundamentals the engineer designs a controller such that any error ϵ will drive the plant output(s) to agree with the criteria, usually called *references* in control theory.

FIGURE 3-4
A block diagram representation of a closed-loop engineering system.

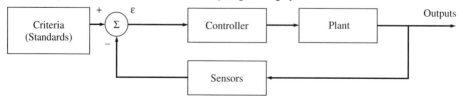

A very simple example of another physical system is the thermostatically controlled home heating (or cooling) system. The thermostat is set for a temperature such as 68°F.* If the house temperature goes below that setting, the bimetallic strip of the thermostat will partially straighten, make contact, and turn on the furnace. The heat added by the furnace causes the house temperature to rise (assuming the house is adequately insulated). The house and its furnace constitute the "plant" for this system. The thermostat in this simple system is its sensor and controller. When the temperature reaches 68°F, the bending of the bimetallic strip will break contact and cut off the furnace.

In many daily activities we use feedback (closed-loop) control techniques: we plan, act, observe results, and modify our actions when it seems that the improvements would justify the extra effort. I am using the technique now. I plan (determine) the information I wish to convey, write, read the material, and rewrite. This concept is developed further in the next chapter on communication.

Managers use these control theory techniques to coordinate and optimize the other three management functions: planning and decision making, organizing, and leading. These techniques are used by management at all levels to maintain organizational progress toward the criteria (i.e., the goals and objectives) set in planning. The concepts are depicted in Figure 3-5.

In an industrial organization, the plant represents the operations system. This system has subsystems, such as purchasing, receiving, transformation, delivery, merchandising, billing, and collection. The first two subsystems, are involved with obtaining feed stock for the core of operations, i.e., for the transformation of goods and materials into products or services, or both. The last two involve receiving money for the products or services. For their effective and efficient operations, managers require continuous information about each subsystem. The management information system (MIS)

---

*Interestingly, concerning the temperature range for human comfort, the lower temperature is approximately 68°F = 20°C; the upper temperature is the transposition of 68 or 86°F = 30°C.

FIGURE 3-5
Controlling in an organizational system.

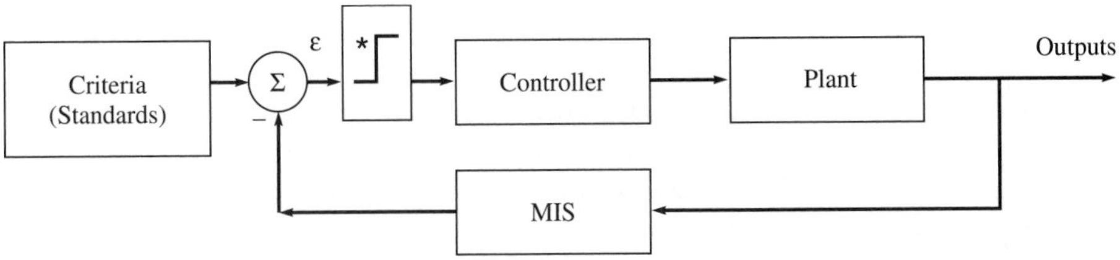

ε = difference between management's standards and the value of the "sensed" outputs.

MIS = firm's management information system.

* Management's discontent threshold for action.

provides managers with that information. It helps management implement the planning, operations, and controlling needed to "stay on course" toward goal achievement. Another MIS is a decision support system (DSS); it provides the "controller" (i.e., management) with the current, relevant data about operations to make the proper decisions. In my youth, I did not understand the paradox "Things have got to get worse before they will get better." If the discrepancy between the outputs and their criteria is not significant, management takes no actions. It is probable that other areas are more demanding of attention and corrective action. The threshold "step" in the model (Figure 3-5) is used to recognize this phenomenon.

Managerial Skills   These are the skills required for a manager to effectively and comfortably carry out the four major phases of the management process. They include technical, interpersonal, conceptual, and diagnostic and analytic skills.

Technical Skills   These skills are concerned with the operations of an organization. Recall that operations are the special activities by which an organization turns inputs (materials, information, human resources, and capital) into outputs of goods or services. These skills are based on thorough understanding of those special activities, especially their theoretical bases. Naturally, your formal engineering education is the first step in developing the required competence. With this knowledge you are in a very advantageous position to benefit from experience and, in turn, further enhance these skills. Technical skills are especially important for first-line managers. To effectively supervise operations, I believe the manager should be able to perform the tasks and fully understand their theoretical bases. It is important to remember, however, that as a first-line manager, your job is to work through others, using interpersonal skills. You cannot and should not try to do all the work yourself. This mistake frequently occurs when an excellent engineer is promoted to a project or first-line managerial position.

Interpersonal Skills   Your technical skills might impress those with whom you work, but to fully achieve your potential, you must interchange pleasantly with others. Probably the greatest factor in interpersonal skills is the ability to understand the position of others. To do this, you must first "put yourself in their shoes" by considering their family situation, cultural background, education, and aspirations. Understanding is the basis of the other two components of interpersonal skills: i.e., the ability to communicate and to motivate both individuals and groups. Interpersonal skills are important at all levels of management. Their importance increases on moving from first-line to middle management and diminishes only slightly at the top level.

Conceptual Skills   These depend on one's ability to think in the abstract. Recall that an *abstract* is the essential qualities or model of the whole. Generally, engineers are strong in these skills, having used mathematical models to both analyze and design systems in their respective fields.

Diagnostic and Analytic Skills   Diagnostic thought is an inferential process. Through it, data are collected from observations and analyzed. From the analysis, the cause and effect relation is inferred. This is a sophisticated way of describing the troubleshooting process used by engineers to guide their technicians. I believe that this skill is best developed from the experience gained in trying to repair devices and equipment. Before graduation, you will have experiments and projects which do not perform as our models indicate they should. Be active in analyzing the problems. Also, as seems reasonable, try repairing autos and various malfunctioning devices as VCR's, stereos, radios and other devices based on engineering principles. Even in failure, you learn.

In the transition from first-line to middle management, the importance of technical skills diminishes and the interpersonal ones become more significant. In middle management, the manager needs (1) to influence people to work as a team and (2) to build coalitions and cooperation. The transition from middle to top management shifts emphasis from interpersonal to conceptual skills. In top management, one needs to think strategically and use the holistic systems approach, i.e., clearly visualize the interdependence and interaction of all components which make up the "big picture."

The engineering curriculum gives the graduate an advantage in all of these skills, except the interpersonal ones. Section 1-4 uses transactional analysis, to present the fundamentals of interacting with others. As stated earlier, engineers, with their high level of intelligence, will recognize the great importance of interpersonal relationships and put forth whatever effort is necessary to develop and master these skills.

3-2 SUMMARY

The activities of management consist of planning and decision making, organizing, leading, and controlling. An organization's goals are set in accordance with its mission. The activities of management keep the organization on course to efficiently and effectively achieve those goals.

There are three levels of management: first-line (supervisory), middle, and top. Managers can also be categorized by areas, such as marketing, finance, and engineering.

There are four skills required by effective managers: technical, interpersonal, conceptual, and diagnostic and analytic. The interpersonal skills are the only ones not strengthened by the formal engineering curriculum. However, all graduate engineers and engineering students should recognize their importance, and, accordingly, exert whatever effort is necessary to gain proficiency in them. Until then, you are oxymoronic (smart, but foolish).

DISCUSSION QUESTIONS AND PROBLEMS

  1 Define a system.
  2 Define an organization as a system.
  3 Define management.
  4 Define a manager.

3-1 LEVELS AND PHASES OF MANAGEMENT

  5 Outline the levels of line management, and give the activities of each level. Which level is largest?
  6 In what areas do staff positions support line managers?

7 Give the responsibilities of the board of directors.
8 List the phases of the management process.
9 Give and discuss the two aspects of planning.
10 Review the six building blocks of organizing.
11 Define leading, and review its four areas.
12 What were Mayo's conclusions from the studies at the Western Electric Hawthorne plant?
13 Outline the need hierarchy proposed by Maslow.
14 Compare Alderfer's need hierarchy with Maslow's.
15 What factors does Hertzberg offer as necessary for worker satisfaction?
16 Discuss attribution theory.
17 Discuss equity theory.
18 Using expectancy theory, evaluate Example 3-1 for your values of E' s, I' s, and V' s. Please state them clearly. You might prefer to consider some alternative other than carpooling.
19 What factor controls the degree to which a group can modify a person's behavior in a given situation?
20 List and discuss the basic kinds of groups.
21 Review each state in a group's "state trajectory."
22 From the 10 conditions which encourage group commitment and involvement by participants, discuss the statement that "positive confrontation is constructive" plus one other.
23 Define leadership.
24 Define power, and discuss the five kinds.
25 Briefly describe McGregor's X and Y theories.
26 Review the managerial grid as an extension of McGregor's theories.
27 Discuss Likert's four management systems as observed in your life, at home, or perhaps in part-time work.
28 In the model developed by McClelland and Burnham, discuss how individual needs influence their effectiveness as managers.
29 Why is participation by those affected important in decision making?
30 Decisions concerning library facilities and hours, student parking, tickets for sporting events, and tuition affect you as a student. Consider one of these or another of importance to you, and use the Vroom-Jago model in Figure 3-3 to decide if the decision-making administrator(s) should include student participation. Justify your conclusions.
31 Review the fundamental difference between an open-loop and a closed-loop system.
32 Review the advantages of closed-loop systems.
33 In the schematic of organization systems (Figure 3-5), the "step function" is used to show that no controlling change is made unless the error ϵ exceeds management's threshold. If the threshold is too small, management might react to trivia ("noise" in the information channel). Present your thoughts on the best threshold level of a "prudent and stable" manager.
34 How does the importance of technical and interpersonal skills change as an individual moves to higher management levels?

REFERENCES

1 Webber, R. A. *Management: Basic elements of managing organizations.* Homewood, IL: Richard D. Irvin, 1979.
2 Griffin, R. W. *Management.* Boston: Houghton Mifflin, 1990.

3 Leap, T. L. & Crino, M. D. *Personnel/human management.* New York: Macmillan, 1989.

4 Napier, R. W. & Gershenfeld, M. K. *Groups.* Boston: Houghton Mifflin, 1989.

5 MacCready, P. B. Reach for the stars. *Technology Review,* MIT, May/June 1981, p. 72.

6 Mescon, M. H., Albert, M.; & Khedouri F. *Management.* New York: Harper & Row, 1985.

7 McClelland, D. C. & D. Burnham. Power is the great motivator. *Harvard Business Review,* March-April 1976, 100–111.

8 Vroom, V. H. & A. J. Jago, *The new leadership: Managing participation in organizations.* Englewood Cliffs, NJ: Prentice-Hall, 1988.

9 Blake, R. R. & Mouton, J. S. *The new managerial grid.* Houston: Gulf, 1978.

BIBLIOGRAPHY

Not out when I started writing, but a very readable treatment of most topics presented in this chapter is

Boone, L. E. and D. L. Kuntz. *Management* (4th ed). New York: McGraw-Hill, Inc., 1992.

Courtesy Fluor Daniel, Inc., Greenville, South Carolina

CHAPTER 4

TECHNICAL COMMUNICATIONS

CHAPTER OUTLINE

Successful business and industrial organizations must gather information and send it to those who need it. Thus, the effectiveness of an organization is dependent on the quality of its reporting, from simple memos to complex technical reports. Regardless of your position, reporting will be a very important part of your work. Without the ability to inform, you will not be successful.

This chapter will be developed as if it were a technical report. Its objectives are to present and demonstrate the techniques of effective communication. The scope is restricted to technical reports.* Whether written or oral, they are used to present *specific* information to *specific* audiences for *specific* purposes. Readers expect to find the information under the appropriate headings and subheadings. While their diligent use helps fulfill these expectations, headings also show the plan and development of the report. In this particular "technical report," acronyms will be used as mnemonics to help you with your writing.

4-1 GUIDES AND ACRONYMS FOR TECHNICAL COMMUNICATION

The development of all projects requires two phases: exploratory and implementation. When these two phases are completed, we can add a third: improvement. This is true in writing technical reports. The usual steps are gathering information, planning your report, writing, and then revising. We immediately recognize that the first two steps, gathering and planning, constitute the exploratory phase; the last two, writing and revising, are (respectively) the implementation and the improvement.

While gathering the information, it is convenient to compile the bibliography, as reviewed in Section 4-2. This practice makes the writing task easier later on.

Exploratory Phase: MAPS

The acronym *MAPS* will help you in the exploratory phase. The letters stand for message, audience, purpose, and scope. They guide us in this creative phase of gathering information and planning the report.

A judgmental attitude squelches creativity; however, we must be judgmental for a short period of time at the very beginning of our endeavor. Technical writing requires considerable time and effort even after you have become proficient. Thus, it is prudent to first determine whether each endeavor will make an adequate contribution to society, to your firm, or to your professional growth, and justify your time and effort. Later, analysis is used to sift the gathered ideas and fit them in a plan designed to achieve your goals and objectives. Toward these goals, we must first determine the message, audience, purpose, and scope of the report.

Message Your message directs the gathering and creative selection of information from three major sources: personal and professional experience, library sources,

---

*Other types are test reports, progress reports, and financial reports.

and original research. [Frequently this last item is your purpose; of course in the report you are reading now, the information comes from the first two sources].*

Here the message is as follows:

- How to write a creditable technical report
- How to refine future work
- How the commonality between written and oral reports can be used to present the fundamentals and effective techniques for oral reports

Audience   This consists of graduate engineers, university students, and others of comparable sophistication who need to know the essence of technical communication (i.e., every professional).

Purpose   The purpose is to give you adequate techniques for creditable work and the motivation to develop high competence and polish in oral and written technical reporting. The greater competence will come from further study of the references; the polish will come with the imagination produced by reflections; and the satisfaction, from using the techniques of this chapter.

Scope   This is limited by the fact that the report is written at the introductory level and considers only technical material; however, it is complete in giving the guides and general information necessary to produce creditable work.

Procedures: PWRR

The procedures for the implementation phase are correlated with the acronym *PWRR*. (Just remember *P*acific *W*estern *R*ail *R*oad.) The letters, however, stand for plan, write, relax, and revise. We will simply review them here; details are presented in Section 4-2.

Plan   This is the single most important aspect of any project and should be considered *first!* The plan really begins in the exploratory phase, but you will be modifying it throughout your writing. Your every step is guided by this plan, and time is not wasted "wondering what to do next." Many successful authors spend as much time planning as they do writing. In the past, they probably developed their plan on handwritten or typed sheets, which were inconvenient to organize and change. The many word processors now available make for an easier, more relaxed approach. They offer extreme ease in modifying the text (for example, inserting sentences or paragraphs anywhere), and allow you to store the entire report on disks.

From the outline and the first draft to the final copy, capturing and retrieving ideas are greatly facilitated by using a word processor. The ease with which the text can be saved and changed tends to remove anxiety about typographical errors, misspelled

---

*One important principle of writing, discussed later, stresses the elimination of unneeded words and phrases; here they are enclosed in brackets [and would be deleted in the final draft].

words, or poor placement of ideas. This lack of anxiety promotes creativity, and the general efficiency of information handling increases productivity.

With the word processor you can control the volatility or permanence of the text. As you type your ideas and organize the gathered information, it is stored in the computer's memory as a "working copy." A portion of the page just before the line you are typing is displayed on the monitor's screen; however, any information stored in the working copy can be easily scrolled into view. Furthermore, by the use of simple key strokes, letters, words, or blocks consisting of phrases or paragraphs can be added, deleted, or moved anywhere in the working copy.

Important: The working copy and any modifications you may have made are lost if the computer is turned off or power is otherwise lost. With a little care, this is no problem. The copy is easily made permanent by storing the working copy on a disk; a filename must be assigned to the information for retrieval. To avoid losing any of the work you have done, frequently "save to disk." Word processors like the IBM Display Write, Word Perfect, and others have a quick save which is very convenient. One last caution: Saving new material under the same name as old material "wipes out the old," although most processors first advise of the impending overwrite and ask you to confirm your intentions. Thus, it is wise to have a "backup" disk and to update it after each change on the manuscript disk. When a disk is nearly full, however, trying to exceed its capacity could result in losing the file you are updating. This is a special problem with Word Perfect, it stores the new before deleting the old. Thus, about twice the document capacity is required on the disk.

Finally, in planning, the basic outline of a paper (introduction, body, and conclusion) is given below in Section 4-2. You might find it best to start with a topical outline. Always begin with the main text, for it is the core of the report; the other parts are designed to support it. In Section 4-2 we consider details of planning as we continue the development of this chapter as a technical report.

Write   This task is usually less stressful than planning; guided by our outline, we simply write a rough draft for each section. Write fast! Get your ideas on paper (or in the computer) while you are in a creative mode. Ideas easily slip away when you are distracted, even in the few seconds needed to check punctuation or synonyms. At this point, give little thought to spelling, grammar, etc. This is easier to say than do, for while recognizing the value of this approach, we catch ourselves picking up a dictionary or wondering about a comma. Avoid this tendency, for writing is dynamic. We make false starts and have sudden inspirations. Remember that the rough draft is mainly for your use only; after completing it, relax.

Relax   Put your manuscript aside for at least 5 or 6 h, preferably a day or a week, if possible. Relaxing after completing a task is both good time management and good mental health, that is, satisfying and motivating. While resting, your mind is recharging; but of greater importance is that you become more objective in making a critical review of your work. In business and industry, most people will have to do something else rather than rest. The important thing is to get away from your draft for an extended period of time.

Revise   Revising is the most important aspect in producing top-quality reports. As implied throughout this book, all endeavors are accomplished easiest and most pleasantly in small increments. Thus, revise in many steps; in each, recycle through *PWRR*. At this point the planning aspect will be minor. Reflect on the objective of technical writing: to present specific information to specific people for a specific purpose.

Relevant to the last statement is a basic goal of technical communications: *Give the audience the most information of value while requiring the minimum time and effort.* Conciseness in writing is an outstanding virtue, greatly appreciated by the busy engineer and manager. But be warned: Writing that is too terse can cause problems for the reader. It can require extra time to analyze and supply what you left out in your attempt to save readers' time. We review this below when we consider your reader.

In revising, think about the following:

- Accuracy, clarity, organization, and logical presentation
- Unity and coherence in paragraphs
- Conciseness in grammar, spelling, and punctuation
- Smooth transitions

These are discussed below, not in order of importance but in the order which seems to facilitate revising. This is a feedback process. All excellent writing involves ever increasing circles of feedback: from sentences, to paragraphs, to sections, to the whole report. While writing we recognize other topics that should be included in our outline. Regarding revisions, only our judgment can guide us in deciding if the expected improvement will justify the effort and time for another cycle.

First Revision   Read each section's rough draft for technical accuracy. Do not be distracted by any other details. Check closely for valid statements; truthfulness is the soul of science and engineering, the essence of professionalism. Avoid misleading statements and other forms of misrepresentation. Scientific honesty reflects on your character as a trustworthy engineer.

Also check for errors in equations and in mathematics. After corrections, have another rough draft typed. If you do your work with a word processor, "run off" a copy of your revised draft. Relax; then we are ready for the second *PWRR*.

Second Revision   Strive for clarity. Write not so that you *can* be understood, but so that you *cannot* be misunderstood! Make all sentences as simple and concrete as possible; avoid implications and jargon not generally known to the intended readers. Sentence length strongly affects the clarity of written material. Here is a good rule: Avoid writing sentences which have more than 20 words. Never form a sentence with more than 30, and never follow a 30-word sentence with one over 20 words. You have probably observed that a line of type has 10 to 12 words. Thus, a proper-length sentence should be about 2 lines. Occasionally, a sentence over 2 lines long is suitable.

The technical report is for a specific audience; using your knowledge or an estimate of their technical level, set the level of your writing slightly lower. You must know your audience to match the report to their interest, need, and technical level. Every sentence should reflect your empathy for the reader. Concerning the use of *he* or *she,* after consulting references [1, 2] and colleagues, I think it seems best to alternate equally between the genders or to use *he or she* where appropriate. Sometimes *they* can be used by making the sentence plural. Following this revision, again run a copy of the new draft, relax and recycle through *PWRR* for your third revision.

Third Revision Good organization and logical development promote ease of reading and assimilation of material. Carefully emphasize the main points. Underlining can be quickly overused. Liberal use of headings and subheadings can lead the reader through the thought processes you wish to present. This revision could indicate the need for new sections. It is important, however, to keep the readers' needs and interest uppermost in your thoughts. Ask yourself if each section, paragraph, sentence (yes, even each word) contributes to these needs.

Fourth Revision The paragraph is a tool in organization; you might say that it is a form of punctuation. It must be unified; that is, all of its sentences must relate to *the* topic of the paragraph. Since the paramount objective of technical writing is to inform with the greatest possible ease, do not obscure the topic of any paragraph. Put the topic sentence *first*. It should immediately present the central idea. All other sentences must support and develop it. In well-organized papers and books, the reader should be able to get a summary by reading the first line of every paragraph.

Fifth Revision Conciseness is necessary but not enough to fulfill the basic rule of giving the reader the most valuable information while requiring the least time and effort. To achieve conciseness, ask yourself how much can be deleted without interfering with the readers' understanding, needs, or ease of assimilation. We humans are constantly seeking recognition by writing to impress more than to inform.

Sixth Revision Errors in grammar, punctuation, or spelling spoil what might otherwise be an excellent report. When in doubt, consult the dictionary. A spelling check routine is another advantage of word processors; such a routine also catches many "typos." Section 4-2 contains a short guide to grammar and punctuation.

Seventh Revision Use of transitional words, phrases, or sentences makes it much easier for the reader to follow your thought pattern. Good writers use transitional words between phrases, sentences, and paragraphs; however, be careful that your transitional sentences do not distract from the topic sentence.

Further Revisions It seems that all writing can be made clearer and presentations more logical with another revision, this time by a competent colleague or perhaps yourself after a week or a month. As in all endeavors, we must use judgment in determining if the expected "gain" is worth the time and effort. In my opinion, the revisions listed above are the minimum for a high-quality report. The effectiveness of a report seems proportional to its quality. Thus, quality should match importance.

The Engineering Approach to Technical Communications

As presented in the Preface, the scientific or engineering approach* is the underpinning of every section of this text.

---

*Since the scientific or engineering approach is widely applicable to interdisciplinary systems, including socioeconomic ones, it is frequently referred to as the *systems approach.*

We discussed this approach in Chapter 1; here we will define it and focus its use in the area of technical communications. It consists of five steps [3]:

1 Formulate the problem; that is, determine the requirements and objectives sought.

2 Using the relevant fundamentals (for example, the physical and socioeconomic cause and effect relations), develop a model of the project or situation. Then, considering the requirements, generate alternative methods to achieve the system's objectives.

3 Develop ways to implement the methods which pass the test of preliminary analysis.

4 Simulate or carry out their implementation, and evaluate each against the requirements and objectives.

5 Use differences between the results of the best alternative and the objectives to decide on modifications for improving the next generation of solution.

We immediately recognize that this approach leads us to the optimum solution: the one at which the gain of further improvement does not justify its cost.

In technical writing, the *message, purpose,* and *scope* in *MAPS* are the objectives and requirements in step (1) of the engineering approach. Assessment of the audience is part of step (2). In *PWRR,* the plan or outline is an additional part of step (2); writing and relaxing are steps (3) and (4). Finally, the revisions are the iterative modifications of step (5). In the last section, we reasoned that seven revisions might be optimum.

4-2 TECHNICAL REPORTS

In this section we will continue to develop Chapter 4 into a technical report, following the guidelines discussed in Section 4-1. I have already reviewed *MAPS,* the results of my exploratory phase. We now need to organize the information we wish to present in a form or format that is easy to read and to access.

Format

Although the format is often set by your firm or the client, the one given below [4] will generally be close to it. In the report, there are three major sections, the front matter, the main text, and the back matter. The subheadings below show this grouping. The first one and the last support the main text, which is the place to start, using *PWRR.* Furthermore, as stressed below, start with the body.

A FORMAT FOR TECHNICAL REPORTS

Front matter
    Letter of transmittal
    Title page
    Preface
    Table of contents
    List of figures
    Abstract or summary

Main text
  Introduction
  Body
  Conclusions
Back matter
  Bibliography and references
  Appendix
  Index

Written Report

Writing the technical report is made easier by an outline. From it, each section and subsection is developed. First, let us review an old principle.

There is a pattern which was given by a superb country preacher; you probably know it. When asked the secret of his success as a speaker, he said, "Tell them what you are going to tell them, tell them, and then tell them what you told them." This pattern can be identified in the *main text* as the introduction, body, and conclusions. The principle is so important that it should be used to check every paragraph, section, and (of course) the whole report.

Sequence of Planning and Writing   Start with the body of the main text, for both the introduction and conclusions depend on the material in the body. Planning and writing are dynamically coupled; we can improve each while doing the other. The first step in planning is an outline. Organize and develop an easy-to-follow order for the presentation of your gathered information. List the most important points, those to be emphasized. Do not hide them under a load of supporting but distracting material. Append that material where comprehension of the main ideas permits. Conveniently, this list of important points serves as a rough draft of both your outline and the table of contents; the latter is discussed below.

There are three types of outline: topic, sentence, and paragraph. In the topical outline, each entry is a phrase or a single word. This type is best for technical writers for it does not require as much detailed thinking as the others.

The fundamental principle of outlining is division. Divide the subject matter into the major concepts, their supporting development, and additional information. Reference [5] shows this is analogous to mathematical equations. Consider the total report as TR and its major parts (the important points referred to above) as MP1, MP2, ...
  Then

$$TR = MP1 + MP2 + \cdots$$

Similarly, the major parts have supporting parts, SP1, SP2, . . .
  Thus

$$MP1 = SP1 + SP2 + \cdots$$

and similarly for MP2, MP3, ... Again a task is made easier by doing it little by little. The table of contents for this chapter and all the others are nearly the same as their respective outlines. The differences are discussed in the section, headings, and subheadings. Now let us return to the report's format.

Formal Reports   In a formal report, all of the parts in the format above should be considered and are usually included. We will now discuss each.

The **front matter** consists of the parts listed earlier in "A Format for Technical Reports": letter of transmittal, title page, preface, table of contents, list of figures, and abstract or summary.

The *letter of transmittal* is generally directed to a single person, the primary recipient of the report. Immediately, the letter should identify the report and give other information such as the relevant work or project. It should include the issue date of the report and be signed by the person responsible for its generation and transmittal. Furthermore, it should include the title of the report; contract, work, or project number; the number of copies accompanying the letter (or sent under separate cover, each with a copy of the letter of transmittal); and distribution, that is, names and addresses of other parties to whom you might wish it to be sent.

Title pages, like report covers, dignify the report; however, more important, they provide identifying information and help orient the readers to their reading task. They should be attractive and well designed, that is, symmetrical and not crowded.

The *preface* is similar to the letter of transmittal but is for general readers. Like the letter of transmittal, it introduces the reader to the report. It should be brief but include a statement of subject and purpose, and acknowledgments. Ethics demand proper credit to colleagues for ideas and assistance, and credit for any material reproduced from other sources. These reproductions also require written permission from the source(s).

A *table of contents* can perform three significant functions. First, it indicates the page number on which each major topic begins; that is, it helps the readers to find information. Second, it shows the extent and nature of the topical coverage. Finally, and perhaps most important, it serves as an outline in the planning and writing of the report. The outline can help emphasize topics by the order of their presentation. Even the final draft should be checked for its conformity with the emphasis of material as designated by the outline.

A *list of figures* is used if a report has 10 or more illustrations. Placed on the page with the table of contents, or separately, it gives their titles and page location.

Since the terms *abstract* and *summary* are often used synonymously, they are treated together. However, they are distinctly different. First, there are two types of abstracts: descriptive and informative. The descriptive merely tells what the report is about and what it contains. This type usually requires only a brief paragraph and perhaps no more than a sentence or two. The informative abstract is truly a summary and is more appropriate for technical writings. It is actually a minireport, a condensation of the whole which gives most of the salient facts and the more important information on which conclusions are based. Naturally it also gives those conclusions. Generally, only the supporting facts and explanations are lacking. I wish to stress that the reader should be able to quickly get the gist of the report through a clearly worded (for the intended reader), quick-moving abstract or summary.

Although written last, the abstract or summary appears at the beginning of the report, sometimes on a separate page but generally on the first page of the main text.

In our "technical report," for the front matter, only the table of contents is needed. The outline for this chapter, and, of course, the others are offered as examples.

The **main text**, as noted in the format above, consists of the introduction, the body, and the conclusion. They are required in all reports.

The *introduction,* like all other parts of the report, (except the bibliography), is designed to orient and help the reader understand the material presented in the body. Accordingly, it should immediately announce four things: (1) the subject, (2) the purpose, (3) the scope, and (4) the plan of development.

1 The subject is announced, with additional background as needed.

2 The purpose tells the readers why the subject is presented. They can immediately determine, "Does this paper have a high probability of giving me information of value?"

3 The scope, by announcing how broad and conversely how limited the treatment, indicates the level of competence required by the reader.

4 The plan of development gives the organization of the paper; in turn, the reader has a guide to follow through the body of the report. Because the psychological principle of reinforcement is operative, comprehension is easier.

The *body* is naturally the longest section of the report. The message and purpose largely determine its form. The usual rhetorical forms are (a) the technical exposition, (b) technical description, (c) narration, and (d) argumentation.

a Exposition is the most common approach in a technical report. Often you will be called upon to explain aspects of your work. Writing good exposition hinges mostly on *clarity* and *completeness.* These two characteristics can best be achieved by mastering five techniques: exemplification, definition, classification and division, comparison and contrast, and causal analysis. You have probably already been using these.

Exemplification consists largely of a series of generalizations supported by examples. There are two common ways to use examples. The first is to give one or two well-developed examples. The second is to give a series of short examples. Examples are *not* proofs, but they give your writing life and make the generalizations more believable.

Definitions are frequently needed in technical writings; use of simpler everyday words for specialized terms could cause the reader more trouble. You should define any term you feel is not in the reader's normal vocabulary.

Definition is usually accomplished by putting the term in a class and then differentiating it from other members of the same class. If you were defining Dachshund, for example, you would say that it is a dog and then give characteristics which distinguish it from other dogs.

Term = genus or class + differentia

An ohmmeter is an indicating instrument that measures the resistance of an electric circuit

A Btu is a unit of heat energy that equals the amount required to raise 1 lb of water 1°F. *Btu* stands for British thermal unit. 1 kWh (kilowatthour) = 3213 Btu

When writing for nonprofessionals you may decide that a less precise, more understandable, but correct definition is desired. For example, you would define *Btu* as a quantity of heat energy.

Classification and division is a writing technique which uses similarities and differences to condense a large group of items; thus, the repetition and labor required are significantly reduced. To a small extent we have been using that technique in this section. First, the class of rhetorical forms was divided into exposition, description, narration, and argumentation. Under exposition we reviewed five subtechniques, of which *classification and division* is the third.

Comparison and contrast are used to show how one thing is like another (comparison) and how they are unalike (contrast). The strength of this approach is that it can be used to relate the unfamiliar, such as a technical phenomenon, to a familiar experience. For example, it requires slightly more than 300 Btu to heat 1 gal of water from room temperature to bathing temperature (110°F). Thus a 2-min shower (8 gal per min from an unrestricted household shower head) or a 4-in tub bath requires 5000 Btu (about 13 cents if the water is heated by electric energy, at 8 cents per kWh).

In causal analysis, a proposition is stated and then defended by inductive or deductive reasoning.

In *in*ductive reasoning, one takes *in* a series of particular facts and, from them, moves to a generalization. For example, for over 20 years, I have observed that second children in a family, more specifically of the same sex, tend to be extroverts; they give more attention to and receive gratification from externalities, that is, people and material things outside the self.

In *de*ductive reasoning, one moves from the general to the particular. From the generalization proposed in the last paragraph, which might be called the *second-child syndrome,* parents could deduce that a second child would probably be extroverted, even to the extent of being aggressive. Consequently, parents might want to give such children more love and attention to counteract their feeling of having to "fend for themselves."

In the above example, induction was used to present a general proposition: second children tend to be more extroverted. Then deduction was used to determine characteristics parents might expect of a second child. Furthermore, deduction was used to show how these characteristics could be kept within desirable limits. Consider the following possible fallacy.

Note: There is danger in the careless use of causal analysis. Do not draw inferences from insufficient evidence. The logicians call this fallacy *post hoc, ergo propter hoc* (i.e., after this, therefore because of this).

A second danger is reverse logic. Consider the syllogism:

1 All cows have four legs.
2 Bessie is a cow.
3 Bessie has four legs.

Converse (reverse) logic is not valid reasoning; although it could incorrectly be applied to a valid statement, it is just the "argument" that is invalid.

1 All cows have four legs.

2 Dan Patch has four legs.

3 Therefore Dan Patch is a cow. (Dan Patch in reality was a famous trotter, a horse.)

Attention to these dangers helps the astute reader to discover these flaws. In turn, scientific honesty requires that writers be careful to show where they are certain and where there is doubt. These dangers should in no way deter one from reporting observations and generalizations, although special caution must be used to employ proper qualifications, such as the "tend to be," "might," and "probably" used above regarding the second-child syndrome. Recall that nonprofessionals want certainty, but the experts are content with probabilities.

The previous presentation on technical exposition is rather detailed because of the importance of explanations in technical writing. Frequently there is also need to describe, to narrate, or to persuade. These will be considered, respectively, using their rhetorical forms: technical description, narration, and argumentation.

b Technical description is used to effect a rational understanding of some mechanism or place, to give significant physical details, and, in the case of mechanisms or devices, to describe their function. The description follows the common pattern:

1 Divide the object described into its component parts.

2 Describe each of the parts. If complex, they in turn may be divided into their component parts.

3 Show how all the parts work together.

4 Use graphic illustrations when they have real worth.

5 Use analogy to simplify the reader's task.

c Narration tells a story—in technical writing, a factual one. It simply relates a series of happenings or actions (such as those in an experiment or process) in chronological sequence. The author gains informality and ease of style through use of the active voice in the indicative mood, using the personal pronoun "I" as appropriate.

d Argumentation is used when your purpose is to convince the reader of the truth of a proposition. Generally, argumentation and persuasion are distinguished by the technique used. The latter appeals to emotions, whereas argumentation appeals to reason. Of course, in technical writing, the appeal is to reason.

The format for argumentation follows:

1 State the major proposition.

2 Support it by two or three minor propositions, which in turn will be supported by verifiable facts and statements from recognized authorities.

The *conclusion* is the last major section of your report. Like the introduction, it should emphasize the most important ideas or the conclusions of the report. Make this section strong and as specific as proper. Its three main functions are to (1) summarize the report's major points, (2) state your conclusions, and, as appropriate, (3) give your recommendations. Finally, *do not* introduce new material in the conclusion!

Back matter consists of the bibliography or list of references, and the appendix. As discussed earlier, the appendix is used to supply information that is necessary for the report's completeness but would distract the readers' train of thought if included in the normal presentation. An appendix is not needed in our "technical report"; however, a bibliography or list of references is included.

Footnotes and a bibliography or list of references have no standards; however, the recommendations below are drawn from current technical sources, and their use is consistent with a polished report [6, 7].

These methods of documenting your report satisfy two requirements: (a) the ethical obligation of indebtedness to nonoriginal sources and (b) the reader's convenience in seeking additional information. Also, the reader can recognize how your work differs from and supersedes other work. The choice between documenting by footnotes with a bibliography or using a list of references is discussed below.

Footnotes meet two needs. First, they can be used to supply special information and definitions which would obstruct the reader's thought process if included in the text. Examples of this informational type are in Section 4-1.

Also, footnotes are used in documenting nonoriginal sources. When used with bibliographies, as suggested in this "report," footnotes can be as brief as possible. On the other hand, when no bibliography is used, first footnote references to sources must be complete. Let us consider first the brief footnotes used with a complete bibliography (reviewed in the next section). For brief notes, the first reference to a book should contain the following:

Author's surname,
Book's title (underlined), and
Page reference.

Commas should be placed after the author's name and the title, with a period at the end of the note. In typed material, the underlining indicates copy to be set in italics. If you reference sources by different authors with the same surname, their initials must be added. The same form is used for references to magazine articles except that the title of the articles is enclosed in quotation marks, not underlined, and the comma is placed inside the last quotation marks. Below are two notes, the first for a book and the other for a magazine:

[1]Mali, *Writing and Word Processing . . . ,* pp. 73–74.
[2]Eisenberg, "Mastering On-the-Job Writing . . . ," pp. 25–28.

Compare these with their bibliographical entries below.

Mali, Paul, and Richard W. Sykes. *Writing and Word Processing for Engineers and Scientists.* New York: McGraw-Hill, 1985.
Eisenberg, Anne. "Mastering On-the-Job Writing Assignments." *Graduating Engineer,* March 1984, 25–28.

Subsequent footnotes for the same sources may be briefer (only name and pages):

[5]Mali, p. 93.
[7]Eisenberg, p. 32.

If subsequent references to the same source immediately follow, the abbreviation *ibid.* (for the Latin word *ibidem,* meaning "in the same place") can be advantageously used. However, if the subsequent reference is different from the immediately preceding one, this abbreviation is usually not advantageous. Use of *op. cit.* ("in the work cited") or *loc. cit.* ("in the place cited") is discouraged. Their use is inconvenient to readers since it forces them to search back through footnotes.

If only a few footnotes are required, as is frequently the case for technical articles, neither a bibliography nor a list of references is necessary. In those cases, the footnotes should be complete. The following rules apply:

For a book:

Name of author as it appears on the title page, with given name first
Chapter title, if needed
Title of book
Edition, if not the first one
City of publication
Name of publisher, exactly as on the title page of the book
Date of publication
Chapter or page of book referred to

Omit *vol.* and *p.* or *pp.* when both are given; for example, without the abbreviations, use I instead of *vol.* and 22–24 for page ranges.

For an article in a magazine or technical/scientific journal, the entries differ from those for books only in that the title of the article is enclosed in quotation marks rather than underlined; however, the title of the magazine or journal is underlined, again so that the copy will be set in italics. For an illustration, please see the earlier bibliographical entry for "Mastering On-the-Job Writing Assignments."

Additional details for the mechanics of footnotes follow:

1 Number footnotes consecutively throughout your report, using arabic numbers raised slightly above the line (superscripts).

2 Type or print a heavy line part way across the page, perhaps 6 to 10 spaces in length, between the last line of text and the first line of the footnote. Allow 2 spaces above and below the line.

3 Place the superscript numbers after the word, paragraph, section, or quotation to which they refer.

4 Indent the first line of each note 5 spaces, but begin additional lines flush with the left margin.

5 Single-space a footnote longer than 1 line, but double-space between notes.

6 In the footnote, the author's name need not be repeated if already given in the text.

The *bibliography* [8] is an alphabetized list, according to the last name of the senior author (if two or more are given). It should include all written sources you have consulted, from books and magazines to pamphlets and bulletins. It is not necessary that they all be cited in the text of the report. As suggested earlier, and stressed again here, developing your bibliography or list of references while gathering information from written sources avoids later hassle and greatly reduces the effort required. It is then easy to put the entries in the format presented in the next paragraphs.

For a book or any independent publication, such as a pamphlet, the following is a proper order of entry:

Name(s) of author(s), with the last name of the senior author in inverted order. Names of any additional authors are entered in normal order, with first name first.

Title of the book, underlined to indicate that it is to be set in italics.

Place of publication.

The publisher's name and the date of publication.

For an article in a magazine or technical/scientific journal, the entries should be as listed below:

Name(s) of author(s), with the last name of the senior author in inverted order. Names of any additional authors are entered in normal order, with first name first.

Title of the article, enclosed in quotation marks.

The title of the periodical, underlined to indicate that it is to be set in italics.

Volume number and/or date of publication of the issue in which the article appeared, and page references.

The bibliographical entries, first for a book and then for an article, are repeated below from the earlier examples.

Mali, Paul, and Richard W. Sykes. *Writing and Word Processing for Engineers and Scientists.* New York: McGraw-Hill, 1985.

Eisenberg, Anne. "Mastering On-the-Job Writing Assignments." *Graduating Engineer,* March 1984, 25–28.

Note that the bibliographic entries differ from the footnote entries in only three ways: (1) The name of the author is inverted, i.e., last name first (first author only if there are more than one). (2) When *book* volume, chapter, or page numbers are given (not the case in the above example), they follow the title instead of appearing at the end. (3) A period follows each main element: authorship, title, and publishing data.

Furthermore, in a bibliography, each entry is single-spaced and the first line begins flush with the left margin. When needed, additional lines are indented 3 spaces; there is double-spacing between entries.

Finally, in referencing an article when no authorship is given, enter it alphabetically according to the first important word in its title. Frequently, magazine articles are written by staff members without indicating an author.

For the *list of references* [9], the works are listed in the order in which they are cited and are numbered accordingly. The citation numbers, corresponding to the numbered reference in the list of references, may be superscripted (elevated a half space above the text) or enclosed in parentheses or brackets (as in this text). As for the bibliography, the citation numbers are placed after the word, paragraph, section, or quotation to which they refer. Unlike the bibliography, in the list of references a given number corresponds to only one reference; thus, after its first citation, that number can be used anywhere in the report to cite that particular reference. To reduce the number of entries in the list of references, for citing different pages in the same reference, I use the reference number followed by the page number, e.g. [6, p. 42]. The "p" may be left off.

Furthermore, in a list of references the respective numeral is placed at the left margin, followed by a period; the entry is then started after 2 spaces. When needed, additional lines are not indented but start directly below the first line.

Should you use footnotes with a bibliography, or a list of references? The differences are small.

Explanatory footnotes will be needed in most technical writings; here, however, we consider the type used for referencing. As noted earlier, when coordinated with a bibliography, these footnotes can be brief, and briefness is important when there are many notes. Furthermore bibliographies include references that were not cited in the report but were of value in your general understanding of the material. Readers may find them valuable when seeking further information.

In a list of references, the citations are numbered in the order in which they appear in the report. This approach is easy; the references do not need to be alphabetized or realphabetized if a reference is added or overlooked. Also references that are not cited in the report can be included by having two lists: one titled "Cited References" and the others "Biography" or "Additional References."

Since the differences between the methods of documentation are small, your personal preference may rule. I chose the list of references.

Informal Reports [10] These are frequently adequate for many technical matters requiring reports of less than 10 pages, such as project progress reports or reports identifying technical problems. The informal report is usually in the form of a letter or memorandum; thus, it has no cover, letter of transmittal, or table of contents. Nevertheless, it is important and justifies your considering *MAP* and *PWRR*. Of course, these should require much less effort than for the formal report.

Letter forms are similar to the business letter:

1 Heading, your address, and date of report.
2 Inside address and the recipient's address.
3 Attention line, which directs the letter (report) to an individual in the addressee's organization who the writer knows *has the authority to deal with the subject report.* Use of an attention line generally expedites the letter's routing to the responsible person. The line is placed first in the inside address, flush with the left margin. This position facilitates envelope addressing when using a word processor.
4 The salutation is located a double space below the inside address and flush with the left-hand margin. Although the comma is suitable for informal letters, in both the business letter and the informal report the salutation should be followed by a colon. A personal salutation is used only when a person's name heads the inside address, not when the letter (report) is merely directed to that person's attention.

5 The body is, of course, the essence of the report. Although unlabeled, it should have clearly recognizable parts: introduction, the message, summary or conclusion, and a closing, perhaps a single line, to avoid abruptness.

6 The complimentary close is a formal way of signaling the end of the letter. It should correspond in formality to the greeting. *Respectfully yours, Very truly yours,* and *Cordially yours* are commonly used.

7 Organization name refers to the organization issuing the letter (report). This is typed in full capitals, a double space below the complimentary close.

8 The typed signature is located below the organization name, with adequate space for the written signature of the report author.

9 The title of the signer appears under the typed name.

10 The identification symbol is a combination of the writer's initials in capitals and those of the typist in lowercase letters. These are separated by a colon (:) or slash (/). This symbol is placed at the left margin, either directly opposite the typed signature or 2 spaces below it. If you do the typing, use the letters *ms,* the abbreviation for manuscript, instead of the typist's initials. When there are enclosures with the letter, the abbreviation *Enc* is typed just below the identification symbol. Also below the enclosure symbol, *cc* with corresponding names is typed if copies are distributed.

The above information is applicable to all business letters. Only two other details might be noted for the informal technical report letter: the reference line and the subject line. Both are located, one below the other, to the right of center between the heading and the inside address.

The reference line shows the request, authorization, or inquiry that prompted the report; it saves the addressee significant time in searching.

The subject line briefly and accurately states the subject. It is a courtesy which saves time for file personnel at both the sending and receiving organizations. It is comparable to the title of a formal report and allows a reader to determine quickly what the report contains and if it concerns the reader.

Memorandum forms, or memos, are usually used only within the author's organization and are more informal than the letter report. Nonetheless, as in all communications, *MAPS* and *PWRR* should be the guide; that is, determine your objectives (the message and purpose) and those of your readers. Memos are limited to communications of small scope. The format is often preprinted, with each item on a separate line (*DATE, TO, FROM,* and *SUBJECT*) flush with the left margin. Remember the following when you write a memo [11]:

Plan it with purpose and reader in mind.

Be sure your subject line clearly announces the subject.

Use an introduction if needed to familiarize the reader with the subject.

Be as brief as possible.

Adjust the tone of your formality, for instance, more formal to those above your organizational level.

Use headings and subheadings to direct attention to the main topics and to topic changes.

Use lists when possible to focus attention on important points.

As practical, include copies of referenced documents, *but* try to avoid requiring extensive reading of the attachments. If you can, save the reader's time by summarizing the documents' main points.

Place your initials beside your name.

As you type your memo, remember to format it properly. Use single spacing for the text and double spaces between paragraphs. Also leave 2 spaces between *DATE, TO, FROM,* and *SUBJECT.* Allow at least 1 in (2.54 cm) on the left. Although less important than *MAP* and *PWRR,* neatness always contributes to the reception of your reports.

Even though informal reports are not as dramatic as formal ones or proposals (see Chapter 5), they project your scientific and engineering competence as well as that of your firm. This reason alone will make its careful preparation a rewarding investment.

Correct Grammar and Punctuation   These are essential to a polished report. The following is a short guide.*

Abbreviations   Abbreviations are widely used in technical writing. Be sure, however, that they will be understood by your readers. When in doubt, spell the complete word. Avoid using *etc.* in formal writing since readers may not know what is included.

Agreement   Make your verb agree with its subject. This problem occurs most often when a subject and verb have intervening words.

Wrong: A *list* of absent *members* are attached.
Right: A list of absent members is attached.

Make your pronouns agree with their antecedents. In the sentence below, *his* or *her* correctly agrees with the word *everyone.*

Wrong: Everyone in the class raised *their* hand.
Right: Everyone in the class raised *his* or *her*[†] hand.

Use the nominative case for subjects, the objective for objects.

Wrong: Prof. Thompson enjoys teaching the computer to *whomever* wants to learn.
Right: Prof. Thompson enjoys teaching the computer to *whoever* wants to learn.

In the sentence above, *whoever* is the subject of the verb, not the object of the preposition *to.* The whole phrase is the object.

Wrong: The task was divided between *he* and Greg.
Right: The task was divided between *him* and Greg.

---

*Except for Polished English, this section on grammar and punctuation, was written by Professor Emeritus Claire Caskey, Clemson University. His contribution is copied with permission from reference 11, page 73.

[†]Use of *his* or *her* can often be avoided by using plurals. For example: All in the class raised *their* hands. As mentioned before, another approach is to alternate and use male and female pronouns equally.

Both *him* and *Greg* are objects of the preposition *between.*

Ambiguity   This word refers to *double* meaning or vagueness.

Wrong: John told James that Bill didn't like him.
Right: John said to James, "Bill doesn't like me."

Capitalization   Capitalize proper names, the first word of every sentence, the first word of each item in an outline, and the first and every important word in titles of books, magazines, and newspapers. Do not capitalize points of the compass, the seasons (unless personified) and names of academic studies, unless they are specific courses: Philosophy 306, psychology, engineering, mathematics.

Choppy Sentences   Too many short sentences can cause problems because they are monotonous to read. The reader must also determine how the choppy sentences are related. The tendency to write in a choppy style is sometimes referred to as the "Dick and Jane" syndrome or the "Jack Webb" (Dragnet) syndrome. It is best to use occasional short sentences for emphasis only.

Coherence   Make your ideas stick together by logical development and by using transitional words carefully. These words can come within sentences as well as between them. *However, moreover, consequently, as a result, nevertheless, first,* and *second* are examples.

Dangling Modifiers   Opening verbal or prepositional phrases will appear to modify the first noun or pronoun following them. Be sure it is the right word.

Wrong: Having rear end trouble, I was driving slowly.
Right: Because my 1988 Plymouth was having rear end trouble, I was driving slowly.

Fragments   Do not punctuate subordinate clauses or phrases as if they were sentences.

Wrong: While the machine was out of order.
Right: While the machine was out of order, production was down by 6 percent.

Paragraphs   Every paragraph should have a topic sentence, which is usually the first one. The other sentences should stick to the topic and develop the main idea by details, examples, logical argument, comparison, and contrast. Paragraph *unity* means that all sentences help develop *one* idea.

Passive Voice   Use the passive voice when the doer of the action is unknown, unimportant, or not to be mentioned for some reason. Laboratory reports are correctly written in the passive voice.

Active: I *heated* the liquid to 100°C.
Passive: The liquid *was heated* to 100°C.

Use the active voice to emphasize the doer of the action.

Active: The architect recommended that steel-reinforced concrete be used.

Punctuation   Punctuation properly used helps say what you mean to say. Improperly used, it slows [down] communication and, at times, stops it. Learn to punctuate by rule, not by intuition. Do not overpunctuate. Overuse of the comma is a common offense.

The Comma

The comma is used in the following ways:

1 Between two independent clauses (complete sentences) when they are joined by a coordinating conjunction—*and, but, or, nor, for, yet:*

Wrong: James operated the movie projector, and spliced film.
Right: James operated the movie projector, but he did not know how to splice film.

2 After long introductory phrases and clauses:

Wrong: When Mary left, he cried for 2 h.
Right: Even though we had bought the television set used and had it for 3 years, the company repaired it without charge.

3 With words, phrases, and clauses in a series:

Wrong: The letter asked students to send name, address, sex and housing requirements.
Right: The letter asked students to send name, address, sex, and housing requirements.

4 To set off nonrestrictive elements (clauses, phrases, appositives):

Wrong: Mr. James McCrimmon who just took his place on the jury teaches Latin at the academy.
Right: Mr. James McCrimmon, who just took his place on the jury, teaches Latin at the academy.

The Semicolon

The semicolon is used as follows:

1 Between two independent clauses not jointed by *and, but, or, nor, for,* or *yet:*

Wrong: James graduated from high school, he attended college in Charleston.
Right: James graduated from high school; he attended college in Charleston.

2 Between two independent clauses joined by a conjunctive adverb, such as *however, moreover, consequently,* and *nevertheless.*

Wrong: Mistakes were made in the calculations, nevertheless, we were given the contract.
Right: Mistakes were made in the calculations; nevertheless, we were given the contract.

3 Between internally punctuated terms in a series. Use semicolons between terms of a series if the terms contain internal commas.

Wrong: He gave her his hat, which was blue, his scarf, of red wool, and a dollar, his last.

Right: He gave her his hat, which was blue; his scarf, of red wool; and a dollar, his last.

The Colon

The colon is used to introduce a long formal list. Be sure to use a complete sentence before the colon.

Wrong: Our reasons for refusing the contract are:

Right: The following men will report to headquarters as soon as possible:

The Apostrophe

The apostrophe is used to form possessives and to indicate contractions. The possessive of personal pronouns, however, has no apostrophe: *his, hers, its.*

Wrong: The cat caught it's tail in the trap.

Right: The cat caught its tail in the trap.

Polished English

When used as adjectives, verbs ending in *ing* are participles; when used as a noun, they are gerunds. The latter case constitutes a gerund phrase, and the modifying pronouns should be possessive.

Wrong: I appreciate you closing the door.

Right: I appreciate *your* closing the door.

Comparisons followed by complete thoughts should use *as if.*

Wrong: You act like you are sick.

Right: You act as if you are sick.

Words such as *like* and *than* can be used as a preposition or conjunction.

Right: He is shorter than me. (preposition)

Right: He is shorter than I (am). (conjunction)

Some verbs can be transitive (take an object) or intransitive (a predicate adjective modifies the subject).

Right: When wearing gloves I feel poorly (the things I touch). (transitive)

Right: I feel bad. (intransitive)

Technical Articles*

Introduction   As an engineer you will have ideas and projects that you wish to share with members of your profession. One way to do this is to write articles for tech-

---

*This section was written by me (see reference 11). It is copied with permission of the Ginn Press, Needham Heigts, MA.

nical journals such as *Civil Engineering*. As a student, you will be encouraged to submit articles for publication; as a practicing engineer, you may be required to write them. If these are done well, they reflect favorably upon both the writer and the company.

Preparation   Normally the article will be of wider interest than the technical report. It is, therefore, written for a more general audience. The usual steps are gathering information (exploratory phase), planning your paper, writing, and revising [*PWRR*].

Gathering Information   In the gathering and creative selection of information, there are three major sources: personal and professional experience, library sources, and original research. In this phase, first consider the message you wish to present, then your audience (readers and their background), and finally the purpose of your article.

Writing Phase   The first step in this phase is outlining. It should give a mental path that leads the reader from the introduction through the body and the conclusion. Take whatever relaxed, exploratory time necessary to develop the information and concepts you wish to present. As you progress, generously use headings and subheadings to guide the readers and keep them oriented. The remainder of this section reviews the major parts of a technical article.

Introduction   It should present four things: the subject with any necessary background, the purpose, the scope, and the plan of development.

Body   The body develops the information to be presented. To do this, you may describe, narrate, and even argue a point. Your main purpose, however, is usually expository—to explain. Visuals (photos, drawings, charts, graphs) are often helpful in supplementing your writing.

Conclusion   Draw whatever conclusions are warranted by the body of the paper. If no definite conclusions are in order, then a simple summary may be all you need. Above all, do not introduce new material here.

Abstract/Summary   As for technical reports, for technical articles, the terms *abstract* and *summary* are often used interchangeably. However, two types of abstracts can be distinguished: descriptive and informative. The descriptive abstract merely tells what the full report contains; it cannot serve as a substitute for the report itself. The informative abstract gives the salient facts and more important information. The first of these usually takes no more than a brief paragraph. Sometimes it is done in a sentence or two. The informative abstract, on the other hand, gives the major facts on which conclusions are based, the conclusions, and any recommendations. This type of abstract can be called a summary. It is appropriate for technical articles and reports. Normally written last, it appears at the beginning of the article or report.

General Guidelines   A basic rule of technical communications is to give the audience the most information of value while requiring the minimum time and effort. The following guidelines will help you do this. Keep them in mind, especially when you are revising.

1 Use headings and subheadings liberally. They will help readers follow your ideas and retrieve information.
2 Be sure the article contains everything the reader needs.
3 Ask yourself how much you can remove without reducing the readers' understanding or their ease of reading. For example, have you put in things to impress them (big words, or five words when two would be better).
4 Strive to be so clear that you cannot be misunderstood.
5 Be sure your grammar and punctuation are correct.
6 Use transitional words and phrases to help your readers.

Documenting Your Sources Naturally, you will use the works of others in your research. Be sure to give credit to these sources by using proper footnotes and a bibliography (see the material in this chapter on back matter). Since methods vary greatly from one periodical to another, follow the method used in the journal to which you are submitting your article. If, however, you copy more than 250 words (even scattered quotes) or more than 5 percent of an original work, you should obtain written permission from the publisher.

Engineering societies such as IEEE and ASME provide authors' guides and offer an author's kit. These include information which will help you prepare your article in the proper format and thus avoid unnecessary rewrite and delay.

Remember As you read the article below, look for the following: an introduction which leads you into the article, a body which develops the main ideas in a logical sequence, and a conclusion which summarizes the major points.

Sample Technical Article
"The Economic Roles of the Engineer"

Abstract

This article reviews the numerous engineering functions required in defining, designing, developing, manufacturing, and marketing a new product.

This is a descriptive abstract

Introduction

Engineers in both private industry and government are instrumental in meeting the needs of mankind. This article describes a product development by private industry but would also apply to a government project if references to a "sales" versus cost were replaced by "public benefit" versus cost. In a private company the engineer's role normally starts after managers have determined the need for a product and the feasibility of manufacturing it. In this brief article we shall look at the engineer's role in research and development, designing a product, developing and testing a prototype, establishing a pilot production line, and eventually operating a full production line. When the product is highly technical, engineers may also work with sales and marketing personnel in defining the need and applying technical support to the finished product.

This introduction forecasts the structure of the article.

Determining the Need

Consider the block diagram in Figure 4-1. It presents a simplified view of the workings of a private company. The economic process begins with a need. Someone or some group needs something. The need can range from the fundamental, such as a source of pure water for drinking, to the somewhat frivolous, such as a dog food with a higher nutritional content. The company itself may have induced the need, perhaps through a clever advertising campaign. A company may even have made a product and then induced a need for it.

Subheadings help the reader follow the idea. If done carefully, these subheadings form an outline.

Feasibility

In any event, once the need (or potential need) is apparent, the company's management must estimate the potential market size, the cost of manufacturing, and the eventual selling price. Recognizing that profits equal total sales revenue minus total cost, management must answer the question: Is this project economically feasible? If the answer is no, the process stops, as indicated in Figure 4-1, because the company has concluded that it cannot satisfy the need profitably. Profit is essential if the company is to grow and provide jobs to support the economy. But if the answer to the question is yes, the company begins a process that may ultimately lead to a new product. (We say "may" because many projects designed to develop new products fail for a variety of reasons.) At this time a large company's research and development department (usually shortened to R&D) enters the picture.

The chief organizing principle in this article is chronological. It progresses in a series of logical steps.

Using an occasional and or but to start a sentence is all right. But not very often.

The Research Team

Once the company decides to proceed, it assigns a research team (in some cases the team has only one member) the task of determining whether the scientific principles needed to develop the proposed product are known. The research team searches the scientific and

Note the use of time order in giving the steps.

engineering literature as well as the patent registry. If the scientific principles are either unknown or inadequately understood, the company must conduct the necessary research to sufficiently understand these principles. It is during this research phase that an engineer generally becomes involved in the overall development and economic justification process. Once the engineer finds a principle that has no serious drawbacks or flaws, this process proceeds to the next phase. If appropriate, the company normally seeks a patent. We should point out that research costs money. Sometimes when a company does significant work without making progress [toward a solution], management decides not to spend any more [money] on the project. Figure 4-1 does not explicitly show this possibility, but it is an important consideration at each step in the process.

Brackets [ ] indicate unnecessary words. You may find this in other places as well.

Development of a Prototype

The next phase is development. A prototype of the final product is assembled and tested to see if it meets all the necessary conditions. These include performing as required, meeting cost targets, and sat-

FIGURE 4-1
Decision-making in the economy—a simplified view. Gajda, Walter J. and William E. Biles, *Engineering: Modeling and Computation*, 4/e. Copyright © 1977 by Houghton Mifflin Company. Used with permission.

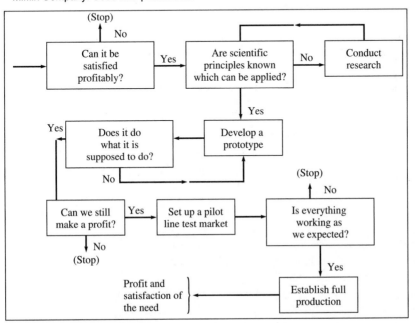

isfying all applicable regulations and laws. If a given prototype is inadequate, another must be developed and tested. Again, if this process is repeated too many times without progress, the project may be terminated. Once a successful prototype has been developed, the company's marketing personnel again investigate and assess the profit potential [of the product]. This reassessment is essential if the research and development phases have consumed significant amounts of time.

Note the use of transitions ("If . . .").

Use small words when available: stopped or ended instead of terminated, but in the last sentence, required might be a better word than consumed.

Pilot Production Line

Next, the company normally establishes a pilot production line, produces limited amounts of the product, and test-markets it. Most of the design details must be determined and documented at this time. If the market tests are positive, the company commits money to the construction or purchase of tools, equipment, and space necessary for full-scale production and national (or even international) sales. Periodically, the company reviews the performance of each product line to ensure that it remains profitable.

The Engineer's Role

Engineers fit into this economic flow at a number of points. Research engineers search for new scientific principles and a more complete understanding of them. Development engineers use these principles to produce prototypes. Product design engineers perform the basic task of determining and documenting product details. Industrial and manufacturing engineers set up the pilot production lines and specify the necessary equipment and space. Plant engineers see that the design and operation of a new production facility are successful. In addition, as stated earlier, highly technical products often require R&D engineers (to help define the need) and application engineers (to assist the user of the product).

Conclusion

From this brief examination of the economic roles of engineers, it can be concluded that they normally enter the process after managers have considered the need for a product and the feasibility of producing it profitably. Engineers initially determine workable principles for producing the product, and then design, produce, and test a prototype. Other engineers document the design and set up a pilot production line. Finally, if full production proceeds, engineers design and oversee the new production facility. The role of the engineer is thus an essential key in the progress of our society.

The article is summarized. No new material is introduced.

Manuscript Mechanics [12]

In technical reports accuracy and clarity are of vital importance; however, first impressions often depend on appearance. Regretfully, it could influence a person's evaluation of the report in all other respects. Thus, it is advantageous for the author to produce a report which is also professional-looking; this includes being free of typographical and grammatical errors and misspelled words.

Although there is no set of rules for manuscript mechanics in a technical report, the following recommendations are good practice and assure a "professional look." As appropriate, they are used in this book and concern the following: paper, margins, headings and subheading, spacing and indenting, and pagination.

Paper   Reports should be typed (or printed) double-spaced on one side of good-quality white sheets, $8\frac{1}{2}$ by 11 in, preferably 20-lb bond. Company policy might require different colored sheets for different types of reports; however, a fresh typewriter (or printer) ribbon used with white paper generally gives a neater copy.

Margins   The left margin should always be either $1\frac{1}{2}$ or 2 in; all others should be 1 in. The greater left margin ensures that your audience will not have difficulty reading the report after it is bound. Naturally, different word lengths will cause the right margin to vary, but it should not be less than $\frac{3}{4}$ in.

Where quotations or other insets are included in the text, an additional 5 spaces of margin should be used on the left side and, if feasible, on the right.

Headings and Subheadings   These signal the main points of the report and should have the same wording as the topical entries in the table of contents. They must be a brief title for the passages they identify.

The main headings name the major divisions of a report. They are in either lower-case (with initial letter in each major word capitalized) or all-capital letters. Main headings are underlined and placed in the center of a new page at least 3 spaces above the first line or subheading. Or they can be numbered (e.g., 2.) and indented 5 spaces from the left margin. In this case, only the initial letter in each major word is capitalized, with 2 spaces above and below. Use of numerals or letters, or both, is optional; however, you should be consistent.

Two levels of subheadings are usually adequate, that is, second-order headings and third-order. Second-order headings should be placed flush with the left margin. Underline each word separately. Use lowercase letters, but capitalize the initial letter of each important word. Double-space above and below the heading, but put no text on the same line and no punctuation after it. The third-order heading is treated exactly as the second-order except for three details: (1) Indent it 5 spaces, (2) put a period after it, and (3) start the text on the same line as the heading (see below).

Again, an alternative is to use numerals; to extend the heading order described above, add decimals (or dashes) for each lower level. For example, with a single numeral for the main heading (e.g., 2) add a decimal (2.1) for the first subheading and another (2.1.1) for the next level.

Spacing and Indenting As noted above, the text should be double-spaced; exceptions are these:

1 Triple- or quadruple-space below center headings.
2 Single-space inset or center listings; if they are numerous, number them.
3 Single-space long quotation (more than 3 lines).
4 Triple-space above and below quotations and listings.
5 Single-space individual footnotes if they are more than a line in length; double-space between notes.
6 Single-space individual entries in the bibliography and double-space between them. For entries greater than 1 line, use hanging indentation; that is, indent the second and succeeding lines 3 spaces.
7 Single-space the abstract if space demands; otherwise, double-space it.
8 Single-space material in the appendix.
9 Double-space above and below side headings.

The beginning of the paragraph is indented 5 spaces; at least 5 more spaces are used at the beginning of a listing. Lists should be numbered if there are more than a few entries.

Pagination Use arabic numerals in the upper right-hand corner; they should be aligned with the right margin, at least 2 spaces above the first line and about $\frac{3}{4}$ in below the top edge. Exceptions are these:

• Prefatory pages (the title page, letter of transmittal, table of contents, and abstract). For the table of contents and the abstract, if on separate pages, use lower-cases roman numerals centered at the bottom of the page, about $\frac{3}{4}$ in from the lower edge. The title page and letter of transmittal are counted, but they have no numerals. Thus, the table of contents would be iii.
• For the first page of the body or for a new division, it is usual to omit the 1; for the other pages, the numeral is placed in the center bottom.

Pages of the appendix are numbered (as in the body) in the upper right-hand corner.

Oral Reports [13,14]

The most diligent person can do only so much; however, through communication, you can enlarge your resources many times over. You can teach others what you know, get others to help you do what you do, and seek understanding for your beliefs. Oral communication is special; it establishes a personal, live relationship between speaker and listener, even when the listener is in a large audience. It is one of the few ways to establish valid, personal communication. Its special effect can be seen even when watching speeches on television. Oral presentation is an effective way to communicate and is a necessity in almost every profession. Therefore, it is worth doing well!

First, we will review the differences between written and oral reports, and then discuss the advantages and disadvantages.

Oral reports have many advantages:

• Inflection of the speaker's voice, that is, changes in loudness and pitch, can be used effectively to stress or subordinate material during presentation.
• The enthusiasm of the speaker is infectious. Close rapport is easily established via smiles and other pleasant facial expressions.
• Eye contact is particularly effective in maintaining listeners' attention and sensing their grasp of the material.
• Flexibility in pace and detail permits immediate response to feedback given by eye contact.
• Oral reports facilitate interchange with others in your field, peer reaction, and earlier exposure to ideas not yet published.

However, there are significant disadvantages:

• The greatest involve distractions, both external and internal. As discussed in Section 1-1 on listening, distractions frequently cause the listener to lose the train of thought or miss a point being stressed. For this reason, in oral reports, ideas need to be repeated by rephrasing the main points.
• There is a heavy dependence on the speaker's articulation. Many listeners will not put forth the extra effort to decipher words that are not clear and distinct.
• Unlike with written reports, listeners cannot orient themselves by scanning headings and subheading, or flip back and reread material. Visuals can be effectively used, however, to reduce this disadvantage.
• Except for a most unusual match between the listener's interest and the subject, the average person's attention span is no greater than about 20 min.
• The learning rate of a listener is much slower than that of a reader. In turn, an oral presentation must be much shorter, more to the point, and less detailed than a written one. Information channels have a characteristic which communication engineers call *bandwidth;* it controls the rate at which information can be transferred. The hearing frequencies are much lower than those for seeing, and the bandwidth is thus much less. This is another reason we learn more by reading (seeing) than by listening.
• Also, it seems it takes a few minutes for listeners to start listening.

The advantages and disadvantages show that the design of an effective oral report requires special attention. Furthermore, the above list shows that for the oral report, delivery is as important as the technical content. Considering substance and then delivery, this section offers methods and formats which ensure a creditable, oral presentation. Again, through your individual creativity and experiences, you will build on these foundations to develop polish.

Report Substance   Here the same acronyms, *MAPS* and *PWRR,* are relevant.

In planning (the *P* of *PWRR*) consider your audience. To whom will you be speaking and where? What is the occasion? Consider the purpose of your talk, and select the material to meet that purpose. Next, organize the material for a clear, complete presentation.

There are four modes of speech: impromptu, memorized, written, and extemporaneous. The occasion usually determines the choice.

Impromptu speeches, by definition, cannot be prepared. You are wise to avoid these unless you know the subject matter particularly well. Even then, an inadvertently chosen word can give an impression you did not wish to convey.

A *memorized* speech is suitable only for tour guides, experienced actors, salespeople or politicians (i.e., those who frequently give the same speech over and over again).

Written speeches are suitable for some occasions, such as when one is seeking political office. The written speech prevents the speaker from making misstatements of facts or emotional statements which may plague him or her later on. Business people use written speeches for the same reason. In both cases, however, the desired spontaneity is lost. Furthermore, in dimly lit rooms, reading is difficult and mistakes are more probable. Nevertheless, papers delivered to professional societies are often written and then read to the group.

The *extemporaneous* speech, unlike the impromptu speech (with which it is often confused), is well planned and practiced. It is the mode most widely used in oral technical reports and is the main subject here. Like the written report, it is planned or organized into three main parts—introduction, body, and conclusion—although the topics, not the words, are presented in a systematic manner. Thus, spontaneity is not lost. The first two of these main parts are considered next. To save time and effort, I will use the conclusion of this report to review the format of conclusions for oral presentations.

Introduction   The opening of a speech should accomplish three goals: (1) create a relaxed, friendly atmosphere; (2) develop interest in the subject; and (3) announce the subject, purpose, scope, and plan of the talk.

1 Mingling and talking informally with your audience beforehand help to establish a relaxed, friendly atmosphere. Prepare an appropriate comment to link you personally to your audience.

2 Suitable anecdotes and jokes are excellent for relaxing the audience, but are suitable only when they are relevant to the subject or the occasion.

3 The subject, purpose, scope, and plan should be the same as for the written report, as presented earlier. However, in the oral report, the plan or organization is more important. Remember that the listener cannot flip back to earlier paragraphs. Organizational guideposts are effective in helping the audience follow your presentation. Numbering is a simple aid: for example, state that "there are three goals," and then refer to them by number. Finally, redundant (but varied) presentations are effective in reinforcing material presented earlier and ensuring its assimilation. For this purpose, the old truism is applicable: Tell them what you are going to tell them, tell them, and then tell them what you told them. This is analogous to the introduction, the body, and the conclusion of your report.

Body   As in the written report, the body of the oral report presents the message. Its main content should consist of concrete information—examples, illustrations, short narratives, analogies, etc.—all supporting a few generalizations (remember, the attention span of your audience is limited). With supporting information, repeat the generalization. Visual aids, reviewed below, are helpful; relating the subject matter to some vital interest of the audience is also effective.

Because of the listeners' limited attention span, cut your subject matter down to the minimum. Use no more than 5 or 6 min to develop each topic. Thus in a 5-min speech you should consider only one topic. Even in an hour-long talk, no more than three or four points should be covered. As noted earlier, it is best to limit your presentation to no more than 20 min.

Considering the purpose of your speech, use the following criteria for its content:

Will it meet the needs and expectations of the audience?
Will it promote the purpose of the talk?
Does the occasion call for it?
Does an accurate presentation require it?

Time and attention span limitations may not permit your meeting all of these criteria; however, your most important points should meet all four.

Organization of the body's subject matter also follows the strategies presented earlier for written communications. A few are reviewed here.

1 The chronological approach is natural for giving the history of an action.

2 When using cause and effect, the basic idea is to show how x has caused y; frequently it is best accomplished in a chronological order.

3 In problem-solution talks, the general form is the obvious: problem, solution, and the effects of the solution. Defining the problem well is crucial; in presenting your choice, again state the problem, the criteria to be applied, and the constraints.

Review each alternative and show how your choice is best, that is, how it most nearly meets the criteria, subject to the constraints.

4 The topical approach is used simply for informing people about a subject. The "newspaper style" could be best. For example, you might wish to tell colleagues about Tau Beta Pi, the national engineering honor society: (a) who Tau Beta Pi members are, (b) what Tau Beta Pi members do, (c) when they meet, (d) where they meet, (e) why Tau Beta Pi exists, and (f) how it achieves its goals.

5 Argumentation is a method addressed to reason; thus you argue about propositions when you have two or more alternatives, which on the first level of comparison seem about equally desirable. As in the written report, argumentation starts by stating the proposition (the major thesis) and then supports it with verifiable facts and statements from recognized authorities. Remember that the argument can be no better than the support presented! Two procedures are useful: (a) Put the strongest minor thesis at the end and the weakest in the middle. (b) When your opponent has a strong point, discuss it and then try to moderate it by focusing the discussion on its weak points in a legitimate, unemotional way. Drama can achieve only short-term success! Recall that emotion (or drama) is more appropriate for persuasion, not argumentation.

Conclusions   We will review the structure of your conclusion at the end of this section, just before the chapter summary.

Visuals   These are often indispensable in all forms of communication. As the saying goes: "A pictures saves thousands of words." The effectiveness, however, depends on their visibility, clarity, and control. Visibility simply means that the audience can see the material. Is it clear, large enough to be read and followed easily by all? To be readable, letters should be at least 1 in high for each 25 ft between them and the reader.

Clarity and simplicity are important because the listener must be able to assimilate the material in only a few minutes.

Control relates to your ability to keep the observer's attention on the material you are discussing. Properly designed, visuals can be used to increase audience concentration; irrelevant materials are distracting. With an overhead projector, an effective plan is to use an opaque overlay. Slide it down to reveal the subject material as it is presented. Small exhibits passed through the audience "kill" concentration, a mistake many professors make when they distribute homework or return quizzes prior to or during lectures.

The chalkboard and the overhead projector are probably the most valuable, convenient, and widely used audiovisual aids. Slide projectors can offer a more professional presentation, but since most comments concerning the overhead projector are also applicable to the slide projector, we will not consider it separately.

In using a chalkboard, write in columns about 2 ft wide; on completing each column, move to the right. Print or write legibly, using large letters; be careful not to write below the area easily seen from the back of the audience. Avoid obstructing their view or "talking to the blackboard."

The overhead projector is convenient for many reasons. First, the transparencies prepared for them can be easily projected on a screen or on a light-colored smooth wall. The transparencies are readily prepared from reproductions, but remember the rule: 1-in letter height is required for every 25 ft to the farthest listener. Notice that a copy of a typewritten page, such as this sheet, can be read easily only at a distance of a few feet. With the projector in the front of the room, it is easy for you to address the audience while referring to the display. Use of a pointer on the screen, rather than a pencil on the transparency, makes it easier to avoid blocking the audience's view.

Probably most important, the overhead projector offers easy control of audience attention. The transparencies can be prepared ahead of time, and if you cover parts and progressively reveal the relevant material, the audience is distracted as little as possible. Remember that the duration of the material's presentation must be adequate for the viewer's assimilation, although this is often difficult to judge. Finally, if appropriate, you can prepare only a portion of the transparency ahead of time and then develop the rest during the presentation using a felt-tipped pen.

In preparing transparencies, *keep them visible, simple, and clear. A complex presentation is discouraging and quickly loses the audience.* In explaining the transparencies, use a pointer, preferably at the screen, standing slightly to one side, so as not to obstruct the view. Transitional statements between visuals save the audience from having to determine what the new material is and greatly facilitate their following your delivery.

Delivery   The effectiveness of your speech greatly depends on your delivery. A slow, relaxed approach to the lectern and an unhurried presentation will have a similar effect on the audience. Greet them in whatever manner is appropriate. Even the most experienced speakers feel the flow of adrenaline. Realize that it can give you the vitality and enthusiasm that come from the presentation of a well-prepared speech, one that is planned and written (but with only the main topics on index cards for easy reference*) with *PWRR* in mind. During delivery, use facial expressions: frown or smile as appropriate, and establish eye contact frequently with a number of people in your audience. Empathy with the audience relaxes you and them; talk *to* (not down or up to) the audience. If feasible, before the presentation learn the names of a few of those in the audience and address them directly. Try immediately to establish an air of mutual concern and admiration; this is enhanced by a pleasing eye contact. Having notes or copies of the report available after your talk shows concern and puts the audience at ease (they don't have to try to record or encode your salient points). Announce this at the very beginning. Naturally, proper, conservative dress will reduce the distractions always present. Your dress has the same effect as the appearance of your written report; it can set prejudices for or against you. Do not start with a handicap! Use a natural posture, stand straight (not stiff), always at ease. Move around if you feel like it, but do not pace. And, of course, avoid meaningless gestures, tapping fingers, or actions such as fiddling with change in your pocket.

Probably of greatest importance in the delivery, beyond eye contact and general rapport reviewed above, are your breathing, enunciation, and speed of speaking. Breathe deeply, slowly, and from your stomach (you can confirm this by using your hand to feel the abdominal movement). This method of breathing gives ease to your speaking. However, since it is not natural, except for experienced speakers and singers, it must be consciously practiced until it can be done subconsciously.

Proper articulation, words pronounced distinctly and clearly, encourages attention and gives a professional aura to the presentation; furthermore, many listeners will not put forth the extra effort to decipher muddled speech. Articulation is affected by anxiety; nervousness affects the pitch of our voices and the speed of our speaking. Nervousness is mainly a nuisance which diminishes with experience and usually lasts only a few minutes at the start of your talk. Control of pitch and speed, however, is very effective in stressing points. A high pitch tends to generate tension in the audience, and increased speeds can arouse interest. Subsequently, use speeds a bit below normal to signal an important point and pause, moving slightly back from the lectern to let your audience reflect on the subject. Be careful: At speeds over about 160 words per minute, articulation loses precision. To pronounce words clearly, the tongue must be free to form the letters *d* and *t,* and the lips must be free for *b*'s and *p*'s.

Suggestions for Easy Improvements   If all of this seems too much, don't be discouraged; delivery techniques are easily developed by practicing oral reading to yourself, with a tape recorder, or to friends. This method is a most pleasant and quickly rewarding approach to becoming an articulate speaker.

Below are suggestions for daily use, whether talking to one person or many:

• Relax and use a pleasant voice; make a special effort to speak not so that you can be understood but so that you cannot be misunderstood! For example, do not use vague expressions such as "Please hand me that thing over there."

*Use of an overhead projector usually makes index cards unnecessary.

- Open your mouth; do not let it act like the mute on a trumpet. A restricted mouth is a common cause of indistinct speech; practice speaking while looking in a mirror. Watch commentators and announcers on television.
- Avoid speaking in monotones; monotony reflects lack of interest in a subject. Variations in pitch and loudness show your enthusiasm. Be alive! Every day!
- Speak smoothly and use simple sentences. In a speech or daily conversation, using words your audience does not understand creates gaps and probably causes them to lose their train of thought.
- Without blasting, practice adjusting the loudness of your voice so that everyone who should hear you can. A loud voice is not necessarily the most effective way of getting attention. Emphasize by leaning forward and dropping your voice slightly; it is natural for others to automatically be more attentive. The pause is similar to an exclamation mark.

Be prepared beyond the material you plan to present. Such preparation acts like "water behind the dam." It gives power to the presentation. Furthermore, you will feel comfortable and at ease if time permits a question and answer period. This will give you an opportunity to get further feedback and also to expand on the major points you wish to stress. Finally, realize that you will often have questioners who want to show what they know. Try to tolerate show-offs and avoid showdowns. It saves time and is less stressful for you and for those in the audience.

To time your speech, practice first using a tape recorder and then, if possible, before colleagues. It gives you the confidence and poise required for the smooth, relaxed delivery which enhances your self-concept as a successful engineer. It also helps prevent you from running overtime.

The conclusion for your talk, like the conclusion for this section, should reflect the purpose and plan as presented in the introduction. Furthermore, it should summarize your major points, reviewing why they are important. Announce the conclusion, and move through it swiftly. If your talk had to do with a proposal, a call to action by a specific time is in order. If it was informational, a memorable quotation or story can be used to close the speech. Here we repeat the format for all technical communications, especially oral ones: Tell them what you are going to tell them, tell them, and then tell them that you told them.

4-3 SUMMARY

Like all excellent communications, technical reports, both oral and written, are the results of an arduous endeavor. Thankfully, the scientific or engineering approach minimizes this effort, mostly because we are guided by a well-thought-out plan. Finally, the iterative modifications of *RR* (relax and then revise) in *PWRR* can produce excellent results, and proper "relaxing" can minimize the arduousness.

The following thoughts by novelist Ernest Hemingway seem appropriate: "There's no rule on how . . . to write. Sometimes it comes easily and perfectly. Sometimes it is like drilling rock and then blasting it out with charges."

Wendell Berry [modern poet and essayist] offered this thought: "What gets my interest is the sense that the writer is speaking honestly and fully of what he knows well." Berry's idea is particularly important in a proposal presentation, which is the second topic of the next chapter.

DISCUSSION QUESTIONS AND EXERCISES

1 What are the advantages of using numerous headings and subheadings?

2 According to the text, what are the phases of all projects?

3 What are the special characteristics of oral communications?

4 Using the fundamentals and mnemonics presented in this chapter, write a brief summary of the following from Chapter 1:

    a "Matching Activities to Capabilities through Time Management"

    b "Self-Discipline" and "Assertiveness"

    c "How to Succeed in Life through and in Engineering"

5 Using the fundamentals and mnemonics presented in this chapter, prepare a written technical report on any nonfictional subject of your choice. Work with a partner, someone from the class who you feel will work diligently with you.

6 Perhaps with the same partner and perhaps on the same topic, prepare and present an oral report.

REFERENCES

1 *The McGraw-Hill author's book.* New York: McGraw-Hill, 1984, p. 14.

2 *West author's guide.* New York: West, 1986, p. 42.

3 Chestnut, H. *System engineering tools.* New York: Wiley, 1966, pp. 24–30.

4 Turner, R. P. *Technical report writing.* San Francisco: Rinehart & Winston, 1971, pp. 17–19.

5 Mills, G. H., & Walter, J. A., *Technical writings* (3rd ed.). New York: Holt, Rinehart & Winston, 1970, p. 78.

6 Mills & Walter. *Technical writings.* pp. 446–452.

7 Mali, P., & Sykes, R. W. *Word processing for engineers and scientists.* New York: McGraw-Hill, 1985, p. 98.

8 *Prentice-Hall author's guide.* pp. 26–31.

9 Turner. *Technical report writing.* p. 27.

10 Turner. *Technical report writing.* p. 29.

11 McAuliff, D., et al. *Effective technical communications* (2nd ed.). Needham Heights, MA: Ginn, 1987, p. 18.

12 Mills & Walter. *Technical writings.* pp. 381–392.

13 Michaelson, H. B. *How to write and publish engineering papers and reports.* Philadelphia: ISI, 1982, pp. 132–140.

14 Mills & Walter. *Technical writings.* pp. 337–344.

BIBLIOGRAPHY

Eisenberg, A. *Effective technical communication* (2d ed.). New York: McGraw-Hill, 1992.

Houp, K. W., & Pearsall, T. E. *Reporting technical information.* Beverly Hills, CA: Glencoe, 1977.

Pearsall, T. E., & Cunningham, D. H. *How to write for the world of work.* New York: Holt, Rhinehart & Winston, 1978.

Courtesy Fluor Daniel, Inc., Greenville, South Carolina

PROJECT MANAGEMENT AND PROPOSAL PRESENTATION

In practice, projects depend on successful proposals. That success, however, requires effective planning and communication. Both of these are greatly facilitated by the use of project networks: CPM (critical path methods) and PERT (program evaluation review techniques). The decision maker to whom a proposal is directed is significantly influenced by the project's time and cost. Effective control of both is possible through project networks. For these reasons CPM and PERT are presented here. Precedence diagraming (PD) is also included. They are followed by proposal presentations in the next section.

PROJECT PLANNING AND MANAGEMENT

Projects vary in size from large ones, such as the SDI ("Star Wars") and the Stealth Bomber, to small ones, such as improving the software of a word processing package. Naturally, for the large projects, many experts or specialists will be involved in planning. Only a few specialists might be assigned to a medium-sized project. Small ones could have only a relatively inexperienced engineer working with a technician who is experienced in the project area. This chapter is intended to aid the engineer in the last category. It should also be a helpful reference to any engineer who has not had recent project planning experience.

Planning is surely one of the most important factors which determine whether a project "goes well" or is plagued by unanticipated misfortunes. Your concerned Parent from Chapter 1 (rather than your critical one) advises that "luck favors those prepared" by a well-organized plan. Planning is work, often unpleasant (for we like "to get on with it"), but the effort is extremely well rewarded.

Preparing and using visual aids are very helpful in developing and organizing our thoughts. The aids are also effective in communicating clearly with others. This is especially true in projects which require the coordination of many activities. Here we present two visual aid–type planning and management techniques: critical path method and program evaluation and review technique. As is the practice in industry, we will refer to them as CPM and PERT, respectively. Both use arrow diagrams or networks to show the interdependence of project activities. Showing this interrelation between tasks is the major advantage of CPM over Gantt or bar charts, from which CPM was developed. The Gantt charts were used as early as World War I. In the late fifties and early sixties, CPM and PERT gained wide acceptance, especially in military contracts such as the Navy's Polaris program. The planning of the Polaris encompassed 23 networks with 3000 activities. Many Air Force programs included PERT as a contractual requirement. Today, many firms, such as IBM, use CPM/PERT diagrams in the early stages of their planning. PERT uses data from CPM-type arrow diagrams to evaluate the randomness due to imperfections in all human and physical systems. Because of this, CPM will be considered first. Examples are developed for CPM (normal and expedited, or crashed) and for PERT. For interruptible activities, precedence diagraming method (PDM) is also presented.

5-1 CPM: CRITICAL PATH METHOD [1]

Benefits and Advantages

CPM provides the following benefits:

• The total project is planned and coordinated by the persons involved.
• The CPM network is a means of communication.* It informs everyone of the overall schedule and individual responsibilities in meeting that schedule.
• This planning technique is based on specific events being achieved at specific times, so that there is accountability and a way of recognizing the need for corrective action.
• The overall progress of the project can be assessed on a periodic basis, such as every week. Thus potential bottlenecks become evident.

Furthermore CPM has these advantages:

• It encourages logical discipline in planning, scheduling, and controlling the cost of the project.
• It promotes both long-range and detailed planning.
• It provides a standard method of documenting and communicating the project plans and scheduling for time and costs.
• It focuses management's attention on the most critical elements (those in the critical path, defined later) and encourages management to work with the engineers in making contingent plans.
• It makes evident the effects of both technical and procedural changes.

These characteristics of CPM apply only when there is little uncertainty concerning the times required for the activities. PERT is a similar network method, but it includes techniques for considering uncertainties. It is presented in Section 5-2.

Mechanics of CPM Activities-on-Arrows (AOA) and Activities-on-Nodes (AON)

1. An activity is any part of a project which requires resources and time to be accomplished. It has a definable beginning and end with no interruptions. Activities may represent labor, machine operations, or negotiations.
2. An event is at a point in time. It represents a state of accomplishment, such as at the beginning and end of an activity. Events may represent a starting time, the completion of a negotiation, or other task.
3. CPM models a project by a network of arrows and blocks (or circles) called nodes. There are two types of arrow diagrams or networks: one type, AOA, represents activities on (by) arrows and events by nodes; this is called the arrow format or scheme. The other type, activity on node (AON), uses nodes to represent activities; arrows are used only to show the interrelationship between activities. An advantage of AON is that it obviates (makes unnecessary) dummy activities, discussed below. It is also more economical in the number of symbols required. Both types are presented.

We will first consider the arrow format. Presently, it is better known, especially by middle and senior management. As expected, the network is often called an arrow diagram. It clearly shows the precedence relationships between activities required for an event and the event required for an activity. The arrows represent activities and indi-

---

*It is suggested that all rational decisions are based on models: when rational people with the same goals differ, they are probably using different models. This problem is reduced by using CPM.

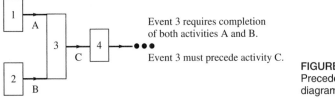

FIGURE 5-1
Precedence relations for arrow
diagrams.

cate the flow of time; nodes, symbolized by blocks or circles, represent events, i.e., an achievement or status at a specific time. The events are like the system state variables used in automatic control system analysis. Because of the symbols available on a word processor, we use blocks instead of circles to indicate events. As seen in Figure 5-1, activities (arrows) begin and end on events (nodes or blocks). By arbitrary convention, events are numbered and activities (also called *tasks* or *jobs*) are identified by capital letters. In practice it is more common to print a few descriptive words or use mnemonics near the arrows to obviate cross-referencing.

4. Dummy activities, represented by dashed-line arrows, are used merely to show dependency of one activity on another. They are always of zero duration. One is used in Example 5-1 (see "G" in Figure 5-2).

As a summary of precedence relations for the arrow method, consider the partial network in Figure 5-1. Sometimes tasks are identified by a pair of numbers, such as might be represented by i and j, corresponding to their beginning and terminal events. For example, in Figure 5-1, activity A may be shown as (1,3). This method facilitates computer programming and analysis.

Procedure

Normally a network is developed moving from the final event to the first. Start with the final box which indicates project completion, and work backward drawing boxes as you think of earlier events needed to achieve subsequent ones. Then draw arrows for the activities required to accomplish the later events. This logical approach is equally applicable to many personal endeavors, such as in the example below.

Example 5-1: A Simple Operation

CRITICAL PATH NETWORK FOR PREPARING A QUART OF TEA

As suggested, we start with the desired state as the final event N; it is a quart of tea ready for heating or cooling. Working backward, let us list this and earlier states from which the later ones can be achieved. The list is shown on p. 137.

Development   Now we can easily draw the CPM model (arrow diagram) for this small project. N-7 will be event 1, of course. The last event (N) will be numbered 8. We will now develop the diagram shown in Figure 5-2.

EVENTS FOR EXAMPLE 5-1

Event number	Event or state
N	1 quart of tea ready for heating or cooling.
N-1	$1\frac{1}{2}$ pints of water with 2 teaspoons of lemon juice in a quart jar; and $\frac{1}{2}$ pint of warm, strong, sweetened tea with two tea bags in a pint jar.
N-2	$\frac{1}{2}$ pint of warm, strong, sweetened tea with two tea bags in a pint jar.
N-3	Quart jar with 2 teaspoons of lemon juice.
N-4	Pint jar containing two tea bags, two $\frac{1}{2}$-grain saccharin tablets, and $\frac{1}{2}$ pint of hot water.
N-5	Same as N-4, except water is at room temperature.
N-6	Empty pint jar on the counter.
N-7	Decide to make tea.

Technological Order and Conventions   In Figure 5-2, the numbers in the boxes identify the corresponding events; those above indicate times that are discussed below. An event has not been achieved (or accomplished) unless all activities terminating on it have been accomplished. This shows that tasks (activities) must be performed in technological order. For instance, a house cannot be wired before the walls are in place. In the figure, the numbers above the arrows show the duration required by the corresponding activity; the letters below identify them. An activity cannot begin until the event from which it originates has been achieved. For this example, the activities are listed in Table 5-1 with their estimated times of duration. Please review it now. For instance, activity A is estimated to require 1 min to get a jar from the cabinet; thus, 1 min is also required to achieve event 2, after event 1 has been achieved. Similarly, 3 min is required to "move" from event 2 to event 3. The finish time (FT) of an activity

FIGURE 5-2
The CPM(AOA) network for preparing a quart of tea.

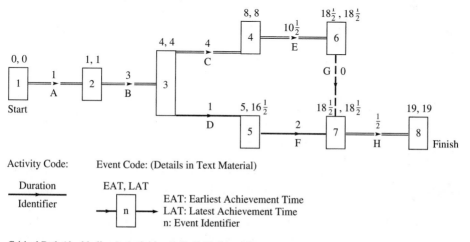

Activity Code: Event Code: (Details in Text Material)

Duration   EAT, LAT
――――――→
Identifier

                    EAT: Earliest Achievement Time
    ―┤ n ├―→   LAT: Latest Achievement Time
                    n: Event Identifier

Critical Path (double lines): Activities A, B, C, E, G, and H
Minimum Time: 19 min

TABLE 5-1
ACTIVITIES WITH PRECEDENCE AND DURATION FOR FIGURE 5-2

Activity	Between Events	Duration	Action
A	1,2	1 min	Get a pint jar from the cabinet.
B	2,3	3 min	From a second cabinet, get two tea bags and two $\frac{1}{2}$-grain saccharin tablets, put them in the pint jar, and add $\frac{1}{2}$ pint of water.
C	3,4	4	Put the pint jar in the microwave oven and set for $3\frac{1}{2}$ min.
D	3,5	1	Get a quart jar from the cabinet near the refrigerator; from the refrigerator, get lemon juice and add 2 teaspoons of juice.
E	4,6	$10\frac{1}{2}$	Let tea steep (in the pint jar).
F	5,7	$\frac{1}{2}$	Add $1\frac{1}{2}$ pints of tap water to the quart jar.
G	6,7	0	Dummy: It only shows that event 7 also requires event 6.
H	7,8	2	Pour contents of the pint jar into the quart jar, and remove the tea bags.

equals its start time (ST) plus the duration (D) of the activity, i.e., FT = ST + D. Although all activities must be finished to "reach" event 8, some are not critical and can be done while other events are being accomplished.

Critical Path and Finish Times   Basic relations for activities:

$$\text{EF (earliest finish time): for activity A,}$$
$$\text{EF}(A) = \text{ES}(A)\ (\text{earliest start time}) + D(A)(\text{activity duration})$$
$$\text{LS(latest start time): for activity A,}$$
$$\text{LS}(A) = \text{LF}(A)(\text{latest finish time}) - D(A)(\text{activity duration})$$

For this simple example (Figure 5-2), we see that there are two paths from event 1 to event 8; we will analyze them using a "forward pass." Conceptually, moving forward (left to right) along a path from event 1, we recognize that the first number above each box is the earliest time that the event represented by the box can be achieved or accomplished. We call it EAT (earliest accomplished time). The EAT(n) of an event n is controlled by the latest EF of any activity terminating on it. In this example, only event 7 has more than one activity terminating on it. From the first "basic relation" above, EF for an activity equals its earliest start time ES plus D, the duration required for the activity. It is the relation used for the "forward pass" through a network. For example, in starting, event 1 obviously cannot start before now: time zero. Event 2 requires only activity A; thus, by the first "basic relation" above, EAT(2) = EF(A) is 1 min (0 + 1) as shown by the first number above event 2. Continuing, adding to it the duration of activity B, the EAT(3), shown at event 3, is 4 min. The EAT's at the other events are determined in a like manner by conceptually continuing to move through all events to the final one. Note, however, that if more than one activity terminates on an event, the one which is completed last controls the EAT of that event. For, as noted above, an event is not achieved until all activities directed to it have been accomplished. Accordingly, event 7 requires the accomplishment of both activity G and F. F has a duration of 2 min and can be started at EAT(5) = 5; hence, EF(F) is 7 min after project start. G is called a *dummy activity;* it requires zero time and is shown by a dashed line. Dummy activities are used to show precedence; that is, their terminating events must be preceded by the originating ones. Thus, in this special case, event 7, besides the accomplishment of the activity F directed to it, also requires the accomplishment of event 6. The dummy activity G (of zero duration) is used to show this. Hence the EAT(7) is controlled by event 6; that is, EAT(7) is EAT (6) plus the duration of activity G. Accordingly, the EAT(7) = $18\frac{1}{2}$ + 0 = $18\frac{1}{2}$ min. Finally, adding the duration for task H gives 19 min for the EAT of event 8 and for the project. The activities which control the project's minimum completion time make up the path called the *critical path,* from which this method gets its name.

A systematic method of identifying the critical path uses the EAT at an event and a second "accomplishment" time, LAT (latest accomplished time). It is the latest time in which an event can be achieved without causing delay of a subsequent activity. If that event is on the critical path (defined below), any delay would delay the total project. The LAT's are determined by a "backward" pass, that is, working backward from the final event. For our example above, we start with the EAT of 19 min for event 8. It is the time required for completing all activities on the project's critical path. As is common practice, take the EAT(8) of 19 min for the project to be management's LAT, that is, the desired latest accomplished time for the project. Enter it as the second value above the box representing the last event. Now, moving backward to event 7, and using the second "basic relation" from above, subtract the $\frac{1}{2}$ min duration time of activity H to get LAT(7) = $18\frac{1}{2}$ at event 7. Likewise, we subtract 0 from it to get an equal value for LAT(6); operating similarly, we get the LAT's for events 4, 3, 2, and 1. At all those events, we note that the LAT's are the same as the EAT's. This identifies the critical activities, which define the critical path. It is shown by double lines in Figure 5-2. At each event on the critical path, LAT = EAT. In large projects there might be many critical paths, that is, ones in which all of their events have LAT = EAT, and they will all be of equal duration. In medium-sized projects there will often be more than one critical path.

Slack: Free and Total Slack times (also called *floats*) show where excess resources could exist. They indicate extra time available for activities not on the critical path. There are two types: total slack and free slack. *Total slack* is the delay possible without delaying the project. In some cases, management (or the project's client) might accept a completion time later than the project's EAT, its earliest achievement time. In those cases, there will be an automatic increase in the total slack for all activities of the project equal to the accepted delay. *Free slack* is the delay possible without delaying any activity. Both types of slack are reviewed below. For arrow networks, slacks are determined from data associated with events, but only the extra time available for activities is usually significant. The required information for evaluating activity slacks is shown above events in Figure 5-2.

In our simple example there are two paths, one noncritical; but for this and larger projects, determining LAT's and the EAT's for all events is advantageous. First, as noted above, their equality defines the critical path(s). Their values shown at events not on the critical path give us special information. At each event, the difference between the LAT and EAT equals the maximum slack of any activity terminating on it. When there is only one terminating activity, the difference is that activity's slack. For two or more activities terminating on the event, their respective slack equals the event's LAT minus that activity's EF. We will see that free slack is always equal to or less than the total slack. Again, the terms *slack* and *float* are often used interchangeably.

In our example, only those activities associated with event 5 have slack; it equals the difference between the LAT(5) and EAT(5): $LAT(5) - EAT(5) = 16\frac{1}{2} - 5 = 11\frac{1}{2}$ min. Let us define the time required for (or duration of) an activity between events i and j as $d(i, j)$. The slack $S(i, j)$ of an activity (i, j) is the difference between the $LAT(j)$, given at event j, and the sum $EAT(i)$ at event i and $d(i, j)$:

$$S(i, j) = LAT(j) - [EAT(i) + d(i, j)]$$

Again, considering our example, activities D and F are not on the critical path and thus have slacks or floats. We will evaluate them, using the data from Figure 5-2, for activity D, $i = 3$ and $j = 5$, and $d(3,5) = 1$. By using the relation above, the slack for activity D (between $i = 3$ and $j = 5$) is given by

$$S(3, 5) = LAT(5) - [EAT(3) + d(3,5)]$$
$$= 16\frac{1}{2} - (4 + 1) = 11\frac{1}{2} \text{ min}$$

If all $11\frac{1}{2}$ min were used for D, the EAT at event 5 would become $16\frac{1}{2}$ min, and there would be no slack for F. Suppose, however, that $5\frac{1}{2}$ were used. Then the $d(3,5)$ of D would become $1 + 5\frac{1}{2} = 6\frac{1}{2}$ and a new EAT(5) would be $5\frac{1}{2}$ greater than originally, or $5 + 5\frac{1}{2} = 10\frac{1}{2}$. Now, the slack for F would be

$$S(5, 7) = LAT(7) - [EAT(5) + d(5, 7)]$$
$$= 18\frac{1}{2} - (10\frac{1}{2} + 2) = 6 \text{ min}$$

We immediately recognize that the sum of the slacks of possible activities (here D and F) equals the slack of the event (here event 5) to and from which the two connect. Furthermore, we realize that the slack of the event can be divided as desired between the incoming and outgoing activities. Thus, calculation of activity slacks is usually unnecessary, and the event's slack is the difference between its LAT and EAT.

Now let us review and evaluate free and total slacks in our example. Recall their definitions. First, free slack is delay possible in an activity without delaying any other activity. Again, the network shows that activities D and F are not on the critical path and, in turn, can have slack. We see that although the calculations in the previous paragraph show that D has slack, it has zero *free* slack; for any delay in D will equally delay event 5 and, in turn, activity F, which begins at event 5. However, from the earlier calculations, F could be delayed $11\frac{1}{2}$ min without delaying the project. Thus, by the definition, the *total* slack for D is $11\frac{1}{2}$ min.

This information on the critical path(s), EAT's, LAT's, and the slacks is invaluable to project managers. It enables them to smooth workloads and recognize potential problems. Also, it shows where extra time can be shifted from an activity without disrupting progress, and where bottlenecks might need special effort and control to stay on schedule. Free slack gives the supervisor flexibility in starting a task. Often time-scaled networks provide managers a clearer view of a project's status.

Time-Scaled Networks   The network diagrams we have reviewed were originally developed from bar charts, also known as Gantt charts. The AOA format is easily adaptable to time scaling. It is very similar to the bar charts. The AOA of Figure 5-2 is shown time-scaled in Figure 5-3. On it the length of the arrows is proportional to the duration of the activities. The slacks are easily recognized, and the critical path is shown by the heavy lines. It is apparent that D has no free slack. As noted earlier, however, the total slack between events 3 and 7 ($18\frac{1}{2} - 7 = 11\frac{1}{2}$) can be divided as the manager pleases between activities D and F.

The present time on this project is the thirteenth day (marked as *Now* on the time line). Accomplished tasks are underlined. The project manager can immediately see which activities are on or behind schedule.

Milestones   In most projects there are events and associated times at which significant progress has been made, e.g., the completion of a phase. These events are called *milestones* and are recognized as events at which many activities have been completed. Milestones are times at which assessments, decisions, and perhaps corrective action must be made; they are a convenient time for project assessment. The small project of our example has only one milestone, event 7. It is not significant since it is the only event before the last.

FIGURE 5-3
Time-scaled network for Example 5-1.

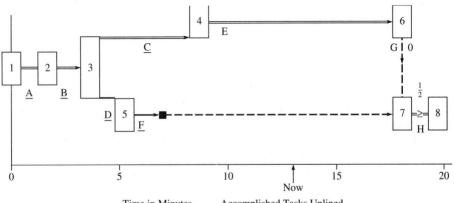

Activity-on-Node Format (AON) [1, p. 30]

As noted earlier, the AON format is presently not as well accepted as AOA. Its advantages, however, are causing it to gain popularity. Its advantages are that it is simpler to explain, more readily understood by nontechnical users, and easier to revise. The arrow format (AOA) of Example 5-1 is repeated in Figure 5-4. We will convert it to AON. Please refer to Figure 5-5. Recall, in this format, that activities are represented by nodes (blocks), and arrows only serve to show the interrelationship between activities. The values below the blocks (nodes) give the durations for the activities they represent. As might be expected for AON, the activity start and finish times are more relevant. They are the ES's (earliest start times) and LS's (latest start times) and are shown above the blocks in that order. They are determined in the same manner as for the AOA (arrow-on-activity) networks: a forward pass for the ES's, a backward pass for the LS's. For the ES values, starting at time zero, the ES(A) is 0. To it add D(A) = 1 to get ES(B)= 1; next ES(C) and ES(D) equal ES(B) + D(B) = 1 + 3 = 4. These ES values and the others are shown in the figure.

For the LS values, let us at the finish node. For LS(H) we subtract $\frac{1}{2}$ from the project's finish time of 19: LS(H) = $19 - \frac{1}{2} = 18\frac{1}{2}$, shown as the second value above H. In turn, LS(F) = LS(H) − d(F) = $18\frac{1}{2} - 2 = 16\frac{1}{2}$. In a similar way the other LS's are determined and are shown as the second value above each activity. LS's = ES's for all activities except D and F. Their inequality shows them to have slack. They have equal total slack: LS's − ES's = $11\frac{1}{2}$. Only F has free slack, for any delay in D delays F. As shown for AOA, the total slack can be divided between them.

Occasionally a task (such as A) needs to be only partially accomplished before another job (such as B) can be started. Both AOA and AON require that an event be completed before a subsequent one can start. A solution is to divide task A into subtasks A(1) and A(2). The division adds complexity to the network. Precedence diagraming (PD) is an extension of AON which makes provisions for this situation. It is also known as the precedence diagraming method (PDM, often just PD) or precedence network (PN) [2, p. 38; 3, p. 134]. PD's provisions for considering variations in the start of subsequent tasks is based on four types of constraints: They are shown in

FIGURE 5-4
The AOA network for preparing a quart of tea.

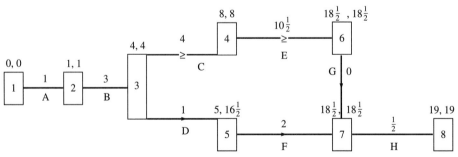

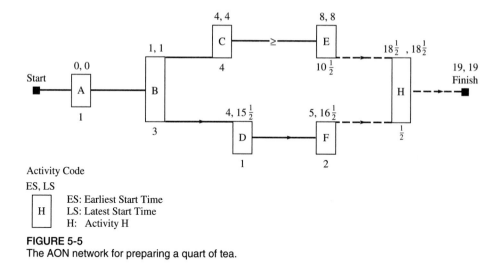

Activity Code

ES, LS

H	ES: Earliest Start Time
	LS: Latest Start Time
	H: Activity H

FIGURE 5-5
The AON network for preparing a quart of tea.

Figure 5-6. The delay n is time units, e.g., hours or days. The first, FS(A-B) $= n = 0$, is the only type possible in AOA and AON.

Example 5-2: Microchip Production

The project we use to demonstrate PN is the largely automated production of bipolar integrated circuit chips. It is greater than necessary for the demonstration but was chosen because of the importance of microelectronics in modern engineering. The activities are presented in Table 5-2 and are diagramed in Figure 5-7.

Circuit chips are built on silicon slices, cut from single crystal. The slice is round; about 15 cm in diameter and millimeters thick. The slices or wafers are processed in batches, typically of about 50. The process has four cycles: for the transistor's collector, base, emitter, and circuit traces. The traces are made by condensing gold vapor; they connect the transistor to the chip pins. At small firms, certain

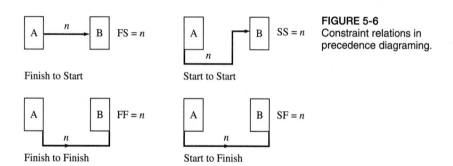

FIGURE 5-6
Constraint relations in precedence diagraming.

stages in the cycle can handle only 10 or 25 wafers at a time; thus, delays for accumulation are required. Delay for cooling after activity A is considered by FS = 4 min. Each cycle has 13 steps (1 through 13), which together consist of 12 activities (A through L). The process is formulated in Table 5-2 [4, Chap. 2]. Please review it now.

A PD (Figure 5-7) for this process appears much simpler than one using an AOA or AON format. There are four cycles: One each, A through I, for collector, base, and emitter; the last cycle is A through H, to J for the traces, and then to K.

To summarize, the total process has four main cycles:

1 For the collector, A through I:

A + I = 3 h

The total of FS(A-B), B, C, SS (D-E), E, F, and SS (G-H), and H = 30.75 min

2 and **3** Similarly, each for base and emitter, I to A to I, equals $2\frac{1}{2}$ h + 34.75 min

4 For the traces, FS (I-A), A through H, J, and FF(J-K): 2 h and 89.75 min or 3 h and 29.75 min

TABLE 5-2
ACTIVITIES IN MICROCHIP PRODUCTION

Step	Activity	Step	Activity
1.	A. The slices are baked in a water vapor environment at 1100°C for 2 h. This covers them with silicon oxide (glass), an insulator.		each. 10 min per minicycle: SS(G-H) = 20 min.
2.	Cooling: FS(A-B) = 4 min.	9.	H. Rinse removes photoresist. Full batch ($\frac{1}{2}$ min).
3.	B. The wafers are then covered (sprayed) with a photosensitive substance called *photoresist* ($1\frac{1}{2}$ min).	10.	I. Diffuse in 1000°C oven; dopant goes where oxide was removed. In the first cycle for C, collectors (deep), the time = 1 h; then in cycles for B (base)
4.	C. They are spun to smooth the coatings ($\frac{1}{2}$ min).		and E (emitter), each $\frac{1}{2}$ h.
5.	D. Proper masks are set up for exposing the wafers to ultraviolet light (UV), one wafer per minicycle to each of 10 human operators using microscopes. Five $\frac{1}{2}$-min minicycles = $2\frac{1}{2}$ min for a batch of 50. Thus, SS(D-E) = $2\frac{1}{2}$ min.	11.	After each, cool 4 min (includes $\frac{1}{2}$ min for system reset after cycles for C and B), FS(I-A) = 4 min.
6.	E. Batch exposed to UV ($\frac{1}{4}$ min).	12.	J. After emitter cycle, A through H, deposit circuit traces: In a vacuum, gold is condensed from gold oxide vapor. Connecting paths are made to the transistor: $\frac{1}{2}$ h.
7.	F. Developed as for regular photos; this removes photoresist from the silicon oxide, where it was protected by the mask ($1\frac{1}{2}$ min).	13.	K. Bond gold traces to chip pins, and package as a finished product. One wafer to each of 10 human operators using microscopes. 5 min per minicycle of 10 wafers: FF (J-K) = 25 min for full batch.
8.	G. Acid (hydrofluoric) bath for etching and removing oxide not covered by photoresist. Two subbatches, 25 min		L. Finished

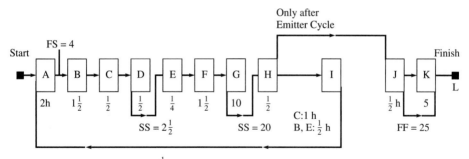

FS = 4, for cooling and reset ($\frac{1}{2}$ min) after C (ollector),
B (ase), and E (mitter) cycle.

FIGURE 5-7
Precedence diagram for micro-chip production.

The total time is the sum of these: 13 h, 10 min. Since there are no parallel paths, all activities are critical.

Crashing a Project

Frequently a project's predicted completion time is too late to meet the client's needs. Or perhaps an earlier completion might be adequately advantageous so that the client would offer a bonus.

Example 5-3: Expediting Installation of an Ice Floor

Consider that the CPM network of Figure 5-2 models the installation in an existing gymnasium of an ice floor for hockey. The numbers above the arrows, however, represent the activities' duration in days (instead of minutes as in the tea making example). Desiring to have the rink ready at the beginning of the season, the client is willing to give an $8000 bonus for completion of the installation in 17 days, 2 days less than presently expected. A crash program is a possible solution and is presented next. A probability analysis using PERT might show the crash to be unnecessary. We will consider it in Example 5-4. For now, let us proceed with "crashing" the project.

Solution: Your firm recognizes that the activities can be expedited in three ways: (1) by purchasing more preassembled refrigeration apparatus, (2) by putting more workers on the job, or (3) by having the present employees work overtime. Naturally, all three ways add to the cost of each expedited activity and to the entire project. After close analysis of the methods most economical for each activity, the firm's engineers developed the crash costs of Table 5-3. Note that since activities D and F have slack ($11\frac{1}{2}$ days between them) far in excess of the 2 days, they were not considered in the cost per day analysis. Only those activities on the critical path need be considered. Had the sum of the slack of D and F been less than 2 days, the

TABLE 5-3
COST FOR THE PROJECT ACTIVITIES

Activity	Normal		Crash		
	Days	Cost	Days	Cost	Cost/day
A	1	$ 500	$\frac{1}{2}$	$ 1,000	$ 1,000
B	3	3,000	2	5,000	2,000
C	4	8,000	3	20,000	12,000
D	1	2,000			
E	$10\frac{1}{2}$	5,500	9	8,500	2,000
F	2	2,500			
G	0	0	0	0	0
H	$\frac{1}{2}$	500	$\frac{1}{2}$	500	N.A.
	Total	$22,000			

2 days' reduction would have made their path critical and they would have had to be "crashed" also.

The cost per day data in Table 5-3 were derived using a linear approximation.* It is given by the equation below:

$$\text{Cost/day} = \frac{\text{crash cost} - \text{normal cost}}{\text{normal duration} - \text{crashed duration}}$$

For example, activity C is already efficiently organized; thus the cost to reduce its duration would be great, for the reduction would require use of all the means of expediting it, as presented above. Its cost per day is determined by

$$\text{Cost / day} = \frac{\$20,000 - \$8000}{4\,\text{days} - 3\,\text{days}} = \frac{\$12,000}{1} = \$12,000\,\text{per day}$$

Naturally, the lowest cost per day would be chosen; it is $1000 per day for activity A. Regretfully, it can be reduced by only $\frac{1}{2}$ day (at a cost of $500). The next lowest cost per day is the same for activities B and E ($2000). As noted above, the smaller the relative change is, the better is the linear approximation. It would be better to reduce E by $1\frac{1}{2}$ days (approximately 15 percent) than to change B by 1 day (33.3 percent).

The total cost for the project is the sum of the costs for all activities; for normal time, the sum from Table 5-3 is $22,000. Crashing A ($\frac{1}{2}$ day) and E ($1\frac{1}{2}$ days) adds a cost of $1000 for A and $3000 for E, for a total crashing cost of $4000. This extra cost must be added to the normal base cost for a new total project cost of $26,000, when completed 2 days early. Investing $4000 to crash the project by 2 days is surely worthwhile. The $8000 bonus far exceeds the cost. However, there could be an easier way.

---

*As shown by a Taylor series, all continuous systems respond in a nearly linear manner if the change of input is adequately small.

Since your firm probably has and will have many clients who will seek earlier completion times, it is wise to consider "the averages"; that is, consider the probability of completing the project early while working on a normal schedule—at no additional cost. When there are considerable uncertainties about the project's activity times, a probability analysis gives valuable information. PERT is a technique for evaluating project probabilities. It is presented next.

5-2 PERT: PROJECT PROBABILITIES [3, p. 41]

Consider a project in which your firm might have had only limited experience. Probably, the supervisory personnel cannot give activity times with much certainty. Often, however, they can give the range of the times with reasonable certainty. Even though the activities' durations do not have a normal or gaussian distribution over these ranges, the normal distribution can be used to estimate the average value of projects' completion times. The rationale for this statement is given in Appendix A of this chapter. In an example below, we use the standard normal distribution to estimate a project's expected or average completion time. We will, however, first review the basic PERT technology.

   This technique uses three time estimates as given by an activity's supervisor: a most likely estimate, an optimistic estimate, and a pessimistic estimate. The first, the most likely estimate, is symbolized by m; it is the most realistic value of the time required for the activity. Statistically, it is the mode (highest magnitude) in the probability distribution. The optimistic time, symbolized by a, is the unlikely but possible time required for the activity if all goes as planned. Finally, the pessimistic estimate, symbolized by b, is the unlikely time required for the activity if "everything goes wrong." These last two, the optimistic and pessimistic times, are considered to be at the extremes of the probability distribution. Recall that for the normal or bell-shaped curve, over 99.7 percent of all occurrences fall within $\pm 3\sigma$, that is, plus or minus three standard deviations. Thus, b at the upper end of the distribution minus a at the lower end equals six standard deviations, that is, $b - a = 6\sigma$.* Hence, the variance for the ith activity can be expressed as

$$\sigma^2(i) = \frac{\left[b(i) - a(i)\right]^2}{36}$$

Furthermore, the variance for the entire critical path equals the sum of the variances of the activities on the critical path.

   Accordingly,

$$\sigma^2 = \sum_{\substack{\text{CP}}} \sigma^2(i)$$

CP i on critical path

---

*As commonly known in statistics, for unimodal distributions, the standard deviation can be approximated as one-sixth of the range of its distribution. This is satisfactory because 89 percent of any (99.7 percent for the normal) distribution lies within three standard deviations of its mean [6, p. 283].

Equally important is the average or expected value of each activity's duration. It has been found (in practice) that an activity's average time is closely given by the following expression: Its symbol is t (exp).

$$t(\exp) = \frac{2m + \frac{1}{2}(a+b)}{3}$$

This equation shows that m, the most likely time, is twice as important (has double the weight) as the average of the optimistic and pessimistic times.

Naturally, the sum of the expected values of the activities on the critical path will be the expected value of the project's completion. You recall that this is true because the average of a sum is the sum of the averages.*

Example 5-4: Probabilities of Earlier Completion of Example 5-3

Consider Example 5-3 concerning the installation of an ice floor to be used for hockey. Let us determine the probability of completing the project in 17 days without crashing any activities. We will need the values for a, b, and m for all activities on the critical path.

Solution: The supervisors for those activities (except for G) gave the estimates in the first three data columns of Table 5-4; the values in the last two columns were determined from the equations above.

Let us calculate $t(\exp)$ and then σ^2 for activities A and E. Using the equations from above,

$$t(\exp) \text{ for A} = \frac{2(1) + \frac{1}{2}\left(2 + \frac{1}{2}\right)}{3} = \frac{13}{12}$$

and

$$t(\exp) \text{ for E} = \frac{2\left(10\frac{1}{2}\right) + \frac{1}{2}(7+12)}{3} = \frac{122}{12}$$

These times and those for the other activities are included in Table 5-4 with the supervisor's estimates. Their sum gives the expected project completion time as $\frac{227}{12} = 18\frac{11}{12}$ = 18.9167. As might be expected, this is close to 19, the deterministic value. We will now evaluate the variances and the probability of a shorter completion time.

Continuing with activities A and E, their data give

$$\text{For A, } \sigma^2 = \frac{\left(2 - \frac{1}{2}\right)^2}{36} = \frac{9}{4 \times 36} = \frac{9}{144}$$

---

*It is often helpful to remember that any linear operation can be commuted; for instance, the integral of a sum is the sum of the integrals. Note, however, that because of the nonlinearity of the operation, the reciprocal of a sum is not the sum of the reciprocals.

TABLE 5-4
ESTIMATED ACTIVITIES DURATION

Activity	a	b (in days)	m	t(exp)	(σ^2)
A	$\frac{1}{2}$	2	1	$\frac{13}{12}$	$\frac{9}{144}$
B	$1\frac{1}{2}$	4	3	$\frac{35}{12}$	$\frac{25}{144}$
C	$2\frac{1}{2}$	6	4	$\frac{49}{12}$	$\frac{49}{144}$
E	7	12	$10\frac{1}{2}$	$\frac{122}{12}$	$\frac{100}{144}$
H	$\frac{1}{2}$	$1\frac{1}{2}$	$\frac{1}{2}$	$\frac{8}{12}$	$\frac{4}{144}$

$$\text{For E, } \sigma^2 = \frac{(12-7)^2}{36} = \frac{100}{4 \times 36} = \frac{100}{144}$$

These and the variances for the other critical activities are listed in the table with the other project data. Their sum gives the variance for the project as $\frac{187}{144}$ or $\sigma = \frac{13.675}{12}$, or $\sigma = 1.14$.

Now, we determine the probability of completing the project in 17 days with no crash efforts. To use the standard normal tables, we need to transform our data to standard form. We define a normalized variable z; it is

$$z = \frac{x - t(\exp)}{\sigma}$$

where x is the completion time for which the probability is sought. For our data, from the table above and recalling that the times are in days,

$$x = 17 \text{ days} \qquad t(\exp) = 18.9167 \qquad \text{and} \qquad \sigma = 1.14$$

$$z = \frac{17 - 18.9167}{1.14} = -1.6813$$

Now using the standard normal distribution in Appendix B of this chapter,

$$\text{Prob}[t(\text{completion}) \leq 17] = \text{prob}[z \leq -1.6813] = .04637$$

Note that the probability value .45363 for positive z is obtained by interpolating between the values for $z = 1.68$ and 1.69. Since z is negative, the probability for $t \leq 17$ is determined by subtracting the .45363 from .5, giving .04637.

This probability indicates that if our firm did many projects like this one, on average it would complete them in 17 days less than 5 percent of the time. Accordingly, the $8000 bonus averaged over those many projects would only average .04637 × $8000 = $370.96. Thus, in this case, any crash cost less than $8000 − $370.96 = $7629.04 would be advantageous. You have probably already realized that even with crashing the project the probability of completing it in 17 days is probably no greater than 95 percent. Using that value for completion 2 days early, on average, the reward for crashing would be

$$\text{Average reward} = .95 \times \$8000 = \$7600$$

Hence, being more realistic, any crash cost less than $\$7600 - \$371 = \$7229$ would be advantageous. The gain would equal the $7600 minus the crash cost, as $\$7600 - \$4000$ (from the original crash calculations) would give a $3600 advantage.

5-3 PROCEDURAL STEPS FOR PLANNING: CPM-PERT

1 Bring together the key supervisors or managers whose units will be involved in the project. This may be only you or could include a fellow engineer or technician.

2 Start with the completion event (e.g., delivery of the product to the customer or completion of a bid).

3 On a large sheet of paper or possibly on a flipchart or chalkboard, list the events needed to achieve completion. Work backward from the last event to the starting event, connect events by the activities needed to achieve them, and continue the process until a skeleton outline of the network has been developed. *Note:* Be sure that all events, except the first and last, are connected by one or more activities to a preceding event(s) and that any outgoing activities are connected to succeeding events.

4 The skeleton network should be redrawn, reproduced, and distributed to each unit or department involved in the project.

5 In these departments or units, division of tasks might be required to develop a complete and detailed network for their activity or activities.

6 Each manager or supervisor who has responsibility for the unit should give an estimate of the time (in hours or workdays) required to achieve the relevant events. If PERT is needed, the three times for optimistic, normal, and pessimistic estimates should be included.

7 After receiving the time estimates from the supervisors, the project coordinator will draw up a final detailed network, with the schedule changed to calendar time. By studying this, each department knows its responsibilities and those of others, as well as the necessary completion times.

8 Complementary to the master plan, the project coordinator also makes assignment sheets for each supervisor or manager with the project events and completion dates for that unit or group.

9 Next, at a meeting with the supervisors of the key units, they and the coordinator should identify possible bottlenecks which could be caused by unanticipated occurrences. Probabilities from PERT can be used to focus the time sensitivities. Building on their mutual creativity, the supervisors and the coordinator should develop contingency plans.

10 After the project begins, each unit supervisor or manager should periodically report (perhaps every week) the status of and potential problems for the unit's activity.

11 From these reports the project coordinator will generate and distribute periodic updates of the project plan, with actual dates of events and predicted project completion.

12 As often happens in "real life," the entire project or specific critical events will get behind schedule. When this occurs, the coordinator and the responsible manager(s) will hold a corrective action meeting to ensure on-schedule completion of the project. If contingency plans are made in advance, this meeting should require only a final review before their implementation.

5-4 SUMMARY

For a project's smooth progress and scheduled completion, hardly any factor is as important as planning. CPM is highly effective in identifying and communicating to all participating units a project's component activities (with responsibilities), their precedence relations, and possible needs for corrective action. It also gives the project coordinator methods to maintain the normal schedule or for crashing critical activities to achieve an earlier completion.

Finally, PERT gives needed information on the probabilities of activities' durations and the project's completion.

CPM, PDM, and PERT systems supply information to the project coordinator that is needed for prudent decisions in project planning and management. These same techniques and data are also needed for significant proposals.

PROPOSAL PRESENTATION

Many industries, particularly computer and other electronic companies, are heavily oriented toward research and development. Proposals are their primary source for obtaining revenue, i.e., income for existence and growth. In our free enterprise system, competition keeps prices to a minimum. Only innovation, however, makes it possible for a company to increase its product's price by establishing that product as having a higher value. Merchandising techniques are employed to do this; however, only *true* innovations, "sold" to management by technical proposals, can keep a company growing. H. J. Wisseman, assistant vice president of Texas Instruments, said, "The technical proposal is our divisions' primary sales tool. . . ." [5] Over time, almost every employee in an engineering firm is affected by the skill of its proposal writers.

The following sections build on the information and concepts of earlier chapters, analyze the "seller-buyer" interaction, and use the information to present methods for developing forceful but ethical proposals. We also review factors which can significantly influence the decision maker(s) who accept or reject the proposal.

5-5 THE ENGINEERING APPROACH

In this approach, we define the "problem" and review the relevant fundamentals. *"Proposal,"* from the French, means to "put forth." As noted in the next paragraph, in our daily lives we put forth many ideas to our friends and, if married, especially to our spouses, who share our efforts to achieve family goals.* In business and industry, almost daily, we offer proposals, ideas, and plans on an informal basis. They might be directed to our immediate (first-line) manager or to a small group. Generally, the "trial balloon approach" is used. For a client or for higher-level management, a formal plan is usually presented. Both professional proposals and the informal daily ones are "seller to buyer" interactions. The same factors influence both. A *proposal* can be defined as an offer to act (e.g., to solve a problem or to provide a product or service) for a specific cost within a specific time.

Definitions and Fundamentals

For the engineering approach, the relevant fundamentals include the logic and emotions which influence this buyer-seller relation. Initially the interchange will probably be person to person. The material in Chapter 1 on interpersonal relations is especially relevant here. Our personal experiences are relevant as well. We recognize that in our youth—often in interchanges with brothers, sisters, and parents—we were frequently "selling" them. Our efforts were to get them to respond in the way we thought best. As adults, evidence of honesty and ethics is strongly influential in proposal acceptance. Although techniques of persuasion are important, history has shown that over the long term, it is wise to be ethical and honest.

Informal Proposals These are often on a personal basis, frequently one to one, you and the DM (decision maker). In these situations you can usually choose a time when you think the DM will be in a favorable "frame of mind." It is said that Benjamin Franklin advised to never seek a decision from someone before lunch. Probably "before lunch" is not applicable today, but still, as Franklin recognized, the DM's frame of mind is very important. Let us review factors which can influence the decision maker's receptivity to your ideas.

A technical or business proposal suggests the existence of a problem, one which your plan can solve. It is very important to determine the person who has the responsibility and authority to implement your solution. Identify the real decision maker. It is also relevant to determine those who influence her. The "selling" process seems to progress through five phases; they can be remembered (encoded) by using the letters *AIETA*. This is not an acronym, for it makes no word; however, it can be pronounced like (and readily associated with) Verdi's opera *Aida;* recall that in Italian, the letter *i* is pronounced the same as our English *e*. Thus, keeping *AIETA* in mind, the first phase of the process is *a*wareness; next *i*nformation is sought about a possible solution. Then the information is *e*valuated. If the results are favorable, a *t*rial solution is made. Finally, if the trial is favorable, the plan or solution is *a*ccepted.

Awareness A problem may not be self-evident. Thus, the DM may or may not be aware of it. If she is aware, the problem manifests itself as a need, want, or desire;

*Including your children in the sharing of your goals enhances their personal growth.

however, she may not have identified it correctly. If the decision maker is unaware of the problem, with tact, the proposing engineer tries to make her aware of it. However, the DM might still feel that the problem is not important enough to warrant attention. Other "fires" could be more pressing. As noted in Chapter 1, often, when planning is inadequate, we give attention to the most pressing situation rather than the most important. In such a case, the engineer might have to wait for another time. Remember the old adage, "Things have to get worse before (action will be taken so that) they can get better." The decision maker could be naturally defensive or perhaps has had a recent promotion. For whatever reason, she might think it best not to take chances on an untried plan. If the DM is definitely unreceptive, then you can only reflect on the failure for future guidance. As stressed below, perseverance has great value, but you do not want to be an annoyance. After a pleasant interchange, however, the DM might later appreciate the value of your plan.

Information   If the decision maker *does* recognize the problem as important and seeks a solution, you can be ready to present it. Knowledge of her background and any prejudices can be helpful in influencing her evaluation. The format for the solution plan is presented below.

Evaluation   The DM studies the solution plan and, as needed, asks for clarifications. The evaluation should be as complete and clear as possible. If part is omitted the DM could not know enough to ask about it. The decision maker weighs the benefits as presented in your plan against the costs and any feelings or emotions she may have toward you or your group.

Trial   To obtain more information, the DM may want to try certain aspects of the solution. Especially important is information indicative of the proposal's feasibility at a reasonable cost.

Acceptance   After evaluating all the information, the DM decides to accept or reject your proposal. In the latter case, she should present her reasons for not accepting. If they can be corrected, you can change your plan and try again; perseverance without antagonism is extremely effective.

For most technical writings, to some degree, the reader is self-motivated to try to understand your work; this is usually less true with proposals, especially when they are unsolicited. For this reason, clarity and conciseness are even more important. These characteristics are easier to develop using an outline. We will review the development of an outline, using the same acronyms we used for technical writings. First, let us consider characteristics peculiar to proposals, particularly formal ones.

Formal Proposals   These are nearly always written and have three significant parts: an offer and plan to solve a technical problem, a specific plan for management, and a specific cost [6]. For large proposals, these three parts are often presented in separate volumes.

The Technical Proposal   This is the major focus of this section. The technical proposal is an offer or approach to solve a problem, often in minute detail, with alternative plans and designs. It is vital that you indicate a thorough understanding of the problem, as understood by the client.

The Plan of Management   This explains in explicit detail how the project will be managed, often with persons responsible for each activity identified by name. Knowledge of those identified assures the customer that the problem will be handled by competent personnel. Lines of responsibility are clearly indicated. A time schedule for the project's completion is usually suggested. (Techniques of project planning and management were presented earlier in this chapter.)

The Cost Proposal   This details the costs in terms of labor and material. These aspects are often closely coordinated with the firm's accounting department. For further details, see Chapter 6 on engineering economics.

Proposals: Solicited and Unsolicited   Formal proposals can be solicited or unsolicited. The methods of preparation, presented in the next section, are applicable to both types. The remainder of this section reviews the special details of the more common, solicited proposal. It is a response to an "invitation to bid" and is often called a *bid of purchase request.* Clients send "invitations to bid" to companies listed in their source file. These are firms that the client knows and respects. Often, however, clients advertise in trade or professional journals.

These invitations are generally called *requests for proposals* (RFP). They present a set of specifications with a general discussion of the client's needs and possibly a formal statement of the work desired; a deadline is also indicated.

Development of Solicited Proposals

When companies receive an invitation to bid, they study its specifications and requirements. From this study, they determine whether or not to submit a proposal. These determinations depend on the company's capabilities, the availability of qualified staff and facilities, and the possibility of profit. If the decision is to bid, the two most important tasks are the development of a suitable design and the preparation of the cost estimate. The first must give the client confidence that the bidder can deliver the services offered. The second makes possible a benefit-cost analysis. Frequently, while studying the invitation, the company confers with representatives of the requesting firm to ensure, as thoroughly as possible, an understanding of the problem. Often the client will hold preproposal briefings for interested firms.

Proposals concerned with large projects are always team efforts, not the work of an individual (although there is usually a coordinator). The team develops the technical solution to the client's problem.

When the client receives the bid, separate teams are used to judge the relative merits of the technical aspects, the plan for management, and the cost details.

Preparation of proposals is important. Few experiences are as frustrating to a firm's engineers and executives as losing a contract to a competitor, especially when the engineers are convinced that they have developed a better design.

The Exploratory Phase   Recall that the acronyms for technical writing are *MAPS* and *PWRR.* For solicited and unsolicited proposals, *MAPS* has a second significance: *make a preliminary study.* This study extends the exploratory analysis of *m*essage, *a*udience, *p*urpose, and *s*cope. Logic and an excellent problem solution, while of greatest importance, are not always adequate to sell your *plan*.* It is also very important that your proposal harmonize with the client's attitudes and values. Specifically seek and address what the customer feels is crucial or important. These characteristics of the client can usually be gleaned from close study of the bid papers and by carefully listening to his representatives. These contacts could reveal the attitudes about alternative approaches. This information is important when (in your proposal) you review the history of previous efforts on the problem.

Study the competency of known or probable competitors. Honestly and truthfully stress your strengths and, as much as proper, note the weaknesses of others. The value of this emphasis is greatest in the introduction of your technical approach.

The Implementation Phase   We are now ready to plan the five parts of the proposal.

Planning   After a careful preliminary study, we can make a logically persuasive and technically strong proposal for solution of the client's problem. *PWRR* shows that the first step is *p*lanning (the proposal).

Surely every activity will be more effective if planned. Just as the proposed project needs a well-developed *plan* to ensure its acceptance, the proposal itself also needs a well-organized plan. Construct at least a rough outline before you start writing. It could be a tentative table of contents with probable headings and subheadings. Base the development of your plan on a mental path which leads the reader from the introduction through the body and the conclusion. Take whatever relaxed exploratory time necessary to clearly develop the information and concepts you wish to present. Generously use headings and subheadings to guide the readers and keep them oriented as they progress.

Elements of a Proposal   Although the "bid request" (or RFP) may set the format for the response, the proposal should have five parts: (1) clear statement of the problem, (2) a history reviewing previous approaches and their results, (3) your firm's (or your) solution and "how it works," with the scheme of implementation, (4) expected results with advantages and disadvantages, and (5) the response desired of the decision maker if the *plan* is accepted. As discussed below, a time frame for the response should be included.

Problem Statement

This statement fits well in the proposal's introduction. The introduction makes your first impression; recall that this part should clearly define the problem and indicate your understanding of it. Keep in mind the letters *SPSP;* that is, as for all technical writing, the introduction presents the *s*ubject and gives the *p*urpose, the *s*cope, and the *p*lan (of the presentation, not of the proposed solution).

---

*To avoid ambiguity, *plan* (italicized) is used in the remainder of this chapter to refer to the proposed problem solution; the word is not italicized when referring to the plan of the proposal.

History

This should be a concise review of other approaches that have been made to the problem. Include alternative solutions, and explain why you chose the one you did.

Your Solution

The specific problem to be solved will set the technical procedure for its solution; however, a convincing *plan* should also include these aspects:

Clear statements that show your thorough understanding of the problem (as the client understands it) and also your competence to solve the problem.

An explanation sufficient to ensure the degree of understanding that the decision maker(s) wants and needs to make a decision. It has been noted that people tend to oppose what they do not understand.

There is often additional material needed by (or of interest to) the reader, such as a justifying mathematical analysis. If including this in the presentation would be awkward or distracting, place it in the appendix.

Expected Results

These, along with costs, are the most important factors affecting the proposal's acceptance. This part should give a summary of the benefits that the decision maker (or her firm) will receive if your (or your firm's) proposal is accepted. Review:

- How it satisfies the agreed needs.
- Supporting proofs.
- Bonus benefits that will accrue to the DM. These might include feelings concerning one's self, one's group, the firm, or environmental responsibility.
- Undesirable features. It seems there are always a few, and not "sweeping them under the rug" enhances your credibility and strengthens your image.

Finally, there is no greater influencer of people than self-interest. It often includes the satisfaction in showing concern for others.

Decision Maker's Response

As considered in the next paragraph, most proposals require that activities begin by a certain time. You should clearly note this at the end of your proposal. Request an appropriate response by a specific date.

A decision by a certain time is desirable even for proposals which have no time limit. It is easy for an excellent proposal to die simply because the DM never got around to making a decision. Most decision makers are accustomed to deadlines. *Note:* Decisions having an indefinite time requirement are usually given low priority.

Scheduling   An important corollary of *planning* is making a schedule. It should enable everyone to meet his responsibilities with minimum stress. Such a schedule should not only consider each person's demands but also allow adequate time for administrative and service functions. Usually someone in management must prepare a letter of transmittal. The department head under whom the proposed work is to be done will surely want to review the finished document. Furthermore the publications department will need adequate time to professionally edit, illustrate, and print your document. Regretfully, first impressions, usually based on appearance, can have a major effect on any endeavor. Because of this, it would be unwise indeed to jeopardize an excellent proposal by poor reproduction and illustrations, often the result of a last-minute "rush job."

PWRR of Your Proposal   We are now ready to put our ideas on paper, keeping *PWRR* in mind.

Planning   Make a first draft of the plan to present the *plan* of your engineers. Evolution of the plan will continue throughout your development of the proposal. To avoid the possible difficulties noted above, be sure to include and schedule all activities, even the supporting ones which run parallel in time. Such a schedule is easiest using the techniques discussed earlier in this chapter: CPM (critical path method) and PERT (program evaluation and review technique). PERT is needed when the time required for project activities cannot be closely estimated. It permits you to consider the randomness of real-world situations.

Writing   A strong proposal combines the rhetorical forms of technical exposition (explanation) and argumentation (objective support for your *plan*) . The personal characteristics of the DM or his firm should not be considered in developing the proposal's technical *plan.* It is, however, suggested that they be considered in the proposal's presentation. Review the discussion of exposition and argumentation in Chapter 4. Then relax, and use the outline for the proposal's plan to write its first draft. Again, do not be distracted by spelling, grammar, or anything else in developing your ideas. Just write, recording your ideas as they come to your creative Child. You are the only one who needs to see the first draft.

Relaxing   Relax for at least 2 or 3 h, preferably overnight. Effective relaxing requires that you put the draft aside and work in other areas as needed.

Revising   Again, it is important that revision be guided by the basic rule of giving the reader the most information of value while requiring the minimum time and effort. Review the revision procedures in Chapter 4. Special attention should be given to transitions, the last revision suggested. They are more important in proposals, where your reader might be less motivated. Well-used transitions make a proposal easy to follow and understand. This ease also makes a favorable impression on the reader. Recognize that as with résumés from job applicants, reviewers look for reasons to eliminate all but one. Frequently the reasons are based on trivial points. Of course, the same can be true when there are two or more nearly equal bidders.

5-6 GENERAL GUIDELINES [7]

1 As with all technical writing, start by analyzing the situation. Reflect on who will be reading the proposal and what they need to know.

2 Consider which organization of the proposal will best present your material. Review the rhetorical options in Chapter 4.

3 Construct at least a rough outline before you start writing. Change it as necessary, but remember that organizing after writing wastes time and effort. Think of your readers and your purpose as you develop your proposal.

4 In the introduction, orient your readers thoroughly. Be sure the introduction identifies the critical problem(s) and your main purpose. Lead the readers into your proposal; present details later.

5 Consider preceding the introduction with an executive summary; it should emphasize the potential benefits and your ability to meet the customer's needs.

6 Use headings and subheadings liberally. Try using statements rather than topics. Your reader should be able to see your organization in these statements.

7 Do not bury important ideas under details. Place nonessential material in the appendix.

8 Concentrate more on the benefits of the proposal than on the method, unless for some reason your method is more important.

9 When you have finished your rough draft, put it aside for a while, preferably a day. Then revise and polish it.

10 Use tables, graphs, drawings, photographs, and other visuals to support and enforce your ideas.

11 Write clearly and as simply as you can, using plain concrete words. Avoid indirect expressions and involved constructions.

12 Write grammatically, punctuate properly, and spell correctly.

13 Remember that transitions help the reader link the ideas in your presentation; this practice facilitates his understanding and makes a good impression on the reader.

14 Also remember that the introduction makes your first impression; because of its importance, it may be best to develop the introduction after the major part of the proposal has been developed.

15 The primary subjects of all proposals are the problem and your proposed solution. In the latter, make absolutely clear your understanding of the problems and stress the major aspects of your solution.

16 Remember that while the proposal's final proof copy does give the author a last chance to make changes, it is less costly to make changes in the earlier drafts.

5-7 PROPOSAL SUMMARY

This title is suggested for the proposal's final section, for it seems that no "conclusions" can be drawn. In a summary, however, a strong final impression can be made, one which emphasizes the benefits of the expected results, discusses the strong points of the plan, and leaves the reader with the belief that the bidder is especially competent and will efficiently carry out the project. Be careful not to introduce any new material or ideas in this final section. If new ideas come to mind while writing the summary, go back and include them in the main sections. It is all part of the value of RR, relaxing and revising.

5-8 CONCLUSION

For proposals, as with other technical writing, follow these essential steps: gather and diligently study all relevant information, be relaxed in this exploratory phase, then try to totally relax for a short period before preparing an outline. Use the outline to write your rough draft; stress the strong points of the plan. For proposals, keep the presentation client-oriented. Liberal use of headings and subheadings makes reading and understanding easier. Finally, promoting the proper attitude of the decision maker is surely important, but most persuasive in selling an idea or plan is its intrinsic value.

May I stress again (from the end of Chapter 4) the thoughts of Wendell Berry, contemporary poet and essayist [8]. They are especially relevant to proposals: "What gets my interest is the sense that the writer is speaking honestly and fully of what he knows well."

DISCUSSION QUESTIONS AND PROBLEMS

1 Give your thoughts concerning the trade-off between the value of a plan and the time and effort required to prepare one.

2 Prepare a CPM network (arrow diagram) for a normal endeavor or project, such as changing an automobile tire or preparing a meal. To show the advantages of CPM, the endeavor should involve two or more people or an automatic device (e.g., a toaster or microwave oven) which permits simultaneous activities. For each activity, determine its EAT and LAT; indicate these on the network (see Figure 5-2). Also, determine the project's critical path with its minimum completion time.

Note: Figure 5-8 is for Problems 3, 4, 5, and 6. It is the CPM/AOA (activities-on-arrows) network for renovating a parts room.

FIGURE 5-8
An arrow network for renovating a parts room.

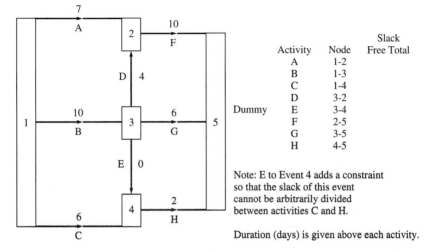

Activity	Node	Slack	
		Free	Total
A	1-2		
B	1-3		
C	1-4		
D	3-2		
E	3-4		
F	2-5		
G	3-5		
H	4-5		

Note: E to Event 4 adds a constraint so that the slack of this event cannot be arbitrarily divided between activities C and H.

Duration (days) is given above each activity.

3 For the project network in Figure 5-8, determine (a) the EAT and LAT for all activities and (b) the project's critical path and minimum time (days).

4 a Determine the free and total slacks of all noncritical activities (Figure 5-8).
   b Which of the slacks is always smaller or equal to the other?

5 For the parts room renovation, by changing the route of a critical electrical conduit, the dummy activity E can be removed. Determine the maximum free slack for (a) activity C and (b) activity H.

6 Draw a time-scaled diagram for the parts room renovation (E can be neglected).

7 Confirm the times given for the precedence diagram in Figure 5-7. Remember FS = 4 min for cooling (in the forward and reset paths).

8 The arrow network for installation of a diesel cogenerator is given in Figure 5-9.
   a Determine its minimum completion time and the slacks, free and total, for the noncritical activities.
   b Times and costs for all activities are given in Table 5-5. Determine the project's total cost to finish it in normal time.
   c Determine the minimum total cost to "crash" the project so as to finish it 1 day earlier than the normal time.

9 In the network for installation of the cogenerator (Problem 8 and Figure 5-9), activity B represents an order for ancillary (Latin for "maid servant," hence, *auxiliary*) equipment. It was discovered that the order includes a valve required in activity C.
   a Redraw the network, and determine the new minimum completion time. *Hint:* A dummy arrow is convenient to show the new requirement in an AOA network.
   b If you did not make an AON network in part a, redraw it as an AON network and determine its critical path. Note, no dummy arrow is required, but it is general practice to add "start" and "finish" modes.

10 Determine the method and total cost to finish the new project in Problem 9 a day early (see Table 5-5).

11 The supervisor for the project in Problem 8 gave the optimistic and pessimistic times (*a* and *b*, respectively) for the activities. The normal times are repeated; for them the symbol is *m*. The set of all activities are given in Table 5-6. The problems are designed to minimize "busy work."

TABLE 5-5
DATA FOR PROBLEMS 8 AND 10

Activity	Normal		Crash	
	Days	Cost	Days	Cost
A	3	$ 600	2	$ 700
B	4	500	3	650
C	5	400	4	600
D	2	1000	1	3000
E	6	600	4	1000
F	3	500	2	700

For activity E, the cost can be considered a linear function of the reduction.

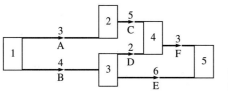

FIGURE 5-9
CPM Arrow Network for Problem 8.

    a Using PERT analysis, determine the probability of normally (no crashing efforts) completing the project 1 day early.

    b If the bonus for the 1 day earlier completion is $5000, determine the expected value of the bonus.

    c Over the long term, the firm has many projects such as this one. Compare the expected value of the bonus (part b) with the extra cost for crashing. Would crashing the project be economically advantageous?

12 Table 5-7 gives the activities for developing a software package. Because of proprietary information, the firm which contracted for the analysis identified the activities by letters only.

    a Develop the CPM/AOA network for the project, and determine its critical path and minimum time.

    b For each event, determine its EAT and LAT; and for each activity, its EF (earliest finish time) and LS (latest start time).

    c Determine the free and total slacks for each noncritical activity. The times may be entered in boxes below each activity as FS/TS, or listed as shown to the right of Figure 5-8.

    d What is the expected value of a $5000 bonus offered for completion of the project 2 days less than the critical path time?

13 In Section 5-5, we summarize the fundamentals of proposal presentation by using the letters *AIETA*. Review them, and give your thoughts on each.

14 What are the principal parts of a formal proposal? Review each of them.

15 Enhancement Project. With a colleague prepare a proposal for a significant project. The project need not be physical; the proposal could be for a change in sports' playoffs, student activities, or a firm's process or policy. In the proposal include a CMP/PERT network. Show the critical path, its minimum time and the slacks, free and total, of the noncritical activities.

TABLE 5-6
SUPERVISOR'S ESTIMATES OF ACTIVITY DURATIONS

	PERT		
Activity	*a*	*m*	*b*
A	3	3	3
B	4	4	4
C	4	5	6
D	2	2	2
E	4	6	7
F	3	3	3

TABLE 5-7
TECHNOLOGICAL ORDER FOR PROJECT ACTIVITIES IN PROBLEM 12

Activity	Preceding* Event	Technological required activities	Supervisor's estimates (weeks) a	m	b
A	1	None	2	3	4
B	1	None	3	4	6
C	2	A	3	5	7
D	3	B	2	2	2
E	4	C & D	1	1	1
F	2	A	3	6	8
G	3	B	3	3	3
H	5	G	4	7	10
J (there is no "I")	7	E, F, & H	3	4	6
K	6	H	3	5	8
L	8	J & K	2	2	2

*These events are not unique, but they should give the answers for part 6 given in the section Answers to Selected Problems.

REFERENCES

1 Peurifoy, R. L. & Ledbetter, W. B. *Construction planning, equipment & methods* (4th ed.). New York: McGraw-Hill, 1985, Chapter 2.

2 Moder, J. J.; Phillips, C. R.; & Davis, E. W. *Project management with CPM, PERT and precedence diagramming.* New York: Van Nostrand Reinhold, 1983, pp. 37–42.

3 Weist, J. D., & Levy, F. K. *A management guide to PERT/CPM* (2nd ed.). Englewood Cliffs, NJ: Prentice-Hall, 1977.

4 Muler, R. S., & Kamins, T. I. *Device electronics for integrated circuits* (2nd ed.). New York: Wiley, 1986, Chapter 2.

5 Mills, G. H., & Walter, J. A. *Technical writings* (3rd ed.). New York: Rinehart & Winston, 1970, p. 298.

6 Mills & Walter. *Technical writings.* pp. 299–300.

7 McAuliff, D., et al. *Effective technical communications* (2nd ed.). Needham Heights, MA: Ginn, 1987.

8 Morrison, C. Editor, *Communications Across the Curriculum News,* Vol. 1, no 3, Apr. 1990, Clemson University, SC.

9 McClave, J. T., & Benson, P. G. *A first course in business statistics.* San Francisco: Dellen, 1989, p. 317.

BIBLIOGRAPHY

Eisenberg, A. *Effective technical communication* (2d ed.). New York: McGraw-Hill, 1992.

Hillier, F. S., & Lieberman, G. J. *Introduction to operation research* (3rd ed.). San Francisco: Holden-Day, 1980.

Kamn, L. J. *Successful engineering, A guide to achieving your career goals.* New York: McGraw-Hill, 1989.

Mali, P., & Sikes, R. W. *Word processing for engineers and scientists.* New York: McGraw-Hill, 1985.

Stech, E. *Flow chart planning and project management.* Kalamazoo, MI: Lindermeyer, 1982.

APPENDIX A: Probability Distribution of a Project's Completion Times

As noted in Section 5-2, the activity times, as estimated by the unit supervisors, do not have gaussian or normal probability distributions. The project's completion time, however, can be represented by a normal distribution. This is true because the completion time is the sum of the average times of the activities on the project's critical path. Each is influenced by many statistically independent factors. Accordingly, by the central limit theorem, the average of the sum of statistically independent values (the average activity times) tends toward a normal distribution. Thus, the average or expected value of the project's completion time can be evaluated using the standard normal distribution [9].

APPENDIX B

CUMULATIVE PROBABILITY FOR Z OR LESS FOR STANDARD NORMAL DISTRIBUTION

Z	0	1	2	3	4	5	6	7	8	9
0.0	0.0000	0.0040	0.0080	0.0120	0.0160	0.0199	0.0239	0.0279	0.0319	0.0359
0.1	0.0398	0.0438	0.0478	0.0517	0.0557	0.0596	0.0636	0.0675	0.0714	0.0754
0.2	0.0793	0.0832	0.0871	0.0910	0.0948	0.0987	0.1026	0.1064	0.1103	0.1141
0.3	0.1179	0.1217	0.1255	0.1293	0.1331	0.1368	0.1406	0.1443	0.1480	0.1517
0.4	0.1554	0.1591	0.1628	0.1664	0.1700	0.1736	0.1772	0.1808	0.1844	0.1879
0.5	0.1915	0.1950	0.1985	0.2019	0.2054	0.2088	0.2123	0.2157	0.2190	0.2224
0.6	0.2258	0.2291	0.2324	0.2357	0.2389	0.2422	0.2454	0.2486	0.2518	0.2549
0.7	0.2580	0.2612	0.2642	0.2673	0.2704	0.2734	0.2764	0.2794	0.2823	0.2852
0.8	0.2881	0.2910	0.2939	0.2967	0.2996	0.3023	0.3051	0.3078	0.3106	0.3133
0.9	0.3159	0.3186	0.3212	0.3238	0.3264	0.3289	0.3315	0.3340	0.3365	0.3389
1.0	0.3413	0.3438	0.3461	0.3485	0.3508	0.3531	0.3554	0.3577	0.3599	0.3621
1.1	0.3643	0.3665	0.3686	0.3708	0.3729	0.3749	0.3770	0.3790	0.3810	0.3830
1.2	0.3849	0.3869	0.3888	0.3907	0.3925	0.3944	0.3962	0.3980	0.3997	0.4015
1.3	0.4032	0.4049	0.4066	0.4082	0.4099	0.4115	0.4131	0.4147	0.4162	0.4177
1.4	0.4192	0.4207	0.4222	0.4236	0.4251	0.4265	0.4279	0.4292	0.4306	0.4319
1.5	0.4332	0.4345	0.4357	0.4370	0.4382	0.4394	0.4406	0.4418	0.4429	0.4441
1.6	0.4452	0.4463	0.4474	0.4484	0.4495	0.4505	0.4515	0.4525	0.4535	0.4545
1.7	0.4544	0.4564	0.4573	0.4582	0.4591	0.4599	0.4608	0.4616	0.4625	0.4633
1.8	0.4641	0.4649	0.4656	0.4664	0.4671	0.4678	0.4686	0.4693	0.4699	0.4706
1.9	0.4713	0.4719	0.4726	0.4732	0.4738	0.4744	0.4750	0.4756	0.4761	0.4767
2.0	0.4772	0.4778	0.4783	0.4788	0.4793	0.4798	0.4803	0.4808	0.4812	0.4817
2.1	0.4821	0.4826	0.4830	0.4834	0.4838	0.4842	0.4846	0.4850	0.4854	0.4857
2.2	0.4861	0.4864	0.4868	0.4871	0.4875	0.4878	0.4881	0.4884	0.4887	0.4890
2.3	0.4893	0.4896	0.4898	0.4901	0.4904	0.4906	0.4909	0.4911	0.4913	0.4916
2.4	0.4918	0.4920	0.4922	0.4925	0.4927	0.4929	0.4931	0.4932	0.4934	0.4936
2.5	0.4938	0.4940	0.4941	0.4943	0.4945	0.4946	0.4948	0.4949	0.4951	0.4952
2.6	0.4953	0.4955	0.4956	0.4957	0.4959	0.4960	0.4961	0.4962	0.4963	0.4964
2.7	0.4965	0.4966	0.4967	0.4968	0.4969	0.4970	0.4971	0.4972	0.4973	0.4974
2.8	0.4974	0.4975	0.4976	0.4977	0.4977	0.4978	0.4979	0.4979	0.4980	0.4981
2.9	0.4981	0.4982	0.4982	0.4983	0.4984	0.4984	0.4985	0.4985	0.4986	0.4986
3.0	0.4987	0.4987	0.4987	0.4988	0.4988	0.4989	0.4989	0.4989	0.4990	0.4990
3.1	0.4990	0.4991	0.4991	0.4991	0.4992	0.4992	0.4992	0.4992	0.4993	0.4993
3.2	0.4993	0.4993	0.4994	0.4994	0.4994	0.4994	0.4994	0.4995	0.4995	0.4995
3.3	0.4995	0.4995	0.4995	0.4996	0.4996	0.4996	0.4996	0.4996	0.4996	0.4997
3.4	0.4997	0.4997	0.4997	0.4997	0.4997	0.4997	0.4997	0.4997	0.4997	0.4998
3.5	0.4998	0.4998	0.4998	0.4998	0.4998	0.4998	0.4998	0.4998	0.4998	0.4998
3.6	0.4998	0.4998	0.4999	0.4999	0.4999	0.4999	0.4999	0.4999	0.4999	0.4999
3.7	0.4999	0.4999	0.4999	0.4999	0.4999	0.4999	0.4999	0.4999	0.4999	0.4999
3.8	0.4999	0.4999	0.4999	0.4999	0.4999	0.4999	0.4999	0.4999	0.4999	0.4999
3.9	0.5000	0.5000	0.5000	0.5000	0.5000	0.5000	0.5000	0.5000	0.5000	0.5000

*If Z is negative, use 0.5000 minus value shown; if Z is positive, use 0.5000 plus value shown.

Courtesy Fluor Daniel, Inc., Greenville, South Carolina

CHAPTER 6

ENGINEERING ECONOMICS

CHAPTER OUTLINE

Most engineers work in industry, or for federal, state, or city governments. In our free enterprise system the primary goal of industry is to maximize profits; in government the primary goal is to maximize the benefit-cost ratio. Both require decisions regarding acquisition and retirement of capital goods.* Many of these decisions are made by managers who rely on information from (and the judgment of) engineers. Engineers have special backgrounds on which to base judgments concerning new technology that is unfamiliar to management. Because of this, the relevant body of financial principles and techniques is called *engineering economics.*** Its primary use is to make quantifiable evaluations of alternatives. The essence of quantifying is accounting. Thus, we first review the elements of accounting. This knowledge improves our interactions with both the accountants and management, since one of management's first questions about a proposed project is, "How much will it cost?"

As alluded to earlier, successful engineers must be happy at home and work, and comfortable socially and financially. Building on the fundamentals of this chapter, Chapter 7 addresses financial planning, i.e., estate creation and personal finances.

6-1 ELEMENTS OF ACCOUNTING [1]

As the term implies, *accounting* has to do with assets and liabilities; however, the recorded information is in terms of monetary evaluations. Common forms of financial organization are the sole proprietorship, the partnership (general and limited), and the corporation. These are discussed later in this chapter. All forms of management use financial reports.

Financial Reports

These reports are based on two fundamental accounting equations. The first equation concerns the balance sheet and is often called the *balance sheet identity.* The second is used to determine profits or losses; thus, it is used for the P&L or income statements.

The Balance Sheet This requires a balance of the basic accounting equation (Equation 6-1):

$$\text{Assets} = \text{liabilities} + \text{equity} \tag{6-1}$$

The balance sheet shows the financial state of the organization at a particular time, usually December 31, in the year for which the statement is issued. The assets consist of two kinds: those considered current and those which are long-term (fixed).

Current assets are (a) cash on hand, (b) cash immediately available from bank accounts, (c) inventory (products finished and those in process), and (d) accounts receivable (usually due within a month). Generally, current assets are either cash or assets that will be (or can be) converted to cash within 1 year. Please review Table 6-1

---

*Capital goods, also known as *real capital,* refer to the buildings, equipment, and other materials used in the production process. They usually have a service life of many years.
**In Financial Management, this material is covered under Corporate Finance.

TABLE 6-1
Balance Sheet for AB Corp.
December 31, 19XX
Assets

Current assets:
Cash.	$ 20,000
Inventory.	80,000

Fixed assets:
Equipment.	130,000
Buildings.	170,000
Total assets	$400,000

Liabilities

Current liabilities:
Accounts payable	$ 15,000
Notes payable	35,000

Long-term liabilities:
SBA note	30,000
Bonds (70 × $1,000 ea.)	70,000

Equity

Capital:
Preferred stock, 400 shares $100 par value	40,000
Common stock, 10,000 shares $10 par value	100,000
Capital surplus.	100,000
Accumulated retained earnings	10,000
Total liabilities & equity	$400,000

for sample entries of assets and liabilities. Often these statements include data from previous years to facilitate comparison and to show growth.

Fixed assets refer to property, plant, and equipment, such as that used for production and maintenance. This entry represents assets not intended for sale but those used over time in the firm's operation.

Liabilities are also divided into current and long-term types. The current liabilities are due within 1 yr: they include accounts payable for goods used for production, e.g., the electricity, gas, and other utility bills; and promissory notes owed to banks for unsecured loans. Also included are long-term liabilities due within 1 yr.

The current assets are the firm's working capital. The net working capital equals current assets minus current liabilities; this is the amount available for projects without having to raise capital by borrowing funds or selling assets. We will recognize its importance in Chapter 7, when we discuss financial planning. Its section, Important Terms, gives an alternative use of "net working capital."

Long-term liabilities are all those due later than 1 yr from the date of issue of the statement. They include loans such as a 5-yr loan from the Federal Small Business Association (SBA), or 3- to 10-yr loans from banks or other financial institutions. After a firm or corporation is well established, it can sell bonds. These are discussed in detail later. For now, bonds are long-term liabilities, loans at a fixed interest rate. The terms are from 10 to 30 yr.

Equity, also called *net worth,* is the portion of the total assets which belongs to the organization's owners. Its value is determined by subtracting all liabilities from the total assets; thus, owners' equity is the variable that constantly changes to balance Equation 6-1. A portion of this net worth is assigned to common stock and is called *common equity.* It is the net worth minus the value of the preferred stock. In our example in Table 6-1, there are 400 shares of preferred stock with a face value of $100. The annual dividend is fixed at $8. The common equity divided by the total number of outstanding shares of common stock plus the common stock equivalents (discussed in Chapter 7) equals the (common) stock's book value. For the small company represented in Table 6-1, the number of common shares is 10,000; thus, the common shares of the company have a book value of $21. This is calculated below:

$$\text{Book value} = \frac{\$210,000}{10,000} = \$21$$

The Income Statement   This is also called the *profit and loss* (P&L) *statement.* It shows the results of business activities which cause the changes in the periods between financial statements. Its fundamental equation is

$$\text{Profits (earnings)} = \text{total revenues} - \text{total costs} \tag{6-2}$$

Of course this is obvious, but the equation is the framework for explaining the component activities (see Table 6-2).

TABLE 6-2
Consolidated Income Statement for (Year) 19XX

Net sales		$45,000
Cost of sales and operating expenses		
Cost of goods sold	20,000	
Depreciation.	2,000	
Selling and administrating expenses.	3,000	
Total costs		$25,000
Operating profits = sales − total costs:		$20,000
Other income		
Dividends and interest		3,000
Total income		$23,000
Less interest on loans		1,500
Income before federal and state income taxes*		$21,500
Provision for federal taxes**		4,300
Net (after taxes) profit for the year.		$17,200

*State taxes were waived for the first three years of operation.
**A federal tax of 20 percent is arbitrarily chosen as representative of the individual and corporate rates for this magnitude of income. Assuming that this book will serve as a text or reference over many years, it seems probable that in struggling to reduce our federal debt, Congress and the administration will change our tax structure several times.

The earnings per share of stock equals the net profit divided by the total number of shares, including the number of stock equivalents discussed in Chapter 7. Our small company only has 10,000 shares; thus, *the earnings per share* is calculated below:

$$\text{Earnings per share} = \frac{\$17,200}{10,000} = \$1.72$$

This value divided into a stock's current market price gives its present P/E ratio, that is, its price to earnings ratio. Thus, if the stock was presently selling for $17, its P/E would be 10.

Break-Even Analysis [2]

We now review the terms in Equation 6-2. Please see Figure 6-1. By Equation 6-2, we see that there is profit only if $N \times S$, the total revenue, exceeds the total cost. It is clear that this condition could never occur unless S, the slope of the revenue line, is greater than V, the slope of the cost per unit line; furthermore, the *break-even point* "○" (a value of N at which the profit is zero) can be defined as that value of N at which $(N \times S) - (N \times V) = FC$, or $N (BE) = FC/(S - V)$. When N is greater than this value, the firm is profitable. An exception could occur when production is excessively above normal. That situation is reviewed below. When the firm is profitable, it has to pay taxes at a rate usually represented by t. This has a decimal value, such as 0.2 for the 20 percent used in calculating the federal tax on income in Table 6-2. Accordingly, for the firm in that table.

$$\text{Net profit} = (1 - t) \times \text{gross profit (or income)}$$
$$\$17,200 = (1 - 0.2) \times \$21,500$$

Again referring to Table 6-2, we note that the fixed costs are depreciation and the selling and administrating expenses. The latter would include building rents and utilities. The costs of "creature comforts" in the production areas would also be classified as fixed.

Cost/Revenue Production Chart

Figure 6-1 is a model using linear approximations. The slope constants are different for production below and above 100 percent. No firm has truly linear operations, but the chart is convenient for visualizing operations, even extended into overproduction. The greater variable costs for production greater than normal are modeled by straight lines of greater slope. In our simple model, revenues do not change with greater production. The greater variable costs in overproduction are due to several factors. Two are most significant: overtime salary rates and reduced productivity in that time. To maintain customer relations by meeting demands, overproduction could be wise, even at reduced profits.

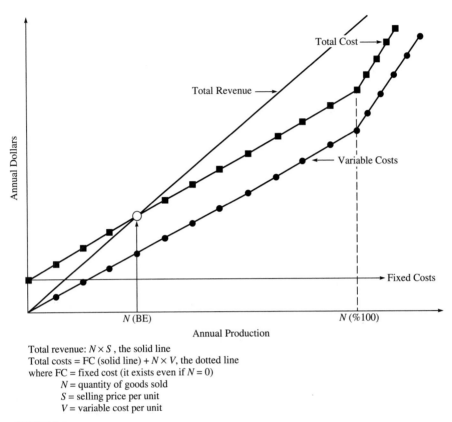

Total revenue: $N \times S$, the solid line
Total costs = FC (solid line) + $N \times V$, the dotted line
where FC = fixed cost (it exists even if $N = 0$)
       N = quantity of goods sold
       S = selling price per unit
       V = variable cost per unit

FIGURE 6-1
Cost/revenue production chart.

Accounting provides information from which both costs and profits can be determined. It is a major part of the management information system (MIS) reviewed in Chapter 3. Profit is the primary objective of private enterprise. Taxes required by the IRS and the legal setting of the enterprise depend on its form of organization.

6-2 FORMS OF FINANCIAL ORGANIZATION [3]

The more common forms of organization are the sole proprietorship, the partnership, and the corporation. We will review each and the limited partnership; however, the focus is mostly on risk and liability.

The Sole Proprietorship

This is a business owned by a single person. That individual receives all the profits and assumes all the risks associated with the enterprise. This form is generally for small organizations, since (at least initially) it is limited in the financial resources available to it. Furthermore, if it is possible to raise large amounts of money, an individual would hesitate to assume the associated risk.

Partnerships

This arrangement is an extension of the proprietorship and involves two or more persons. There two types: general and limited.

General Partnerships   Depending on their individual commitments and involvements, general partners share the enterprise's profits, but the risks are unlimited. The burden of unlimited liability becomes serious when the personal fortune of a particular partner far exceeds the business investment. In addition to the ordinary risks of business, unusual liabilities may be incurred because of an unscrupulous or improvident partner. The liabilities could also result from litigations, or accidental injuries to customers or employees that are not adequately covered by insurance. Any unpaid business creditors may levy upon the personal estate of any partner, if a balance over the partner's personal debts remains. This right of creditors to seize satisfaction from whatever partner has means, subject only to *rule of marshaling,** results in what is known as *joint and several liability (see Chapter 8).* Thus, a partner cannot limit his or her liability to outside creditors by mere agreement with other partners. The only such arrangement that has legal value is the limited partnership; under it, creditors would have been given notice of limited liability.

Limited Partnerships   This type, abbreviated LP, is formed by an agreement among two or more limited partners and the ordinary partners, called the *general partners.* Unlike the general partnership, the limited partnership must file a certificate disclosing special information.** This requirement of filing and publication is designed to protect the position of creditors. It gives them full knowledge of how much they can rely on capital supplied by the limited partners. The limited partner has no voice in management, but unlike the general partnership, the limited partnership can be assigned or sold without causing it to dissolve. Furthermore, there can be incapacitation or death of a limited partner without requiring dissolution of the partnership.

Since the limited partner has no control of management, it is more like an investment. But it is probably less attractive than that offered by a corporation, particularly the S corporation (see below). This is especially true because of the tax advantages removed in 1986. The absence of federal corporate income tax for the limited partnership, however, does make possible higher return than an investment in a regular corpo-

---

*The *rule of marshaling* provides that business creditors have first claim upon business assets, and personal creditors have first claim upon separate personal assets.

**This information includes the name of the partnership, the character of the business, the place of the business, the name and residence of each member, the duration of the partnership, a careful description of the original capital contribution and liability for any additional contributions of the limited partners, the compensation of the limited partners, and the terms of their withdrawal.

ration. The greater return, however, depends on the honesty and competence of the general partner.

Corporations

This form of business is a legal personality; its equity is divided into shares, usually over a million in number. Ownership is divided among thousands of people; thus, the corporation is truly an autonomous, legal unit. It can, however, accomplish through its representatives or agents many things that an individual or partnership normally cannot. Most significant is its capability to raise and own large amounts of capital, such as moneys, property, and equipment. This is possible without shareholders incurring liabilities beyond the value of the shares. The permanence of the corporation is not affected by the death of any shareholder and is generally not significantly affected by the death of any individual manager.

But every joy has its price. For the corporation, that price is extra taxes, both state and federal; these taxes are on income before any distribution to shareholders and, combined, total about 50 percent. The S corporation, however, has tax advantages [3, p. 6].

Small, privately held corporations may elect to be treated as S corporations. The maximum number of shareholders is 35. For tax purposes, this is like a partnership in that earnings are passed through (without being taxed) to the shareholders. Even if not distributed, the earnings are taxed at the personal rate of the shareholders.

All forms of viable financial organizations require objective analyses of potentially profitable projects. In Section 6-1 on accounting, Equation 6-2 shows that profits equal total revenue minus total cost. In Section 6-3 below, we review the techniques needed for evaluating costs and develop the techniques for determining the profitability of projects.

6-3 ENGINEERING ECONOMICS [4]

The viability of industrial and governmental programs is dependent on evaluating existing projects and choosing advantageous new ones. Although the decision probably involves intangibles, the first approach is usually based on an economic evaluation, using principles and techniques of engineering economics.

Basic Concepts

Interest and the Time Value of Money An economic evaluation involves analyzing cash flows (revenues received and costs paid); however, these usually occur at different times. In turn, interest is charged until costs are paid and is collected until revenues, such as accounts receivable, are received. Interest is rent paid by one party for use of money borrowed from another party. For instance, a firm might borrow $10,000 from a commercial bank for a period of 1 yr. If the rent on the $10,000 is $800, the rent, called *interest,* is $8 per $100 and is expressed as 8 percent. *Cent* is short for *centum,* a Latin word meaning "100." The term *percent* also implies for a period of 1 yr; to be explicit, often the term *APR* is used. It stands for annual percentage rate. Later, we will discuss effective (annual) rates for cases in which the rent or interest is paid for periods less than 1 yr. In the loan above, after 1 yr the firm is to pay back $10,800, or $800 more than the $10,000 it received. Because of interest, especially when compounded as considered below, we recognize that money changes value with time. The change is called the *time value of money.* Furthermore, we see that (the change of) value depends on the interest rate. With a 10 percent rate, the $10,000 would have increased to $11,000, instead of the $10,800 for an 8 percent rate. Of course, the firm would borrow the money only if it could make more in a year than the "rent" for use of the money. Often firms borrow money to finance one or more projects. For these projects, the firm would use the borrowed money to buy goods, both capital and materials, and to pay salaries. An economic analysis of each proposed project is required to determine the wisdom of implementing it. Using the techniques of engineering economics, the analysis indicates the probable magnitude of the project's return. It must adequately exceed the project's cost to pay the interest and give the minimum profit which management sets. The minimum profit set by management is discussed later, but naturally it would have to considerably exceed the market interest rate, that is, the rate the firm would have to pay on borrowed moneys. If it had excess cash, instead of using it for a project, the firm could buy a near-risk-free federal note or short-term bond. These financial securities (reviewed in Chapter 7) would pay about the market rate. This rate is the basis on which management sets the firm's minimum attractive rate of return (MARR), which we will discuss later. We will now build on these interest concepts to develop factors. They permit comparing moneys received and dispensed at various times over the life of the project, including those paid before its start-up.

Life Cycle Cost   Interest factors permit life cycle cost analyses. They show that first costs, such as perhaps those from a low bid, are grossly inadequate when comparing projects. Specifically, the total of costs and incomes over the life of the project or piece of equipment must be closely assessed. Interest must be considered to discount (reduce) these monetary quantities to a single amount corresponding to a reference time. Typically, this time is taken to be time zero at the beginning of the project or the time of purchase of the equipment. For the project analyses, we next develop the needed interest factors.

Interest Factors of Compound Interest   The original amount borrowed is called the *principal.* For a specified period, the borrower pays interest as "rent" for use of the money. After that period, if acceptable to the loaning agent, the borrower could elect to make no payments. Interest in the second period, however, would have to be paid on the sum of the principal and the interest due for the first period. This type of plan is known as *compounding.* In it, interest is charged on earlier interest and the principal. Compounding is magic for the investor. We stress its value in Chapter 7. If interest is paid when due (with or without payment on the principal), the arrangement is called *simple interest.* Returning to compound interest, we observed above that after 1 yr, a firm would owe $10,800 if it had borrowed $10,000 at an interest rate of 8 percent. It is understandable that the firm might wish to wait another year before repaying any money to the bank or other financial institution. In that case, it would then have to pay interest on the new debt of $10,800. To determine how much will be due at the end of the second year, let us first set up an equation for determining the amount due after the first year. It is the future cost of the debt, from the firm's viewpoint. It is the future value or worth of the investment, from the bank's viewpoint.

Future Value   Consider the amount of interest as I. Recall that for the first year it is calculated by multiplying the original amount (the principal) by the interest rate i expressed as a decimal. Namely,

$$\text{Interest} = \text{principal} \times i$$

Using P for the principal, the present or original amount,

$$I = P \times i = \$10,000 \times 0.08 = \$800$$

Now, adding this to the original amount, we get

$$\text{Future (amount)} = \text{present (amount)} + \text{interest}$$

or, after 1 yr,

$$
\begin{array}{ccccc}
F(1) & = & P & + & I \\
\$10,800 & = & \$10,000 & + & \$800
\end{array}
$$

Thus, using $I = P \times i$ and $F(1)$ as the future amount after 1 yr,

$$F(1) = P + I = P + P \times i = P \times (1 + i)$$

In turn,

$$\$10,800 = 10,000 \times (1 + 0.08)$$

If we now deferred paying the bank this amount, interest on the $10,800 would be charged for the second year. Thus, interest from the first year is compounded. The principal for the second year is also symbolized as P. Then $P = F(1)$ and is the present value at the beginning of the second year, namely, $10,800. Accordingly, the future amount due to the bank after 2 yr would be

$$F(2) = F(1) \times (1 + 0.08) = 10,800 \times 1.08 = \$11,664$$

Since

$$F(1) = P \times (1.08)$$

then

$$F(2) = F(1) \times (1 + 0.08) = P \times (1 + 0.08) \times (1 + 0.08)$$
$$= P \times (1.08)^2$$

From this, we engineers can generalize to n periods (usually years) and for any interest rate i to arrive at the equation for the future value of the principal:

$$F(i,n) = P \times (1 + i)^n$$

From it, we can define a *transfer factor* for transferring from P to F :

$$F(i,n)/P = (1 + i)^n$$

This term is called the *single payment, compound interest factor,* or just the *future worth factor.* With others developed below, it is given in Appendix A of this chapter. Table A-1 is for $n = 1$ to 50 periods, usually years, but is only for $i = 0.1$, that is, 10 percent. It is felt that except for discussing examples, determining the factors for any i and n using a hand calculator is as easy as using tables. If you prefer, however, the references for this chapter have tables for a wide range of i and n.

Now it is easy to get the future value $F(i,n)$ of any present amount P invested for n periods at interest rate i. It only requires multiplying the amount P by the factor $F(i,n)/P$. As noted above, this factor can be determined using a calculator or by referring to a table of interest factors. Again consider the example above with $i = 0.08$, $P = \$10,000$, and the factor $F(0.08,2)/P = (1.08)^2 = 1.1664$. After 2 yr, its future value $F(0.08,2) = P \times F(0.08,2)/P$. We can consider that P in the numerator cancels the P in the denominator of $F(0.081,2)/P$, and in turn we have

$$F(0.08,2) = \$10,000 \times 1.1664 = \$11,664$$

Present Value   As you have probably already recognized, it is just as easy to determine what present value (amount) invested now is required to realize a certain amount or value at a future time. Consider a family which will need $\$10,000$ in 10 yr, when their child is to enter a university. There is a special kind of bond called a *zero coupon bond;* it is discussed in more detail in Chapter 7. These "zeros" (as the bonds are popularly called) pay no dividend. The interest is reinvested each year; that is, the interest is compounded each year until its maturity in 5 or 10 yr.*

Thus, if P is paid for the bond, its future value in n years is $F(i,n) = P \times F(i,n)/P$, and the present amount needed now to grow to $F(i,n)$ is determined by

$$P = F(i, n) \times \frac{1}{F(i, n)/P}$$

---

*U.S. savings bonds (Series EE) are "zeros". With minor restrictions, their interest (minimum 6 percent) is tax-free if used for qualified educational expenses.

We recognize that the last term is the reciprocal of the future worth factor derived above. The new factor, called the *present worth factor,* is defined as

$$P/F(i, n) = \frac{1}{(1+i)^n} = (1+i)^{-n}$$

Accordingly, if a "zero" is paying 8 percent and has a maturity equal to 10 yr or more,* its $F(i,n)$ for any n less than or equal to the time to its maturity is

$$F(i,n) = P \times F(i,n)/P$$

or

$$P = F(i,n) \times P/F(i,n)$$

To determine the required present worth (amount) per dollar of future worth, let us evaluate $P/F(0.08,10)$; it is

$$P/F(0.08,10) = (1 + 0.08)^{-10}$$
$$= 0.46319$$

Thus, the amount needed to invest now (at 8 percent) in order to have $10,000 in 10 yr is**

$$P = F \times P/F$$
$$= \$10,000 \times 0.46319 = \$4631.90$$

The cash flow diagram in Figure 6-2 helps visualize these operations. Note that we use down arrows to indicate "money in" and up arrows for "money out."

The power of compound interest is surely wonderful; it enables you to more than double your money in 10 yr. It is even more wonderful if you start earlier; for instance, if the family invests at the child's birth, after 18 yr the money at 8 percent interest would increase by a factor

$$F(0.08,18) = (1 + 0.08)^{18} = 3.996$$

and only about $10,000/4 = \$2500$ would be required. Sometimes, as in the early 1980s, Treasury bonds or AA- or AAA-rated corporate bonds pay higher rates; for instance, at 10 percent, $F(0.1,18) = 5.56$; of course, $10,000/5.56 is only $1798.59.

---

*Any maturity of 10 yr or greater is suitable, since the bond can be sold at any time for its value at that time. Note, however, by the equation for $F(i,n)/P$, that an increase (decrease) in the market rate, i, decreases (increases) P, the bond's present value. This is presented in detail in Chapter 7.

**For convenience, where there is no ambiguity, we will drop the arguments of the factors, that is, F/P instead of $F(i,n)/P$.

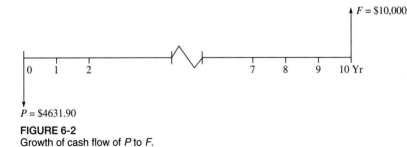

FIGURE 6-2
Growth of cash flow of P to F.

Most commercial and federal bonds are issued with a face value (its redemption value shown on its face) of $1000. Thus, the family in the example above would have to purchase 10 bonds to have a future redemption value of $10,000.

Often investors seek short-term investments by buying long-term bonds which were issued earlier. Consider a 10 percent 10-yr "zero" issued 8 yr ago. It will be redeemed in 2 yr for $1000. At issue, its cost was $385.54. This value is determined similarly as above:

$$P = F(i,n) \times P/F(i,n) = 1000 \times (1.1)^{-10} = 1000 \times 0.38554$$
$$= \$385.54$$

The bond's value after 8 yr is determined similarly:

$$F(0.1,8) = P \times F(i,n)/P = \$385.54 \times (1.1)^8 = 385.54 \times 2.1436$$
$$= \$826.44$$

An investor can purchase the bond for about this amount and, in turn, redeem it for $1000 in 2 yr, realizing the 10 percent rate.

As with other activities, however, the task of saving is easier when accomplished little by little. Thus, let us now consider investing an amount A each year. We can consider A as an annuity.

Annuities An annuity is a series of equal amounts paid periodically over n periods. These series have future and present values.

Future Value

We first consider the future value of a series of equal amounts A, invested at the end of each period. This can be determined using a series of future worth factors. The periods can be for any length of time; we will consider the period as 1 yr. For the annuity each year, a corresponding future worth factor would be used. Each would be for the number of years remaining from the time of that particular annuity A until the future time, n-years, being considered. Each annuity is then multiplied by its respective factor. The total future value is the sum of all the products. So much for the hard way; in

Appendix B, we develop a routine to put the series in a compact, closed form. First, for a clearer perspective, please review Figure 6-3. As in Figure 6-2, down arrows indicate investments, "money in(vested)"; up arrows indicate disinvestments, or "money out." In this example, we will use $i = 8$ percent and $n = 10$. Thus, for A at the end of the first year, there are $k = 9$ yr remaining in which to gain interest. Accordingly, $k = 8$ after the second A, etc. Recall the future worth factor from above, and letting $n = k$ for the number of remaining years for gaining interest after each A, the future worth factor for the kth annuity is

$$F(i,k)/P = (1 + i)^k$$

Hence, for $A = P(k)$ for all the k's $= 9$ to 0.

$$F(\text{total}) = A \times F(0.08,9)/P + A \times F(0.08,8)/P + \cdots + A \times F(0.08,0)/P$$

Taking the A's outside the brackets and then substituting for the F/P's,

$$
\begin{aligned}
F(\text{total}) &= A \times [F(0.08, 9)/P + F(0.08, 8)/P + \cdots + F(0.08, 0)/P] \\
&= A \times [(1 + 0.08)^9 + (1 + 0.08)^8 + \cdots + (1 + 0.08)^0] \\
&= A \times [1.999 + 1.851 + \cdots + 1] = A \times 14.487
\end{aligned}
$$

Upon reflection, we see that the sum of the future worth factors enclosed by the brackets can be defined as $[F(0.08,10)/A]$, an annuity-to-future-worth factor. On multiplying it by A, the future value of the annuity is determined. In Appendix B, we show that, for any i and n, this factor equals

$$[F(i,n/A] = [(1 + i)^n - 1]/i$$

or

$$
\begin{aligned}
F(0.08,10)/A &= [(1 + 0.08)^{10} - 1]/0.08 = [2.1589 - 1]/0.08 \\
&= 1.1589/0.08 = 14.487
\end{aligned}
$$

Now, for this type of investment, there is no difficulty in determining A, the amount the family needs to save each year. See Figure 6-3. In order to have $F(0.08,10) = F(\text{total}) = \$10,000$ after 10 yr,

FIGURE 6-3
Growth of annuities.

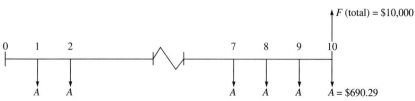

$$A = F(0.08,10) \div [F(0.08,10)/A] = \$10,000/14.487$$
$$= \$690.29$$

In Chapter 7 we review the possibility of using annual investments as an alternative to ordinary or whole life insurance.

You have probably noticed that an annuity is paid at the end of the tenth year; at the same time that F, the full future value, would be received. This approach is merely to facilitate the mathematics, since this plan is applicable to another type of evaluation which we consider next. It is the present worth of payments A for n years in the future.

Present Worth of an Annuity

In developing this value, let us consider a lottery.

Example 6-1

A lottery might pay \$1000 per month for a 10-yr period. What is the present worth of the winnings? If 8 percent interest is typical over those 10 yr, then the interest per month is the 8 percent divided by 12. Furthermore, over 10 yr there would be 120 payments, one each month.

Solution: To get the present value of these 120 uniform payments, for an interest rate per month of $0.08/12 = 0.006667$, we could calculate $F(0.0066667,120)/A$ and then multiply it by $P/F(0.006667,120)$; the multiplication cancels the F's to get $P/A(0.006667,120)$. To save us from this extra multiplication, we can get the equation for $P/A(i,n)$ by simply multiplying the equations for $F(i,n)/A$ and $P/F(i,n)$; the F's cancel, leaving

$$P/A(i,n) = [F(i,n/A] \times [P/F(i,n)]$$

Using the equations from above, and dropping the (i,n)'s, we have

$$F/A = [(1 + i)^n - 1]/i \qquad \text{and} \qquad P/F = (1 + i)^{-n}$$

Accordingly, on substitution

$$P/A(i, n) = \{[(1 + i)^n - 1]/i\} \times (1 + i)^{-n}$$
$$= [(1 + i)^n - 1]/i \times (1 + i)^n$$

Taking the reciprocal gives

$$A(i,n)/P = i \times (1 + i)^n/[(1 + i)^n - 1]$$

Returning to our example, with $i = 0.006667$ and $n = 120$ (months), again dropping the (i,n)'s

$$P/A = [(1 + 0.006667)^{120} - 1]/0.006667 \times (1 + 0.006667)^{120}$$

The convenient calculator gives

$$P/A = [2.21973 - 1]/0.006667 \times 2.21973$$
$$= 1.21973/0.0148 = 82.42$$

Thus, the present value of the 120 payments of $1000 is

$$P = A \times P/A = \$1000 \times 82.42 = \$82,420$$

While this is a nice win, it is worth only about two-thirds of the $120,000 usually advertised.

Nominal, Actual, and Effective Interest   Banks and other financial institutions seemingly play games in stating interest rates. Financially, it is a definite advantage to understand their games.

The *nominal* rate is the rate we have been indicating by i. It is the charge per unit of $100 to use that quantity of money for 1 yr, when there is no compounding for any periods shorter than a year. It is the rate usually given in the financial press, even though there could be compounding on a quarterly, monthly, or even daily basis. If the period for using the money is a fraction of a year, such that there are m periods in 1 yr, then the nominal rate is $m \times i$(actual). The *actual* rate is the rate per period, such as the 1.5 percent per month charged by credit card companies and firms extending credit for "installment" purchases. For an actual rate of 1.5 percent per month, the nominal rate is 18 percent:

$$i(\text{nominal}) = m \times i(\text{actual})$$
$$18\% = 12 \times 1.5\%$$

Consider someone borrowing $1000 for 6 months ($1/2$ yr) at a nominal rate of 8 percent. Since there are two 6-month periods in a year, $m = 2$, and i(actual per 6 months) $= \frac{8}{2} = 4$ percent. After 6 months, the borrower pays F [0.04,1(for 1 period)]:

$$F(0.04,1) = P(1 + i/m)$$
$$= \$1000(1 + 0.08/2) = \$1000(1 + 0.04) = \$1040$$

This charge is said to be at an actual rate of 4 percent per 6 months. In this example there are no intermediate periods for compounding. With a value of $m > 1$, compounding over the year gives an effective rate greater than the nominal rate.

The effective rate equals the equivalent nominal rate which gives the same return in 1 yr as that charged per $100. Consider the $1\frac{1}{2}$ percent (0.015) per month charged by many credit cards companies, banks, and department stores. The nominal rate of 12 $\times$ $1\frac{1}{2} = 18$ percent is great enough, but if the balance is unpaid for 1 yr, the charge is compounded monthly. That policy gives an effective rate of 19.6 percent. It is calculated by the compound interest factor given above. The year-end payment per $100 is

$$\$100 \times (1 + 0.015)^{12} = \$100 \times 1.196 = \$119.60$$

We immediately recognize that this is the same as a nominal rate of 19.6 percent per year. It is called the *effective rate* (of return, to the lender); for any number of periods *m* per year, it can be determined in terms of *i*, the nominal rate per year, by the expression

$$\text{Eff. rate} = (1 + i/m)^m - 1$$

As considered above, for the nominal rate, $i = 18$ percent, compounded for 12 months, $m = 12$,

$$\text{Eff. rate} = (1 + 0.18/12)^{12} - 1 = (1 + 0.015)^{12} - 1$$
$$= 1.196 - 1 = 0.196 \rightarrow 19.6\% \approx 20\%$$

Sometimes savings and loan institutions and money market funds call the effective rate the *yield.* Consider an advertisement for nominal rate $i = 8$ percent, yield = 8.33 percent. In money market funds, the interest is usually compounded daily, that is, 365 times per year. Let us apply our expression for $i = 0.08$ and $m = 365$.

$$\text{Yield} = (1 + 0.08/365)^{365} - 1 = (1 + 0.00022)^{365} - 1$$
$$= 1.08328 - 1 = 0.08328 \approx 8.33\%$$

Continuous Interest [5]   This rate results when the number of compounding periods is increased without limit, i.e., to infinity. Because of the 80-20 rule presented in Section 1-2, I would like to stress that more is not always (or even usually) better. But it is usually worth considering. Increasing *m* in the expression for effective interest leads to an easier way to handle all compound interest factors. Only a few of us, not I, might recall the definition of the napierian logarithm base *e*; it is

$$\lim_{m \to \infty} (1 + i/m)^m = e^i$$

Consider the compound interest factor for *m* compounding periods per year. For a nominal rate *i*, the interest per period is *i/m*. Thus, from the future value factor above,

$$F(i,n)/P = [(1 + i/m)^m]^n$$

Now for compounded continuously—that is, letting *m* in *i/m* go to infinity, as defined above—the factor becomes

$$F(i,n)/P = e^{(i \times n)}$$

And the continuously compounded effective interest, using $n = 1$, that is, for 1 yr

$$i(\text{eff}) = e^i - 1$$

When confusion seems possible, we will call the *i*, used previously, the discrete interest, such as for discrete periods of months or years. Below are comparisons between the nominal (per year) rate i and i(eff) of continuous compounding. For i less than 15 percent, the differences are small. In comparing projects, the small error produced in using the more convenient continuous interest would be common to all evaluations. In turn, the error is of no significance.*

Per year:	Nominal i (percent)	Continuous i(eff) (percent)
	5	5.127
	10	10.517
	15	16.183
	20	22.140

As noted earlier, Table A-1 in Appendix A is only for the discrete rate of 10 percent. Other tables are needed for other interest rates. By using continuous interest, only Table A-3 is needed for all i's and n's. This is true because the exponent in all the factors is $i \times n$, the product of i and n. Thus, in using the table, we first evaluate the product for the i and n in the particular analysis. The future worth factor for i and n using continuous interest is developed above. It is

$$F(\text{cont})/P = e^{(i \times n)}$$

This factor shows an equality which makes conversion easy from the discrete factors to the continuous ones: $(1 + i)^n$ in the discrete factors becomes $e^{(i \times n)}$ in the continuous factors. We indicate the distributed annuity as $\mathring{A}$; the integral (sum) of these distributed flows over that period equals the discrete amount, that is $A = n \times \mathring{A}$. Unlike discrete periods, $\mathring{A}$ used with continuous interest is not at a single instance, such as A is at the end of the interest period. $\mathring{A}$ is distributed uniformly throughout that period. In turn, the total amount over n periods, usually in years, is $\mathring{A}$ multiplied by n. We will show the product as $n\mathring{A}$. Thus, in converting from the discrete to the continuous factors, change all $(1+i)^n$ to $e^{(i \times n)}$ and any A in the discrete factors is replaced by $n\mathring{A}$, and i by $(i \times n)$. These relationships are evident in Table 6-3.

The last factor $P/n\mathring{A}$ in Table 6-3 is determined easiest by multiplying $F(i,n)/n\mathring{A}$ by $P/F(i,n)$. If determined by the substitutions first suggested, multiplication of the numerator and denominator by $e^{-(i \times n)}$ is required to get the more compact form.

Example 6-2

For comparison, we will use continuous interest to determine the present worth of the lottery in Example 6-1. Recall that the lottery pays $1000 per month for 10 yr;

---

*In comparing alternatives, any cost common to both need not be considered. For example, when comparing two routes on a road map, we do not consider the part of the routes which is the same for both.

TABLE 6-3

Factor	Discrete	Continuous
$F(i,n)/P$	$(1 + i)^n$	$e^{(i \times n)}$
$P/F(i,n)$	$(1 + i)^{-n}$	$e^{-(i \times n)}$
$F(i,n)/A$	$[(1 + i)^n - 1]/i$	
$F(i,n)/n\mathring{A}$		$[e^{(i \times n)} - 1]/(i \times n)$
$P/A(i,n)$	$[(1 + i)^n - 1]/i(1 + i)^n$	
$P/n\mathring{A}$		$[1 - e^{-(i \times n)}]/(i \times n)$

thus there are 120 periods (12×10), and the product $n\mathring{A} = \$120,000$. The interest rate per month is $i = 0.08/12 = 0.00667$; $i \times n = 0.00667 \times 120 = 0.8004$.

Solution: Using the last factor from Table 6-3

$$P = n\mathring{A} \times P/n\mathring{A} = n\mathring{A} \times [1 - e^{-(i \times n)}]/(i \times n)$$
$$= \$120 \times 1000 \times [1 - e^{-0.8004}]/0.8004$$

Again using our calculator, $e^{-0.8004} = 0.44915$ and

$$P = \$120,000 \times [1 - 0.44915]/0.8004$$
$$= \$120,000 \times 0.68822$$
$$= \$82,586.32$$

which differs only 0.2 percent from the value of $82,420 determined by using discrete factors in Example 6-1.

The small error is significantly affected by the per month payments; that is, they are much closer to "continuous" flow than per annual payments. Of greater significance, however, is that in practice, most money flows are either random or monthly; and, in turn, continuous interest factors are often more appropriate than the discrete annual ones.

Depreciation and Taxes

Capital goods are buildings and equipment used in the production process. They have a service life in excess of 1 yr, usually many years. Their value can be lost by simple wear or obsolescence; for both, the devaluation is a cost of production. Thus, it reduces the firm's earnings and, in turn, the taxes which the firm would pay.

Depreciation is an important cost needed in evaluating a firm's true earnings. In making this evaluation the accounting department usually uses straight-line depreciation. As discussed below, accelerated depreciation schedules allowed by the IRS permit a firm's quicker recapture of investments. Thus, the accelerated schedules are generally used for tax purposes. After presenting straight-line depreciation, we will review the most prevalent types of accelerated depreciation: the sum-of-digits (also referred to as *sum-of-years*) and declining-balance. As reviewed later, the latest tax laws (1981 and 1986) have established two new schedules or depreciation systems: the accelerated capital recovery system (ACRS) and the modified accelerated capital recovery system (MACRS). ACRS applies to capital property placed in service between 1981 and 1986. MACRS applies to capital property placed in service after 1986. Embedded in both systems is a combination of 200 percent and 150 percent declining-balance and straight-line schedules. We discuss these next. Since established firms could have property placed in service before 1981, we also review the then often used schedule, the sum-of-digits.

Straight-Line Depreciation   The initials *SL* also identify this method. As the name suggests, the amount of this depreciation is the same each year, a constant fraction of the equipment's total depreciation. We will see that the book value of the equipment decreases in a straight-line manner. The book value $BV(n)$ is the asset's value at a time n years into its total life of N years. It is defined below. The IRS publishes guidelines for appropriate service lives for the different types of fixed assets. Consider a piece of equipment with data given below:

$$\text{Initial or first cost, } C(f) = \$23,000$$
$$\text{Service life, } N = 10 \text{ yr}$$
$$\text{Salvage value, } C(\text{sal}), \text{ after its service life of 10 yr} = \$3000$$

We quickly recognize that in 10 yr the equipment loses $20,000 in value. This is called its depreciable first cost $C(d)$. Or

$$C(d) = C(f) - C(\text{sal})$$
$$\$20,000 = \$23,000 - \$3000$$

For SL, the depreciation is the same for each year. It is the depreciable cost divided by *N,* the asset's service life (henceforth referred to as its *life*). Thus, $D(k)$, the depreciation for year *k,* is

$$D(k) = [C(f) - C(\text{sal})] \times d(\text{SL}k)$$
$$= C(d) \times d(\text{SL}k) = C(d)/N$$

where $d(\text{SL}k)$ is the fraction for the *k*th year. For a SL schedule, it is a constant equal to $1/N$. Thus,

$$D(k) = C(d)/N \qquad \text{same for every year}$$

For our example,

$$D(k) = \$20,000/10 = \$2000 \text{ every year}$$

The book value at the end of year *n* is $BV(n)$. It is defined as the asset's original or first cost, $C(f)$, minus the total of all the depreciations taken through the *n*th year.

$$BV(n) = C(f) - \sum_{k=1}^{n} D(k)$$

Note that substituting $D(k) = [C(d)/N] \times 1$, the same each year, the book value after N years, at the end of its service life, is

$$BV(n) = C(f) - [C(d)/N] \times \sum_{k=1}^{N} 1$$
$$= C(f) - [C(d)/N] \times N = C(f) - C(d)$$

And, as expected,

$$BV(N) = C(\mathrm{f}) - C(\mathrm{d}) = C(\mathrm{sal})$$

For all depreciation schedules,

$$C(d) = \sum_{k=1}^{N} D(k)$$

$D(k)$ is not the same in each year for accelerated schedules.

Next we review two accelerated depreciation schedules: sum-of-digits (SD) and double-declining-balance (DDB). SD is needed for long-life capital goods placed in service before 1981. As noted above, DDB is used in both ACRS and MACRS.

Sum-of-Digits Schedule   SD is often known as sum-of-years; unlike $d(\mathrm{SL}k)$, the depreciation fraction for the SL method, $d(\mathrm{SD}k)$ changes each year. It is determined by

$$d(\mathrm{SD}k) = [N - k + 1]/0.5N(N + 1)$$

The numerator equals the number of service years remaining at the beginning of the kth year. The denominator equals the sum of the digits from 1 through N. As for our example, with $N = 10$, $0.5 \times 10(10 + 1) = 1 + 2 + 3 + \cdots + 10 = 55$. Accordingly, in the first year, $k = 1$,

$$d(\mathrm{SD}1) = [10 - 1 + 1]/55$$
$$= 10/55 = 0.18182$$

and in the second year,

$$d(\mathrm{SD}2) = [10 - 2 + 1]/55$$
$$= 9/55 = 0.16364$$

Their sum is 0.34546; thus, in only 2 yr, over one-third of the depreciable cost can be recovered from the profits tax-free. For our example, in the first 2 yr, $0.34546 \times \$20,000 = \$6909.20$ would be recovered. By the SL schedule, the amount free from taxes would be only $2 \times \frac{1}{10} \times \$20,000 = \$4000$. Because of this difference, firms frequently use SL to determine true earnings and SD for tax purposes. Later we review how taxes affect the net cost of capital assets, but first let us consider a second accelerated schedule.

Double-Declining-Balance Schedule   This type of accelerated depreciation has the notation *DDB*. The yearly depreciation is a fixed fraction, $F(DDB)$, of the preceding year's book value. $F(DDB) = 2/N$ in the double-declining schedule. Since it is twice that for the straight-line schedule, it is also noted as 200 percent declining balance. In discussing straight-line depreciation, we defined book value. At the end of any time n years into an asset's life, for any type of depreciation schedule, the book value is

$$BV(n) = C(f) - \sum_{k=1}^{n} D(k)$$

For DDB, defined earlier,

$$D(k) = F(DDB) \times BV(k-1)$$

Developing from $n = k =$ yr 1, $BV(1-1) = C(f)$,

$$D(1) = F(DDB) \times C(f)$$
$$BV(1) = C(f) - D(1) = C(f) - F(DDB) \times C(f)$$
$$= C(f)[1 - F(DDB)]$$

Similarly for $k = 2$,

$$D(2) = F(DDB) \times BV(1) = F(DDB) \times C(f)[1 - F(DDB)]$$
$$= C(f) \times F(DDB) \times [1 - F(DDB)]$$
$$BV(2) = BV(1) - D(2)$$
$$= C(f)[1 - F(DDB)] - C(f) \times F(DDB) \times [1 - F(DDB)]$$

On factoring $C(f) [1 - F(DDB)]$ from the right side of the last equality,

$$BV(2) = C(f)[1 - F(DDB)] \times [1 - F(DDB)]$$
$$= C(f)[1 - F(DDB)]^2$$

From these we generalize to

$$BV(k) = C(f)[1 - F(DDB)]^k$$

and

$$D(k) = C(f) \times F(DDB) \times [1 - F(DDB)]^{k-1}$$

On reflection, it is noted that the largest value of $F(\text{DDB})$ is desired. The IRS limits its value to $2/N$, where N is the service life of the asset; the life for tax purposes is also set by the IRS. A factor $F(\text{DB})$ may be chosen less than the maximum allowed. $F(\text{DB}) = 1.5/N$ is 150 percent declining balance.

Since the DDB schedule is based on a fraction of book value, if pursued without care, the total depreciation would exceed $C(\text{d}) = C(\text{f}) - C(\text{sal})$. As you would guess, the IRS does not permit that. In the year in which the book value, $C(\text{f})$ minus the depreciation to that year, would be less than $C(\text{sal})$, the IRS requires that the firm switch to SL. Furthermore, this method gives only slightly faster depreciation than SD.

Net Costs with Taxes   The total cost of capital assets is the sum of its depreciable part $C(\text{d})$ and its salvage value $C(\text{sal})$, i.e., $C(\text{f}) = C(\text{d}) + C(\text{sal})$. Since the depreciable part is a cost which reduces profits, it also reduces the income tax. This reduction in taxes is, in effect, an "income." The IRS, however, requires that this cost be apportioned on a per year basis over the service life of the asset. The "income" is symbolized by $I(t,k)$. Its amount for the kth year equals the tax savings, and it is calculated by multiplying the tax rate t by $D(k)$, the depreciation for the kth year.

$$I(t, k) = t \times D(k) = t \times d(\text{XX},k) \times C(\text{d})$$

where $D(k) = d(\text{XX},k) \times C(\text{d})$, and XX in $d(\text{XX},k)$ is for XX = SL, SD, or other types of depreciation. Recent changes in the tax laws have introduced two other types: the accelerated capital recovery system (ACRS) was established in 1981, and the modified accelerated capital recovery system (MACRS) was introduced in 1986. These are reviewed later.

When using either discrete or continuous interest, with or without taxes, the factors developed in this section are very helpful. They can be conveniently used to compare costs of fixed assets considered for purchase. In all the factors, i (the interest rate before taxes) must be changed to r (the interest rate after taxes). The value of r is developed later.

Comparison of Alternatives

Comparison without Taxes   Because money has time value, all of the costs of a project or equipment over its life need to be transferred to a specific time. Transferring all costs to present value is a common approach. Typical costs are the initial or first cost, $C(\text{f})$, and the annual operating and maintenance costs, O&M; Example 6-3 considers these and others as it shows this approach to compare the costs of machines A and B.

Example 6-3: Comparison of Machines A and B

The following data apply:

Machine	A	B
C(f), initial cost	$50,000	$70,000
O&M, annual operating & maintenance	3,000	1,000
Q, annual savings from fewer rejects	–	500
C(sal), salvage	5,000	10,000
N, service life	10 yr	10 yr
Cost of money	10%	10%

Please see Figures 6-4 and 6-5 for the corresponding cash-flow diagrams for machines A and B, respectively. Note that the annual O&M costs and Q, quality (fewer rejects) savings, are (respectively) end-of-year expenses and "incomes." All amounts are in dollars.

Solution: We will first get the present value of each cash flow for machine A. There are only three: C(sal), O&M, and the C(f) of $50,000. This last one is already at the present value, that is, at time zero. Note that C(sal) is a distant "refund" or a negative cost. To bring it from the 10-yr end-of-life time, we need only to multiply it by $P/F(0.1,10)$. From Table A-1 in Appendix A, this factor's value is 0.38554. Hence, the present worth of salvage is

$$P(\text{sal}) = C(\text{sal}) \times P/F(0.1,10)$$
$$= -5000 \times 0.38554 = -1927.7$$

This value is negative because we are considering costs. As noted above, C(sal) refunds part of C(f), the first cost. Thus, it is a negative cost.

The present value of the annual O&M costs is also easily determined. For it, use the factor for uniform costs. Again, using Table A-1, $P/A(0.1,10)$ is 6.1446.

$$P(\text{O\&M}) = \text{O\&M} \times P/A(0.1,10)$$
$$= 3000 \times 6.1446 = 18,433.80$$

Now algebraically adding these values to C(f) = $50,000, we have the total cost to be

FIGURE 6-4
Cash flows for machine A.

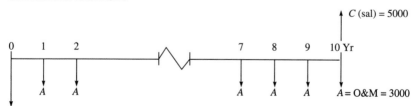

C(f) = $50,000 (Other Amounts Also in Dollars)

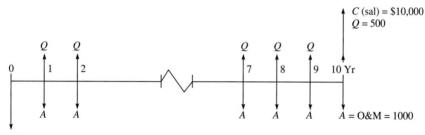

$C(f) = \$70,000$ (Other Amounts Also in Dollars)

FIGURE 6-5
Cash flows for machine B.

$$P(\text{mach A}) = \$50,000 + 18{,}433.8 - 1927.7 = \$66{,}506.1$$

The evaluation can be done by first putting $C(f)$ in a calculator's memory. Then calculate the present worth of each of the other terms and add it to or subtract it from the memory, as appropriate. When all have been determined, the net present value for the problem can be recalled from the calculator's memory.

Evaluation of machine B is similar; the method differs only in the savings that result from better quality control. The cash flows are shown in Figure 6-5. It shows that A for O&M is reduced by Q, the quality savings. Accordingly, the net uniform annual costs are

$$A(\text{net}) = A - Q = 1000 - 500 = 500$$

The corresponding present worth is $A(\text{net}) \times P/A(0.1,10)$; and using the value of $P/A(0.1,10) = 6.1446$, from above,

$$P[A \text{ (net)}] = 500 \times 6.1446 = 3072.30$$

Now, let us bring $C(\text{sal})$ to present value; again using the value of the corresponding factor used for machine A,

$$
\begin{aligned}
P(\text{sal}) &= C(\text{sal}) \times P/F(0.1,10) \\
&= -10{,}000 \times 0.38554 = -3855.40
\end{aligned}
$$

The net cost then is

$$
\begin{aligned}
P(\text{mach B}) &= C(f) + P[A(\text{net})] - P(\text{sal}) \\
&= 70{,}000 + 3072.30 - 3855.4 \\
&= 69{,}216.90
\end{aligned}
$$

Comparing this present value cost with that of machine A shows that purchasing machine A would give a savings of $2710.80, based on a money cost of 10 percent.

Suppose two machines have different service lives; it would no longer be correct to compare them on a present value basis. That is true because it would be difficult to give a correct value to the extra years of the machine with the longer life. We will now consider machines of different lives.

Example 6-4

The data for the machines of this example are given below:

Machine	A	B
$C(f)$, initial cost	$50,000	$70,000
O&M, annual operating & maintenance	3,000	1,000
Q, savings from fewer rejects	–	500
$C(\text{sal})$, salvage	5,000	10,000
N, service life	10 yr	15 yr
Cost of money	10%	10%

Correct comparison using present value requires that the time period for both machines be the same. In this example in which A has a life of 10 yr and B 15 yr, a suitable time period would be 30 yr. In that time we assume the salvaging and purchasing of machine A three times and that of machine B twice. But there is an easier way.

Solution: We will analyze each machine on a per year basis; that is, we will determine $A(A)$, an annual cost for machine A, and $A(B)$, an annual cost for machine B. To make our task easier, notice that except for machine B having a 15-yr life, the data for this example are the same as those for Example 6-3. Thus, getting the annual cost of machine A only requires multiplying its present value $P(\text{mach A})$ by $A(0.1,10)/P$. Both values can be obtained from Example 6-3. $P(\text{mach A})$ is $66,506.10; A(0.1,10)/P$ is the reciprocal of $P/A(0.1,10) = 6.1446$, or

$$A(0.1,10)/P = 1/6.1446$$
$$= 0.16274$$

Now,

$$A(\text{mach A}) = P(\text{mach A}) \times A(1.0,10)/P$$
$$= \$66,506.10 \times 0.16274$$
$$= \$10,823.20$$

Next we will get the annual cost of machine B; its cash flows are shown in Figure 6-6. Determining the net annual costs A for Q and O&M, both occurring every year, only requires subtracting

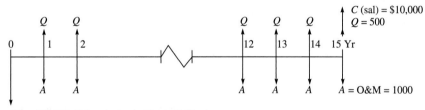

$C(f) = \$70,000$ (Other Amounts Also in Dollars)

FIGURE 6-6
Cash flows for machine B, Example 6-4.

$$A(O\&M/Q) = O\&M - Q$$
$$= 1000 - 500 = 500$$

For $C(f)$, the 15-year annual equivalent $A[C(f)]$ is

$$A[C(f)] = C(f) \times A(0.1,15)/P$$
$$= \$70,000 \times 0.13147$$
$$= 9202.90$$

where the value of $A(0.1,15)/P$ is obtained from Table A-1 in Appendix A.

Finally, the 15-year annual equivalent for the future value $C(\text{sal})$, a negative cost, is

$$A[C(\text{sal})] = C(\text{sal}) \times [A(0.1,15)/P \times P/F(0.1,15)$$
$$= -10,000 \times 0.13147 \times 0.23939$$
$$= -314.73$$

The value of $A(0.1,15)/P$ is the same as above, and the value for $P/F(0.1,15)$ is from Table A-1.

Note that for machines of longer life, the salvage value is small and can often be neglected.

The total annualized cost of this machine is the algebraic sum of all the A's we have evaluated, that is,

$$A(\text{mach B}) = A(O\&M/Q + A[C(f)] - A[C(\text{sal})]$$
$$= 500 + 9202.90 - 314.73$$
$$= 9388.17$$

Comparing this value with $A(\text{mach A})$ shows that for a money cost of 10 percent, purchase of machine B would give a per year savings of

$$\text{Savings per year, } A(\text{sav}) = A(\text{mach A}) - A(\text{mach B})$$
$$= \$10,823.20 - 9388.17$$
$$= 1435.00$$

The cents were dropped as a reminder that even results implying accuracy to tens of dollars are hardly justified. This is true because, as usual, the data are insufficiently accurate.

Incremental Rate of Return (IROR) [6]   As the name implies, IROR is the annual percent gain produced by an incremental increase in an investment. In Example 6-4 above, machine B required an incremental increase of investment of $20,000 in the initial cost: $70,000 − $50,000. This example shows that the greater investment, however, produces a per year savings of $1400. Thus, purchase of machine B provides an annual return of $1400. The incremental increase in investment required is $20,000, and in turn the

$$IROR = 1400/20{,}000 = 0.07 \twoheadrightarrow 7\%$$

Is this rate of return adequate to justify the additional investment? We consider that question next.

Minimum Attractive Rate of Return (MARR)   This rate is based on alternative outside investment opportunities. All investments must compete with the return possible from nearly risk-free federal notes and short-term bonds. Accordingly, from their rates and considering the probable risk associated with a project, management sets the firm's MARR.

Before considering taxes, let us return to IROR and consider its application to a personal investment.

Example 6-5

Your 5-year-old car needs a new battery. From reading this chapter (or some other source), you realize the wisdom of caring for and keeping an auto for 10, 15, or even perhaps 20 yr. On getting prices from several merchants and from calculations given below, you have the following data:

Option	Guaranteed battery life	Cost including installation	Cost per yr	Incremental Cost	Life	IROR
A	2 yr	$30	$15			
B	3	40	13.33	$10	1 yr	150%
C	5	50	10	10	2	133%

Solution: With option A, the cost per year is $15. Option B offers an additional year of life (a $15 value based on A) for an incremental cost increase of $10. The ratio of gain to cost gives the IROR of 150 percent. Similarly, option C gives a

$13.33 gain per year (based on B) for an incremental cost of $10. This ratio gives the IROR of 133 percent for C over B. Although the IROR for C is less than that for B, it is probably much greater than the MARR available from other opportunities.

Comment: A similar analysis of paying for our local newspaper gives an IROR of 48 percent for paying semiannually over quarterly, and an IROR of only 15 percent for paying annually over semiannually. As for the battery purchase above, the newspaper's IRORs are effectively after taxes. As we will see in the next section, the equivalent before-tax rate equals the after-tax rate divided by $(1 - t)$, where t is the decimal tax rate. As an example, the IROR of 48 percent for paying for the newspaper semiannually has an equivalent before-tax rate (as for all investments, except for tax-exempt securities reviewed in Chapter 7) of

$$\text{IROR (before taxes)} = \text{IROR (after taxes)}/(1 - t)$$

It is probable that your combined marginal state and federal tax rate t will be at least 33.33 percent. Accordingly, for IROR(after taxes) $= 48$ percent; IROR(before taxes) $= 48/(1 - 0.33) = 72$ percent. Similarly, the interest paid on a home mortgage is deductible from one's taxable income. Again, the effective income from reduced taxes gives an after-tax interest rate $r = i(1 - t)$. We will now consider how taxes effectively reduce the depreciation costs of capital investments.

Comparisons Considering Taxes   We have showed how tax reduction on an asset's depreciable cost effectively produces an "income"; it is symbolized by $I(t, k)$ for the kth year and at tax rate t. Here we will review the effects of taxes for the more common corporations whose earnings exceed $100,000. For them, the tax rate is in the magnitude of 40 percent, when federal, state, and local taxes are combined. Thus, we will use $t = 0.4$ for the decimal tax rate. The after-tax rate r must be used.

First Cost with Taxes   From the discussion on depreciation, we note that $C(d)$, the loss of value of an asset due to obsolescence or wear, is a cost of production. This cost reduces a firm's earnings and hence its income taxes. Furthermore, as opposed to normal expenses, the IRS requires that the depreciable part of an asset's cost be apportioned over its service life of N years. Let us review how the tax reduction affects the net cost of the equipment and its present value $P[\text{net } C(f)]$. Before tax considerations, recall that

$$C(f) = C(d) + C(\text{sal})$$

Considering taxes, the tax reductions act as "incomes" $I(k,t) = t \times D(k)$, which can be subtracted from $C(d)$. Let us modify the equation for $C(f)$ in two steps: (1) include these "incomes," and (2) discount the future cash flows of the tax reductions and $C(\text{sal})$. These steps give the present value of all cash flows associated with the asset, that is, $P[\text{net } C(f)]$.

Step 1. As presented earlier,

$$I(k,t) = t \times D(k) = t \times d(XX, k) \times C(d)$$

The term $d(XX,k)$ is the fraction of $C(d)$ allowed in the kth year as $D(k)$ by the depreciation schedule XX, such as SL, SD, or DDB.

Thus, on subtracting the $I(k,t)$'s and substituting from above, the new expression for the first cost becomes

$$C(f) = C(d) - \sum_{k=1}^{N} t \times D(k) + C(\text{sal})$$

This equation does not account for the time value of money. It is considered next, in step (2).

Step 2. The "incomes," subtracted from the total cost of the asset, are "paid" over the asset's life and must be discounted (brought back to the present) by multiplying each $I(k,t)$ in year k by the factor $P/F(r,k)$. $C(\text{sal})$, received at the end of the asset's life, cancels part of $C(f)$, but it must be discounted by $P/F(r, N)$. Recall "N" is the service life of the capital asset. Accordingly, the asset's present value after taxes, with the time value of money included, is

$$P[C(f)] = C(d) - \sum_{k=1}^{N} t \times D(k) \times P/F(r, k) + C(\text{sal}) \times [1 - P/F(r, n)]$$

Using $D(k) = d(XX,k) \times C(d)$, where $d(XX,k)$ is the fraction of $C(d)$ depreciated in the kth year as allowed by depreciation schedule XX,

$$P[C(f)] = C(d) - \sum_{k=1}^{N} t \times d(XX, k) \times C(d) \times P/F(r, k) + C(\text{sal}) \times [1 - P/F(r, N)]$$

Since $C(d)$ and t do not change with k, they can be brought before the summation. In two steps, we have

$$P[C(f)] = C(d) - C(d) \times \left[\sum_{k=1}^{N} t \times d(XX, k) \times P/F(r, k) \right] + C(\text{sal}) \times [1 - P/F(r, N)]$$

or $\quad P[C(f)] = C(d)\left\{ 1 - t \times \left[\sum_{k=1}^{N} d(XX, k) \times P/F(r, k) \right] \right\} + C(\text{sal}) \times [1 - P/F(r, N)]$

$$\tag{6-3}$$

For depreciation schedule XX, the terms in the last equation enclosed by the braces constitute the present value per dollar of depreciable cost $C(d)$. For the terms in the brackets, let us use the symbol $P/XX(r,n)$; values for $P/SL(0.1,n)$ and $P/SD(0.1,n)$ are given in Table A-1 of Appendix A. These values for other i's and n's are available in the references. Now, using $P/XX(r,n)$,

$$P[C(f)] = C(d)\{1 - t \times P/XX(r,n)\} + C(\text{sal}) \times [1 - P/F(r,n)] \tag{6-4}$$

$C(d)$ times 1 in the first term plus $C(\text{sal})$ times 1 in the second term equals $C(f)$. The first negative sign considers tax refunds due to depreciation costs. They are discounted because of lost interest. The second negative sign considers the salvage value refund, also discounted for lost interest. Note that this equation considers only the effects of taxes on the equipment's first cost. Next we review how taxes effect a reduction of other costs. In Example 6-6, we show the effects of taxes on the total cost.

Interest and Operation & Maintenance with Taxes   Both of these are operating expenses, and effectively they are reduced by the factor $(1 - t)$; the term *expense* indicates that these costs can be deducted from revenues in the year they are incurred. Repairs to and overhauling of a machine would also be expenses. Recall that interest after taxes is $r = i(1 - t)$.

Example 6-6

Considering taxes, evaluate the total annual cost of machine B in Example 6-4. As is typical of moderate- to large-sized firms, use a tax rate $t = 0.4$. The data for machine B are repeated below. Make evaluations for both SL and SD.

Machine	B
$C(f)$, initial cost	$70,000
O&M, annual operating & maintenance	1,000
Q, annual savings from fewer rejects	500
$C(\text{sal})$, salvage	10,000
N, service life	15 yr
Cost of money, before taxes	10%
Tax rate	0.4

Solution: The cash-flow diagram, including tax "incomes" due to depreciation $t \times D$, for considering taxes is shown in Figure 6-7.

Taxes affect the present value of $C(f)$ in two ways: first, because of the "incomes" from tax reductions produced by depreciation; and second, by the cost

FIGURE 6-7
Cash flows for machine B, Example 6-6.

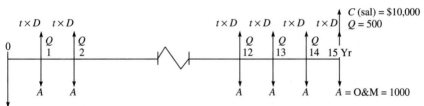

$C(f) = \$70,000$ (Other Amounts Also in Dollars)

of money, that is, the interest rate being reduced by $(1 - t)$. In this example, the before-tax interest rate is 10.0 percent; r, the *after-tax* rate,

$$r = i(1 - t) = 10.0(1 - 0.4) = 6\%$$

The net annual expenses of O&M and Q are also reduced by the factor $(1 - t)$; their present value and that of $C(sal)$ are also affected by the reduced cost of money.

We now use Equation 6-4 to determine the after-tax present value of $C(f)$, that is $P[C(f)]$. The "incomes" produced by depreciation depend on the schedule. SL is used in the first part of this example; SD, in the second part.

Depreciation Schedule, SL We first consider the effect of taxes on the equipment's first cost, that is, $P[C(f)]$. Substituting accordingly in Equation 6-4, and after getting the factors for $r = 6$ percent and $n = 15$ from Table A-2 in Appendix A, and using $C(d) = C(f) - C(sal) = \$70,000 - \$10,000 = \$60,000$,

$$P[C(f)] = C(d)\{1 - t \times P/SL(r, n)\} + C(sal) \times [1 - P/F(r, n)]$$
$$= \$60,000 \{1 - 0.4 \times 0.64748\} + \$10,000 \times [1 - 0.41727]$$
$$= \$60,000 \times 0.7410 + \$10,000 \times 0.58273$$
$$= 44,460.48 + 5827.30 = \$50,287.78$$

If taxes were zero, the value of $C(d)$ would be unchanged and the discounting factor of $C(sal)$ would be 0.23939 (at 10 percent) instead of 0.41727 above (at 6 percent). Accordingly, without taxes,

$$P[C(f)] = \$60,000 + 10,000 \times [1 - 0.23939]$$
$$= 67,606.10$$

Again, the calculations should be rounded to at least hundreds of dollars. We note that the tax reductions for depreciation expenses and in the cost of money effect a savings in the present value cost of capital goods, about \$17,000 in this example.

Let us continue by evaluating the effects of taxes on the O&M costs and the quality savings. As noted earlier, taxes effectively reduce both by $(1 - t)$, or $1 - 0.4 = 0.6$. In turn,

$$A(O\&M\backslash Q) = [1000 - 500] \times 0.6 = \$300$$

The O&M\Q costs are already on an annual end-of-year basis. It is easy to convert $P[C(f)]$, from above, to an annual basis. We only need to multiply it by $A/P(r,n)$; from Table A-2 in Appendix A for 6 percent and 15 yr, it is 0.10296. Hence,

$$A\{P[C(f)]\} = P[C(f)] \times A/P(r,n) = \$50,287.78 \times 0.10296$$
$$= \$5177.63$$

The total annual cost is the sum of $A\{P[C(f)]\}$ and A (O&M/Q). Thus, considering taxes with an SL depreciation schedule, we have

$$A[\text{machB}] = A\{P[C(f)]\} + A(\text{O\&M}/Q)$$
$$= \$5177.63 + \$300 = \$5477.63$$

The annual cost, not considering taxes, as determined for machine B in Example 6-4 is $9388. Of course, a firm does not save by paying taxes, but part of the asset's cost is paid for by reduced taxes. As noted earlier, regarding personal expenses, if you are in a 28 percent marginal federal tax bracket, 10 percent paid on a home mortgage is effectively reduced to $10(1 - 0.28) = 7.2$ percent. This is true because interest paid on a home mortgage is deductible from your taxable income. State income tax would further reduce the effective home mortgage rate.

The significant point is that taxes should be considered in comparing the costs of assets. Let us now evaluate machine B using the SD depreciation schedule.

Depreciation Schedule, SD   Only $P[C(d)]$ is different from the SL evaluation. On substituting the SD schedule in Equation 6-4 and getting the value for $P/SD(r,n)$ from Table A-2, we have

$$P[C(f)] = C(d)\{1 - t \times P/SD(r,n)\} + C(\text{sal}) \times [1 - P/F(r,n)]$$
$$= 60{,}000\{1 - 0.4 \times 0.73441\} + 10{,}000 \times [1 - 0.41727]$$
$$= 60{,}000 \times 0.7062 + 10{,}000 \times 0.58273$$
$$= 42{,}374.16 + 5827.30 = \$48{,}201.46$$

As in the first part of the example, this present value is changed to the annual cost by multiplying it by $A(0.06,15)/P = 0.10296$. The results are

$$P[C(f)] = \$48{,}201.46 \times 0.10296 = \$4962.82$$

To this we add the same annual net cost from the first part of the example for O&M\$Q$ = \$300; the total annual cost using the SD depreciation schedule is

$$A[\text{machB}] = A\{P[C(f)]\} + A(\text{O\&M}\backslash Q)$$
$$= \$4962.82 + \$300 = \$5263$$

From the first part, the annual cost for machine B is $5477.63 when the SL depreciation schedule is used. Thus, use of the SD schedule produces a per year savings of about $200. Saving this much on many machines, such as 1000 automated welders, would be significant.

In the comparison of alternatives just presented, we used the "classic" depreciation schedules. They must be used for equipment placed in service before 1981. In 1981, a new schedule was set up. In 1986, a modification of that schedule was established. As noted above, both use a combination of the declining-balance and straight-line systems. Accordingly, the concepts for comparing alternatives are the same as for the classic schedules. The change in implementations is presented next.

ACRS: Accelerated Capital Recovery System [7,8] This method for depreciation was established by the Tax Reform Act of 1981. It has two major advantages over the pre-1981 schedules: (1) The computations use *property class lives;* they are less than the actual useful lives (called *class life* before 1981). (2) Salvage values are assumed to be zero. Relevant to (1), the classes are divided into eight groups. The first six are for personal property (i.e., all property except real estate). The last two (the seventh and eighth groups) are for real property, i.e., real estate. The seventh group is for residential rental property; the eighth is for nonresidential property. Each of the eight property classes has its corresponding property life, as opposed to the "service life" for the pre-1981 schedules set by the IRS. Also, for each property life, there is a depreciation for each year. It is determined by using the corresponding percent for that year, taken from IRS tables. The year's depreciation is that corresponding percentage of the property's original cost. For all of these, the half-year convention is used.

The half-year convention is the arrangement which allows only one-half the normal schedule depreciation in the first year. The other half is deferred to the year after the property life. This convention is apparent in Table 6-4 (for MACRS).

MACRS: Modified Accelerated Capital Recovery System [9,10] This method differs from the ACRS in only two small ways. The first concerns changes in the equipment included in the eight property classes. Second, there are slight changes in the percentages in a few of the years for the various classes. These percentages are presented in Table 6-4 and can be calculated. Their determination uses double-declining (200 percent) balance for property with recovery lives of 3, 5, 7, and 10 yr. For property with recovery lives of 15 and 20 yr, 150 percent declining balance is used.

The property classes are presented next. They apply only to tangible property. The straight-line schedule still applies to the intangible type, such as patents and franchises.

Guidelines 6-1: MACRS Property Classes Depreciable tangible property may be assigned to one of the following eight classes [11]:

1 *3-yr property.* With a few important exceptions, this class includes qualifying property with a class life (pre-1981 service life) of 4 yr or less. This involves tractor units for over-the-road use; special handling devices for the manufacture of food and beverages; special tools for the manufacture of rubber, finished plastic, glass, and fabricated metal products; and special tools for the manufacture of motor vehicles. Automobiles and light general-purpose trucks, which had been 3-yr property under the ACRS, are excluded here. They are now 5-yr property.

2 *5-yr property.* Included is property with a class life of more than 4 yr but less than 10 yr, plus automobiles; light general-purpose trucks; computer-based telephone central office switching equipment; research and experimentation property; certain thermal-, solar-, and wind-energy property; and certain biomass property. Some properties included in this class life interval are computers and peripheral equipment; office machinery (typewriters, calculators, copiers, and so on); heavy general-purpose trucks; motor transport assets; offshore oil and gas well-drilling assets; construction assets; timber-cutting and sawing assets; and many assets used for the manufacture of knitted goods, carpets, medical and dental supplies, chemicals, and electronic components.

TABLE 6-4
MACRS DEDUCTION RATES, $d(k)$, IN kTH YEAR OF RECOVERY

Percentages for 3, 5, 7, and 10-yr Property Are 200 Percent DB with the Half-Year Convention. For 15- and 20-yr Property, the Percentages Are 150 Percent DB with Half-Year Convention. EOY Is the Abbreviation for End of Year.

EOY	3-yr	5-yr	7-yr	Property 10-yr	15-yr	20-yr
1	33.33%	20.00%	14.29%	10.00%	5.00%	3.750%
2	44.45	32.00	24.49	18.00	9.50	7.219
3	14.81	19.20	17.49	14.40	8.55	6.677
4	7.41	11.52	12.49	11.52	7.70	6.177
5		11.52	8.93	9.22	6.93	5.713
6		5.76	8.92	7.37	6.23	5.285
7			8.93	6.55	5.90	4.888
8			4.46	6.55	5.90	4.522
9				6.56	5.91	4.462
10				6.55	5.90	4.461
11				3.28	5.91	4.462
12					5.90	4.461
13					5.91	4.462
14					5.90	4.461
15					5.91	4.462
16					2.95	4.461
17						4.462
18						4.461
19						4.462
20						4.461
21						2.231

Source: U.S. Internal Revenue Publication 534, Published yearly.

3 *7-yr property.* Included is property with a class life of 10 yr or more but less than 16 yr, property without any class life and not specifically included in the 27.5- and 31.5-yr categories to follow, and several specifically named items. Some assets in this interval include office furniture, fixtures, and equipment; oil and gas exploration assets; theme and amusement park assets; and most assets used for manufacturing such things as food products, spun yarn, wood products and furniture, pulp and paper, rubber products, finished plastic products, leather products, glass products, foundry products, fabricated metal products, motor vehicles, aerospace products, athletic goods, and jewelry.

4 *10-yr property.* This class includes property with a class life of 16 yr or more but less than 20 yr. Some assets included are vessels, tugs, and similar water transportation equipment; assets used in petroleum refining; and assets used in the manufacture of grain, sugar, and vegetable oil products.

5 *15-yr property.* This class includes property with a class life of 20 yr or more but less than 25 yr, plus sewage treatment plants, and telephone distribution plants and comparable equipment used for the two-way exchange of voice and data communications. Some assets included in this interval are land improvements such as sidewalks, roads, drainage facilities, sewers, bridges, fencing, and landscaping; assets used in the manufacture of cement; some pipeline transportation assets; some liquefied natural gas equipment; and certain utility property.

6 *20-yr property.* Property with a class life of 25 yr or more, other than real property with a class life of 27.5 yr or more, plus municipal sewers is included. Such assets are farm buildings; some railroad electric generating equipment; certain transmission lines, pole lines, buried cable, and repeaters; and most other utility property.

7 *27.5-yr residential rental property.* Included is a rental building or structure for which 80 percent or more of the gross rental income for the tax year is rental income from dwelling units. A *dwelling unit* is a house or an apartment used to provide living accommodations, but not a unit in a hotel, motel, inn, or other establishment in which more than one-half of the units are used on a transient basis.

8 *31.5-yr nonresidential real property.* Depreciable property that has a class life of 27.5 yr or more and is not 27.5-yr residential rental property generally falls in this class.

In Example 6-6, machine B from Example 6-4 is evaluated using SL and SD. In Example 6-7, we evaluate its cost using MACRS.

Example 6-7

Evaluate the annual cost of machine B from Example 6-4 using MACRS. First, we must decide the property class. It sets the number of years which must be used for recovery of the total capital. Also, it sets the percentages (of the total cost) recovered each year. From Guidelines 6-1, for a machine with a 15-yr service life (also referred to as *class life* in pre-1981 guidelines), the property class life is 7 yr. For that class, Table 6-4 gives the percentages used below. Recall that MACRS considers the salvage value as zero, so $C(f) = \$70,000$ is used instead of $C(d) = \$60,000$ in determining the depreciation. Furthermore, because of the half-year convention, the recovery time is 1 yr greater than the property class life. The tax $t = 0.4$, and the after-tax cost of money is 6 percent.

Solution: For MACRS, the present value of the total cost is given by Equation 6-3 with $C(d) = C(f)$, $C(\mathrm{sal}) = 0$,

$$d(\mathrm{XX}, k) = d(k) \qquad \text{and} \qquad P/F(r, k) = 1/(1+r)^k$$

$$P[C(f)] = C(f)\left[1 - t\sum_{k=1}^{N+1} d(k)/(1+r)^k\right]$$

$$= \$70,000\left\{1 - 0.4\left[0.1429/(1.06) + 0.2449/(1.06)^2\right.\right.$$

$$+ 0.1749/(1.06)^3 + 0.1249/(1.06)^4 + 0.0893/(1.06)^5$$

$$\left.\left. + 0.0892/(1.06)^6 + 0.0893/(1.06)^7 + 0.0446/(1.06)^8\right]\right\}$$

(6-5)

The terms inside the rectangular brackets could be defined as P/MACRS(0.06,7), that is, the present value of each dollar of original cost for 7-yr property and after-tax cost of money of 6 percent, using MACRS. Using a hand calculator, it is 0.7916618. Multiplying it by $t = 0.4$ and subtracting the results from 1 gives 0.683335.

Thus, by Equation 6-3,

$$P[C(\mathrm{f})] = \$70{,}000 \times 0.683335 = \$47{,}833.47$$

Since this machine has a service life of 15 yr, the per year cost should be determined on a 15-yr basis. Accordingly, to get the annual cost of its first cost, we multiply the present value by A/P(0.06,15). From Table A-2 in Appendix A, it is 0.10296. The results are

$$A\{P[C(\mathrm{f})/\mathrm{MACRS}]\} = \$47{,}833.47 \times 0.10296 = \$4924.93$$

Now adding the after-tax cost for operating and maintenance reduced by the quality savings (from Example 6-6),

$$A[\mathrm{machB/MACRS}] = A\{P/[C(\mathrm{f})/\mathrm{MACRS}]\} + A(\mathrm{O\&M}/Q)$$
$$= \$4924.93 + \$300 = \$5224.93$$

This is not significantly different from the costs using the SD schedule. The difference increases rapidly with higher after-tax cost of money.

Having developed the techniques for comparing the cost of projects, we will now focus on methods to evaluate their profitability.

Profitability: Six Methods [12]

We have analyzed methods for determining the costs of projects. Here we will present techniques for determining their profitability. Furthermore, the techniques will be extended to the evaluation of a project's financial advantage per dollar invested. *Cash flow* is often used as the name implies, to indicate a flow of money. It is, however, defined as the money from earnings that is available to a firm for discretionary use. In a given year, it is equal to the after-tax profits plus the depreciation, and amortization less preferred dividends (if any), all for that year.

The six profitability methods progress from easy and approximate to more detailed and accurate. They are (1) payout (or payback) time without interest, (2) payout time with interest, (3) return on original investment, (4) return on average investment, (5) net present value (or venture worth), and finally, the most sophisticated, (6) discount cash flow rate of return, also known as return on investment (ROI).

A simple project will be used to develop and illustrate these methods. Data for this 4-yr enterprise is presented in Table 6-5. Recall that in a given year, cash flow equals after-tax profits plus depreciation. Also, in a given year, the cumulative cash flow equals its amount from the previous year plus the cash flow for the given year. The first simplification is that, unrealistically, the project's total investment is considered to

TABLE 6-5
CASH FLOWS OF A SIMPLE PROJECT

Time, end of year	After-tax profits	Depreciation	Cash flow	Cumulative cash flow
0	−$1200*	$000	−$1200	−$1200
1	400	200	600	−600
2	300	200	500	−100
3	250	200	450	350
4	200	200	400	750

*All amounts are in 000's.

be made at a single time, zero. Cash flows determined as end-of-year expenses, how-ever, are the usual practice. Straight-line depreciation is used; it is easy, without detracting from the presentations. $C(\text{sal}) = \$400,000$; thus, for $C(f) = \$1,200,000$, $C(d) = \$800,000$. In turn, the SL depreciation is $200,000 in each of the 4 yr.

Payout Time without Interest   Although simple and very approximate, we will see that this quick technique can be valuable as a first test. Consider the data in Table 6-5. Investments are considered as negative cash flows. Figure 6-8 shows the data from Table 6-5. From this and the plot of the cumulative cash flow, extrapolation gives $T(\text{payback}) = 2 + 100/450 = 2.2$ yr, the time of zero cumulative cash flow. Please note that this is a linear approximation.

We show in Appendix B of this chapter that for projects with economic lives much greater than the payout time, there is a particular situation. The reciprocal of the pay-out time (without interest) is approximately equal to ROI, the return on investment. This ROI (rate of return) is used for project comparisons by the discounted cash flow rate of return method, discussed below. For instance, consider an energy-saving refrig-erator with a typical 20-yr life. It costs $60 more than the less efficient refrigerator, but saves $20 per year on operating costs. Its payout time (for the cost of the better effi-ciency) would be $\frac{60}{20} = 3$ yr. The return on the extra investment of $60 would be

FIGURE 6-8
Cash flows for payout of a simple project

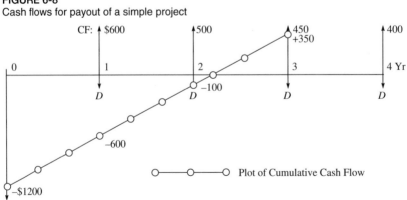

approximately 33 percent, over the 20-yr life of the refrigerator. In Chapter 7, we recognize that this effective after-tax rate is an excellent investment.

Payout Time with Interest   To facilitate the presentation, 10 percent interest is used. It is a per year expense equal to one-tenth (10 percent) of the remaining investment (cumulative cash flow) in that year. The interest cost is deducted from the cash flow for that year. Table 6-6 shows the data. Depreciation is included in the cash flow and is not listed separately. This cash flow is the value before deducting interest. We will review development for yr 1. As shown, the investment for yr 1 is its original cost, $C(f) = \$1,200,000$. The interest at 10 percent is $120,000. We subtract it from the cash flow of $600,000 for yr 1. The subtraction gives the after (interest) deduction amount of $480,000 (see Table 6-6). Now adding it to the initial value of cumulative cash flow gives the value after yr $1 = -\$720,000$. Continuing the method, the cumulative cash flow becomes zero in the third year. Interpolation gives the payout time as $2 + 292/421 = 2.7$ yr.

Return on Original Investment   This is the first direct method for estimating the important information: return on investment.

$$\text{Ret. on orig. invest.} = \frac{\text{average profit over the project's life}}{\text{first cost + working capital}}$$

Working capital is defined in Section 6-1 on accounting. For this simple project, it is zero. The other data for this calculation are obtained from Table 6-5. Thus,

$$\text{Ret. on orig. invest.} = \frac{[400 + 300 + 250 + 200]/4}{1200 + 0}$$

$$= \frac{287.5}{1200} = 0.2396, \text{ or } 24\%$$

TABLE 6-6
CASH FLOWS OF A SIMPLE PROJECT: INTEREST DEDUCTED

Time, end of year	Investment for year	Interest for year	Cash flow interest		Cumulative cash flow
			Before	After deduction	
0	$0000*	$000	-$1200		-$1200
1	1,200	120	600	$480	-720
2	720	72	500	428	-292
3	292	29	450	421	+129
4	129	13	400	387	+516

*All amounts are in 000's.

Return on Average Investment   This is the next step to improve analysis. It requires only a little more effort. The return is the average profit divided by the average investment, both over the project's life.

$$\text{Ret. on ave. invest.} = \frac{\text{average profit over the project's life}}{\text{average invest. over the project's life}}$$

The average investment is determined by summing the net investments for each year over the life of the project and then dividing that summation by the number of years of the project. Note, the net investment in the kth year equals its value in the (kth $-$ 1) year minus the depreciation for year k. The net investment for year 1 is $1200(000); the depreciation is $200(000) taken at the end of each year. The process is shown below.

Year		Net investment
1		$1200
2	1200 − 200	1000
3	1000 − 200	800
4	800 − 200	600
	Ave.	900

Using the average profit of $287.50, as calculated above,

$$\text{Ret. on ave. invest.} = \frac{287.50}{900.00} = 0.319, \text{ or } 32\%$$

Before proceeding with the last two methods, we consider a reasoned approach for determining a firm's cost of capital. It will give a reference for comparing rates of return and in evaluating the worth of a project.

We will consider the average cost of capital (ACC). The financial statement of a firm, as presented in Section 6-1 on accounting, shows the firm's capital structure. Basically, capital consists of debts (liabilities) and equity (common and preferred stock); these are the primary sources of its funds. Consider a firm with a structure of 60 percent common stock, 30 percent preferred stock, and 10 percent debt (bonds are the usual long-term debt). Let us first review the cost of funds from each of these sources.

The P/E ratio, also reviewed in Section 6-1 on accounting, indicates the cost of money by sale of equities (common stock). Its reciprocal multiplied by 100 is the rate in percent. By sale of preferred stock, a firm promises a definite rate of return and first claim to earnings. After a firm is well established, it is able to sell long-term bonds. They usually have a term of 10 to 30 yr. Their interest rate is normally 6 to 10 percent, one-half the usual equity rate.

The rates a firm might have to pay for funds are these: 8 percent for debt, i.e., commercial bank loans or bonds; 12 percent for equity; and 6 percent for preferred stock. For our firm with the capitalization given above, the *a*verage *c*ost of *c*apital (ACC) is determined below. For convenience, we round it off to 10 percent.

$$\text{ACC} = 0.6 \times 12\% + 0.3 \times 6\% + 0.1 \times 8\% = 9.8\%$$

Net Present Value (Venture Worth)   In this method, the present values of all cash flows, positive or negative, are summed algebraically over the n years of the project's life. The present value of each, from the kth year, is determined using the present worth factors $P/F(i,k)$. The interest rate i depends on the current financial environment. The rate is usually the ACC rate. For our example, we will use $i = 10$ percent, as determined above. The cash flow in the kth year is given in both Tables 6-5 and 6-6. The k's in the P/F's correspond, respectively, to the number of years in the future of each cash flow. First, we develop the general equation for the net present value (NPV) using the equation for $P/F(k) = (1 + i)^{-k}, k = 0, . . ., n$.

$$\text{NPV} = - C(\text{f})/(1 + i)^0 + CF(1)/(1 + i)^1 + \cdot \cdot \cdot + CF(n)/(1 + i)^n$$

Now, for example,

$$\text{NPV} = -1200 + 600/(1 + 0.1) + 500/(1 + 0.1)^2 + 450/(1 + 0.1)^3 + 400/(1 + 0.1)^4$$
$$= -1200 + 545.45 + 413.22 + 338.09 + 273.21 = 369.97$$

Recall that the amounts are in thousands; thus this project for money worth 10 percent has a present worth of about \$370,000. This evaluation gives us two significant data. First, if the money for the investment was borrowed at 10 percent, the net profit on this project or venture is approximately \$370,000 minus the present value of any other expenses required to implement it. Second, this venture is proportionally better than any other project of smaller NPV, or vice versa for one of larger NPV.

In the NPV method, the interest rate is set and the present value is determined. If the first cost in the example was −\$1,570,000 (that is, −1,200,000 − 370,000), the venture's NPV would be zero. The firm, however, would have a rate of return equal to its ACC rate, and it could pay all of its investors at their proper rate of return (ROR). If the first cost was less than the \$1,570,000, the ROR would be greater than the ACC rate. The next method, called the *discount cash flow ROR* (rate of return), can be used to find that rate.

The Discounted Cash Flow Rate of Return   This approach follows nicely from the net present value. Instead of determining the NPV for a given interest rate, now the NPV is set equal to zero and the corresponding interest rate is determined. That rate is the *discount flow rate of return* (DCFROR). Again we first set up the general equation.

$$\text{NPV} = - C(\text{f})/(1 + i)^0 + CF(1)/(1 + i)^1 + \cdot \cdot \cdot + CF(n)/(1 + i)^n$$

Recall that an equation is nonlinear if its coefficients are a function of the dependent variable. That is the situation in this equation, since i is the dependent variable. It has been my experience, though with no proof, that there are only three ways of solving nonlinear equations. (1) Use linear approximations; this technique is suitable only for small variations about a point, as in the small signal analysis of electronic amplifiers. (2) Use graphs; this method is suitable for large signal analyses of electronic amplifiers. (3) Use a series of trials with proper modifications after each trial. Techniques (1) and (3) can be used in a complementary way in solving the equation for NPV set to zero. Our previous solution for $i = 10$ percent gave NPV $= 370$. For $i = 0$, the NPV $= -1200 + 600 + 500 + 450 + 400 = 750$. Plotting these two values (Figure 6-9) gives a line which, projected to zero NPV, indicates the next guess for i; it is 23 percent.* Using this,

$$NPV = -1200 + 600/(1.23) + 500/(1.23)^2 + 450/(1.23)^3$$
$$+ 400/(1.23)^4$$
$$= -1200 + 487.8 + 330.49 + 241.82 + 174.76 = 34.87$$

A value of $i = 25$ percent gives NPV $= -5.76$; extrapolation to NPV $= 0$ gives the DCFROR $= 23.7$, or 24 percent.

This value is close to the one obtained by the return on original investment method. That is true because the project's life is short, and thus the effects of discounting future cash flows are small. The short life was chosen to simplify the presentation. For more realistic projects or for machinery with lives of 10 to 30 yr, the DCFROR method, discussed here, is a far better model of the true situation.

The life and growth of a firm depend on controlling costs and implementing profitable projects. The fundamentals and techniques of this chapter are the essences of these endeavors. Furthermore, hardly any other aspect of engineering is as widely applicable to our personal lives. Financial security gives us a mental reserve. Often the confidence it gives is a definite asset in interacting pleasantly with others. It is equally

---

*You probably recognize this as the secant method for determining a real root of a polynomial.

FIGURE 6-9
DCFROR estimates for discounted cash flow.

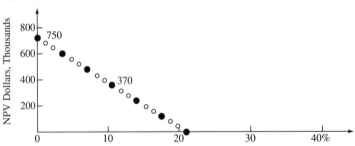

helpful in succeeding at work. Chapter 7 presents fundamentals and techniques required to build an estate and achieve financial security, that is financial planning and the results of carrying out the plans.

DISCUSSION QUESTIONS AND PROBLEMS

6-1 ELEMENTS OF ACCOUNTING

1 a What basic equation is often called the *balance sheet identity?*
   b What variable continuously changes to keep this equation (identically) balanced?
2 From the data in Table 6-7:
   a For the Enterprising Corporation, calculate the owners' equity.
   b Determine the firm's net working capital.
   c The firm has 400,000 shares of preferred stock ($100/sh) and 100 million shares of common stock; what is the book value of the common stock?

TABLE 6-7
Enterprising Corporation Partial Financial
Reports—December 19XX
Balance Sheet
(in millions)

Assets		Liabilities and owners' equity	
Current assets	.$226	Current liabilities	$ 88
Fixed assets	754	Long-term debt	478

Income Statement

Sales	$1437
Costs	882
Depreciation	164
Interest paid	48

3 For the Enterprising Corporation:
   a Determine its before-taxes earnings.
   b If taxes are 34 percent, determine the after-tax earnings per share of common stock; the dividend per share of preferred stock is $8.
4 The Sharp Pencil Corporation has a lease payment on the production building and equipment of $4000 per month. The variable cost is $0.40 per thin-lead mechanical pencil.
   a If the pencils have a selling price of $1.00 each, how many pencils must Sharp produce annually to reach the break-even level?
   b What will be the firm's annual earnings before taxes if it normally produces and sells 150,000 pencils for $1.00 each?
   c The firm is given a one-time order for 100,000 pencils at $0.80 per pencil. Filling the order is estimated to require 1 yr. For all production above 200,000 (in 1 yr), the average variable cost is estimated to be $0.70 per pencil; would the order be profitable? This special order would be over and above the firm's normal production of 150,000 per year. *Hint:* The firm's annual fixed cost is already paid by the normal production.

6-2 FORMS OF FINANCIAL ORGANIZATION

5 a What are the three basic forms of business organization?
   b Review the differences between a general and limited partnership.
   c What is the primary advantage of an S corporation?

6-3 ENGINEERING ECONOMICS

6 Immediately after marriage, a couple decide to save a substantial amount for a down payment on a house.

  a Each spouse (both are engineers) puts $1000 in a 5-yr CD that pays 6 percent interest compounded annually. What will be the total value of their CD after the 5 yr?

  b They decide that an additional $3000 would put them in a better position when they negotiate a mortgage. A low-risk short-term bond fund pays 8 percent, compounded annually. What additional amount would they have to put in the short-term bond fund for it to have a value of $3000 in 5 yr?

  c To keep more funds available for the "unexpecteds," rather than putting the total required amount in the bond fund, they decide to put in only the minimum initial amount of $1000 and add $50 per month. These deposits compound monthly at $\frac{8}{12}$ percent per month. In 60 months (5 yr) what will be the total value of these monthly payments plus the $1000 compounded annually at 8%? *Note:* The odd interest ($\frac{8}{12}$ percent) requires use of a calculator and the equation developed in Section 6-3 on future values of an annuity, that is,

$$[F(i,n)/A] = [(1 + i)^n - 1]/i$$

7 a When inflation is low, money markets require a lower daily nominal rate (no need to add 4 to 5 percent to give a "real" rate between 3.5 and 4.5 percent). If in a period of low inflation the money market rate is 4.5 percent, compounded daily, calculate the effective (per annual) rate.

  b Determine the effective rate if the 4.5 percent is compounded continuously.

8 The subscription for a newspaper may be paid in advance:

Months	Amount
1	$4.35
6	25.00

For each payment, consider that blank checks cost $0.05, postage costs $0.30, and time and effort (perhaps 15 min) are worth $0.50. What APR (annual percentage rate) is earned by paying 6 months in advance? *Hint:* By paying for 6 months in advance, on the excess over the 1-month cost, one gets an annuity-type income of $4.35 + 0.05 + 0.30 + 0.50 at the beginning of each of the next 5 months. A cash-flow diagram would be helpful. Also, some trial and correction are required. Start with 8 percent per month.

9 A manufacturer plans to buy a 250-hp electric motor for a chiller (air conditioning). In Table 6-8, the cost data are given in the column for motor A (the next problem uses data for motor B). For motor A, draw a cash-flow diagram and determine the present value of its total cost. The cost of money is 10 percent.

10 Using data for motor B given in Table 6-8 and considering that the cost of money is 10 percent, (a) draw its cash-flow diagram and determine its annual cost. Hint: Since O&M is already on an annual basis, economy of effort suggests converting the other two individually to their annual costs. For the salvage value, the A/F factor is needed but is not included in the Appendix. That's no problem. As you have probably realized, it is readily obtained by

TABLE 6-8
COST DATA ON ELECTRIC MOTORS FOR PROBLEMS 9, 10, 11, AND 12

Motor	A	B
First cost	$1500	$1200
O&M (operation & maintenance)	200 per year	300 per year
Salvage	300	200
Service life	8 yr	5 yr

multiplying factors A/P and P/F. (b) By converting the present value of the cost of motor A to an annual cost, determine which motor is more economical and by how much.

11 Using data for motor A (Table 6-8), determine (a) the straight-line per year depreciation of motor A and its book value at the end of yr 3 and (b) the depreciation of motor A in the second year using the sum-of-digits depreciation schedule.

12 If the firm in Problem 9 has a tax rate of 34 percent, determine the after-tax annual cost of each motor (a) using the straight-line depreciation schedule and (b) using the sum-of-digits schedule. As above, the after-tax cost of money $r = 6$ percent. Tables in the Appendix give depreciation data.

13 The firm in Problem 9 has a tax rate of 34 percent; determine the after-tax annual cost of motor B using the MACRS. The after-tax cost of money is 6 percent. To reduce your unproductive efforts, without reducing the learning experience in the problem, the shorter-life motor B is used; however, this problem does not show the greater cost savings which MACRS gives on property with longer service life.

The following information is for Problems 14 through 19.

A professional realty group is considering the purchase of a 40-unit apartment complex. It is to be sold after 4 yr. To shorten the period for recovery of capital, they use the sum-of-digits depreciation schedule. After reviewing records for the complex and making the needed calculations, they distilled the data given in Table 6-9.

14 Determine the payout time (without interest) for the apartment project.

15 Determine the payout time for the apartment investment if the interest after taxes is 10 percent.

16 Determine the return on the original investment for the apartment investment.

TABLE 6-9
CASH FLOWS FOR AN APARTMENT COMPLEX

Time, end of year	After-tax profits	Depreciation (sum-of-digits)	Cash flow	Cumulative cash flow
0	−$1200*	$000	−$1200	−$1200
1	400	400	800	−400
2	300	300	600	200
3	250	200	450	650
4	200	100	300	950

*All amounts are in 1000's.

17 Determine the return on the average investment for the apartment investment.

18 Determine the net present value (venture worth) of the apartment investment. The cost of money, after taxes, is 20 percent.

19 Determine the discount cash flow rate of return (DCFROR) of the apartment project.

REFERENCES

1 Curran, W. S. *Principles of corporate finance.* New York: Harcourt Brace Jovanovich, 1988, p. 671.

2 Humphreys, K. K. *Jelen's cost and optimization engineering* (3rd ed.). New York: McGraw-Hill, 1991, p. 149.

3 Moyer, R. C., McGuigan, J. R., & Kretlow, W. J. *Contemporary financial management* (4th ed.). St. Paul, MN: West, 1990, p. 15.

4 Humphreys. *Jelen's cost.* p. 21.

5 Humphreys. *Jelen's cost.* p. 84.

6 Collier, C. A., & Ledbetter, W. B. *Engineering economic and cost analysis* (2nd ed.). Philadelphia: Harper & Row, 1988, p. 248.

7 Newman, D. G. *Engineering economic analysis* (3rd ed.). San Jose, CA: Engineering Press, 1983, pp. 281–286.

8 Humphreys. *Jelen's cost.* p. 65.

9 White, J. A., Agee, M. H., & Case, K. E. *Principles of engineering economic analysis* (3rd ed.). New York: Wiley, 1989, p. 261.

10 Humphreys. *Jelen's cost.* p. 64.

11 Commerce Clearing House. *1987 depreciation guide.* June 18, 1987.

12 Humphreys. *Jelen's cost.* pp. 104–110.

APPENDIX A

TABLE A-1
TABLE OF FACTORS FOR 10%
DISCRETE COMPOUND INTEREST = 10%

	Single payment		Uniform annual series		Depreciation series	
	Compound-interest factor $(1+i)^n$	Present-worth factor $\frac{1}{(1+i)^n}$	Unacost present-worth factor $\frac{(1+i)^n - 1}{i(1+i)^n}$	Capital-recovery factor $\frac{(1+i)^n}{(1+i)^n - 1}$	Sum-of-digits present-worth factor $\frac{n - F_{PP}}{0.5n(n+1)i}$	Straight-line present worth factor $\frac{1}{niF_{PK}}$
	P to F	F to P	A to P	P to A	SD to P	SL to P
n	F/P	P/F	P/A	A/P	P/SD	R/SL
1	1.1000E 00	9.0909E-01	9.0909E-01	1.1000E 00	9.0909E-01	9.0909E-01
2	1.2100E 00	8.2645E-01	1.7355E 00	5.7619E-01	8.8154E-01	8.6777E-01
3	1.3310E 00	7.5131E-01	2.4869E 00	4.0211E-01	8.5525E-01	8.2895E-01
4	1.4641E 00	6.8301E-01	3.1699E 00	3.1547E-01	8.3013E-01	7.9247E-01
5	1.6105E 00	6.2092E-01	3.7908E 00	2.6380E-01	8.0614E-01	7.5816E-01
6	1.7716E 00	5.6447E-01	4.3553E 00	2.2961E-01	7.8321E-01	7.2588E-01
7	1.9487E 00	5.1316E-01	4.8684E 00	2.0541E-01	7.6128E-01	6.9549E-01
8	2.1436E 00	4.6651E-01	5.3349E 00	1.8744E-01	7.4030E-01	6.6687E-01
9	2.3579E 00	4.2410E-01	5.7590E 00	1.7364E-01	7.2022E-01	6.3989E-01
10	2.5937E 00	3.8554E-01	6.1446E 00	1.6275E-01	7.0099E-01	6.1446E-01
11	2.8531E 00	3.5049E-01	6.4951E 00	1.5396E-01	6.8257E-01	5.9046E-01
12	3.1384E 00	3.1863E-01	6.8137E 00	1.4676E-01	6.6491E-01	5.6781E-01
13	3.4523E 00	2.8966E-01	7.1034E 00	1.4078E-01	6.4798E-01	5.4641E-01
14	3.7975E 00	2.6333E-01	7.3667E 00	1.3575E-01	6.3174E-01	5.2619E-01
15	4.1772E 00	2.3939E-01	7.6061E 00	1.3147E-01	6.1616E-01	5.0707E-01
16	4.5950E 00	2.1763E-01	7.8237E 00	1.2782E-01	6.0120E-01	4.8898E-01
18	5.5599E 00	1.7986E-01	8.2014E 00	1.2193E-01	5.7302E-01	4.5563E-01
20	6.7275E 00	1.4864E-01	8.5136E 00	1.1746E-01	5.4697E-01	4.2568E-01
25	1.0835E 01	9.2296E-02	9.0770E 00	1.1017E-01	4.8994E-01	3.6308E-01
30	1.7449E 01	5.7309E-02	9.4269E 00	1.0608E-01	4.4243E-01	3.1423E-01
35	2.8102E 01	3.5584E-02	9.6442E 00	1.0369E-01	4.0247E-01	2.7555E-01
40	4.5259E 01	2.2095E-02	9.7791E 00	1.0226E-01	3.6855E-01	2.4448E-01
45	7.2890E 01	1.3719E-02	9.8628E 00	1.0139E-01	3.3949E-01	2.1917E-01
50	1.1739E 02	8.5186E-03	9.9148E 00	1.0086E-01	3.1439E-01	1.9830E-01

Source: This table is abridged, and modified from Table A1.1 in *Jelen's Cost and Optimization Engineering*, 3d ed. by Kenneth K. Humphreys, 1991 with the permission of the publisher.

213

TABLE A-2
TABLE OF FACTORS FOR 6%
DISCRETE COMPOUND INTEREST = 6%

	Single payment		Uniform annual series		Depreciation series	
	Compound-interest factor	Present-worth factor	Unacost present-worth factor	Capital-recovery factor	Sum-of-digits present-worth factor	Straight-line present worth factor
	$(1+i)^n$	$\dfrac{1}{(1+i)^n}$	$\dfrac{(1+i)^n - 1}{i(1+i)^n}$	$\dfrac{(1+i)^n}{(1+i)^n - 1}$	$\dfrac{n - F_{RP}}{0.5n(n+1)}i$	$\dfrac{1}{niF_{PK}}$
	P to F	F to P	A to P	P to A	SD to P	SL to P
n	F/P	P/F	P/A	A/P	P/SD	R/SL
1	1.0600E 00	9.4340E-01	9.4340E-01	1.0600E 00	9.4340E-01	9.4340E-01
2	1.1236E 00	8.9000E-01	1.8334E 00	5.4544E-01	9.2560E-01	9.1670E-01
3	1.1910E 00	8.3962E-01	2.6730E 00	3.7411E-01	9.0830E-01	8.9100E-01
4	1.2625E 00	7.9209E-01	3.4651E 00	2.8859E-01	8.9149E-01	8.6628E-01
5	1.3382E 00	7.4726E-01	4.2124E 00	2.3740E-01	8.7515E-01	8.4247E-01
6	1.4185E 00	7.0496E-01	4.9173E 00	2.0336E-01	8.5927E-01	8.1955E-01
7	1.5036E 00	6.6506E-01	5.5824E 00	1.7914E-01	8.4382E-01	7.9748E-01
8	1.5938E 00	6.2741E-01	6.2098E 00	1.6104E-01	8.2880E-01	7.7622E-01
9	1.6895E 00	5.9190E-01	6.8017E 00	1.4702E-01	8.1419E-01	7.5574E-01
10	1.7908E 00	5.5839E-01	7.3601E 00	1.3587E-01	7.9997E-01	7.3601E-01
11	1.8983E 00	5.2679E-01	7.8869E 00	1.2679E-01	7.8614E-01	7.1699E-01
12	2.0122E 00	4.9697E-01	8.3838E 00	1.1928E-01	7.7268E-01	6.9865E-01
13	2.1329E 00	4.6884E-01	8.8527E 00	1.1296E-01	7.5958E-01	6.8098E-01
14	2.2609E 00	4.4230E-01	9.2950E 00	1.0758E-01	7.4683E-01	6.6393E-01
15	2.3966E 00	4.1727E-01	9.7122E 00	1.0296E-01	7.3441E-01	6.4748E-01
16	2.5404E 00	3.9365E-01	1.0106E 01	9.8952E-02	7.2232E-01	6.3162E-01
18	2.8543E 00	3.5034E-01	1.0828E 01	9.2357E-02	6.9906E-01	6.0153E-01
20	3.2071E 00	3.1180E-01	1.1470E 01	8.7185E-02	6.7699E-01	5.7350E-01
25	4.2919E 00	2.3300E-01	1.2783E 01	7.8227E-02	6.2649E-01	5.1133E-01
30	5.7435E 00	1.7411E-01	1.3765E 01	7.2649E-02	5.8191E-01	4.5883E-01
35	7.6861E 00	1.3011E-01	1.4498E 01	6.8974E-02	5.4237E-01	4.1424E-01
40	1.0286E 01	9.7222E-02	1.5046E 01	6.6462E-02	5.0719E-01	3.7616E-01
45	1.3765E 01	7.2650E-02	1.5456E 01	6.4700E-02	4.7575E-01	3.4346E-01
50	1.8420E 01	5.4288E-02	1.5762E 01	6.3444E-02	4.4756E-01	3.1524E-01

Source: This table is abridged, and modified from Table A1.1 in *Jelen's Cost and Optimization Engineering*, 3d ed. by Kenneth K. Humphreys, 1991 with the permission of the publisher.

TABLE A-3

TABLE OF FACTORS FOR CONTINUOUS COMPOUND INTEREST: VALUE OF

$F/P = e^{+i \times n}$

i\n →	0.00	0.01	0.02	0.03	0.04	0.05	0.06	0.07	0.08	0.09
0	1.0000E 00	1.0101E 00	1.0202E 00	1.0305E 00	1.0408E 00	1.0513E 00	1.0618E 00	1.0725E 00	1.0833E 00	1.0942E 00
0.1	1.1052E 00	1.1163E 00	1.1275E 00	1.1388E 00	1.1503E 00	1.1618E 00	1.1735E 00	1.1853E 00	1.1972E 00	1.2092E 00
0.2	1.2214E 00	1.2337E 00	1.2461E 00	1.2586E 00	1.2712E 00	1.2840E 00	1.2969E 00	1.3100E 00	1.3231E 00	1.3364E 00
0.3	1.3499E 00	1.3634E 00	1.3771E 00	1.3910E 00	1.4049E 00	1.4191E 00	1.4333E 00	1.4477E 00	1.4623E 00	1.4770E 00
0.4	1.4918E 00	1.5068E 00	1.5220E 00	1.5373E 00	1.5527E 00	1.5683E 00	1.5841E 00	1.6000E 00	1.6161E 00	1.6323E 00
0.5	1.6487E 00	1.6653E 00	1.6820E 00	1.6989E 00	1.7160E 00	1.7333E 00	1.7507E 00	1.7683E 00	1.7860E 00	1.8040E 00
0.6	1.8221E 00	1.8404E 00	1.8589E 00	1.8776E 00	1.8965E 00	1.9155E 00	1.9348E 00	1.9542E 00	1.9739E 00	1.9937E 00
0.7	2.0138E 00	2.0340E 00	2.0544E 00	2.0751E 00	2.0959E 00	2.1170E 00	2.1383E 00	2.1598E 00	2.1815E 00	2.2034E 00
0.8	2.2255E 00	2.2479E 00	2.2705E 00	2.2933E 00	2.3164E 00	2.3396E 00	2.3632E 00	2.3869E 00	2.4109E 00	2.4351E 00
0.9	2.4596E 00	2.4843E 00	2.5093E 00	2.5345E 00	2.5600E 00	2.5857E 00	2.6117E 00	2.6379E 00	2.6645E 00	2.6912E 00
1.0	2.7183E 00	2.7456E 00	2.7732E 00	2.8011E 00	2.8292E 00	2.8577E 00	2.8864E 00	2.9154E 00	2.9447E 00	2.9743E 00
1.1	3.0042E 00	3.0344E 00	3.0649E 00	3.0957E 00	3.1268E 00	3.1582E 00	3.1899E 00	3.2220E 00	3.2544E 00	3.2871E 00
1.2	3.3201E 00	3.3535E 00	3.3872E 00	3.4212E 00	3.4556E 00	3.4903E 00	3.5254E 00	3.5609E 00	3.5966E 00	3.6328E 00
1.3	3.6693E 00	3.7062E 00	3.7434E 00	3.7810E 00	3.8190E 00	3.8574E 00	3.8962E 00	3.9354E 00	3.9749E 00	4.0149E 00
1.4	4.0552E 00	4.0960E 00	4.1371E 00	4.1787E 00	4.2207E 00	4.2631E 00	4.3060E 00	4.3492E 00	4.3929E 00	4.4371E 00
1.5	4.4817E 00	4.5267E 00	4.5722E 00	4.6182E 00	4.6646E 00	4.7115E 00	4.7588E 00	4.8066E 00	4.8550E 00	4.9037E 00
1.6	4.9530E 00	5.0028E 00	5.0531E 00	5.1039E 00	5.1552E 00	5.2070E 00	5.2592E 00	5.3122E 00	5.3656E 00	5.4195E 00
1.7	5.4739E 00	5.5290E 00	5.5845E 00	5.6407E 00	5.6973E 00	5.7546E 00	5.8124E 00	5.8709E 00	5.9299E 00	5.9895E 00
1.8	6.0496E 00	6.1104E 00	6.1719E 00	6.2339E 00	6.2965E 00	6.3598E 00	6.4237E 00	6.4883E 00	6.5535E 00	6.6194E 00
1.9	6.6859E 00	6.7531E 00	6.8210E 00	6.8895E 00	6.9588E 00	7.0287E 00	7.0993E 00	7.1707E 00	7.2427E 00	7.3155E 00
2.0	7.3891E 00	7.4633E 00	7.5383E 00	7.6141E 00	7.6906E 00	7.7679E 00	7.8460E 00	7.9248E 00	8.0045E 00	8.0849E 00

Source: This table is abridged and modified from Table A2.1 in *Jelen's Cost and Optimization Engineering,* 3d ed. by Kenneth K. Humphreys, 1991 with the permission of the publisher.

TABLE A-3

TABLE OF FACTORS FOR CONTINUOUS COMPOUND INTEREST: VALUE OF
$P/F = e^{-i \times n}$

in → ↓	0.00	0.01	0.02	0.03	0.04	0.05	0.06	0.07	0.08	0.09
0	1.0000E 00	9.9005E-01	9.8020E-01	9.7045E-01	9.6079E-01	9.5123E-01	9.4176E-01	9.3239E-01	9.2312E-01	9.1393E-01
0.1	9.0484E-01	8.9583E-01	8.8692E-01	8.7810E-01	8.6936E-01	8.6071E-01	8.5214E-01	8.4366E-01	8.3527E-01	8.2696E-01
0.2	8.1873E-01	8.1058E-01	8.0252E-01	7.9453E-01	7.8663E-01	7.7880E-01	7.7105E-01	7.6338E-01	7.5578E-01	7.4826E-01
0.3	7.4082E-01	7.3345E-01	7.2615E-01	7.1892E-01	7.1177E-01	7.0469E-01	6.9768E-01	6.9073E-01	6.8386E-01	6.7706E-01
0.4	6.7032E-01	6.6365E-01	6.5705E-01	6.5051E-01	6.4404E-01	6.3763E-01	6.3128E-01	6.2500E-01	6.1878E-01	6.1263E-01
0.5	6.0653E-01	6.0050E-01	5.9452E-01	5.8860E-01	5.8275E-01	5.7695E-01	5.7121E-01	5.6553E-01	5.5990E-01	5.5433E-01
0.6	5.4881E-01	5.4335E-01	5.3794E-01	5.3259E-01	5.2729E-01	5.2205E-01	5.1685E-01	5.1171E-01	5.0662E-01	5.0158E-01
0.7	4.9659E-01	4.9164E-01	4.8675E-01	4.8191E-01	4.7711E-01	4.7237E-01	4.6767E-01	4.6301E-01	4.5841E-01	4.5384E-01
0.8	4.4933E-01	4.4486E-01	4.4043E-01	4.3605E-01	4.3171E-01	4.2741E-01	4.2316E-01	4.1895E-01	4.1478E-01	4.1066E-01
0.9	4.0657E-01	4.0252E-01	3.9852E-01	3.9455E-01	3.9063E-01	3.8674E-01	3.8289E-01	3.7908E-01	3.7531E-01	3.7158E-01
1.0	3.6788E-01	3.6422E-01	3.6059E-01	3.5701E-01	3.5345E-01	3.4994E-01	3.4646E-01	3.4301E-01	3.3960E-01	3.3622E-01
1.1	3.3287E-01	3.2956E-01	3.2628E-01	3.2303E-01	3.1982E-01	3.1664E-01	3.1349E-01	3.1037E-01	3.0728E-01	3.0422E-01
1.2	3.0119E-01	2.9820E-01	2.9523E-01	2.9229E-01	2.8938E-01	2.8650E-01	2.8635E-01	2.8083E-01	2.7804E-01	2.7527E-01
1.3	2.7253E-01	2.6982E-01	2.6714E-01	2.6448E-01	2.6185E-01	2.5924E-01	2.5666E-01	2.5411E-01	2.5158E-01	2.4908E-01
1.4	2.4660E-01	2.4414E-01	2.4171E-01	2.3931E-01	2.3693E-01	2.3457E-01	2.3224E-01	2.2993E-01	2.2764E-01	2.2537E-01
1.5	2.2313E-01	2.2091E-01	2.1871E-01	2.1654E-01	2.1438E-01	2.1225E-01	2.1014E-01	2.0805E-01	2.0598E-01	2.0393E-01
1.6	2.0190E-01	1.9989E-01	1.9790E-01	1.9593E-01	1.9398E-01	1.9205E-01	1.9014E-01	1.8825E-01	1.8637E-01	1.8452E-01
1.7	1.8268E-01	1.8087E-01	1.7907E-01	1.7728E-01	1.7552E-01	1.7377E-01	1.7204E-01	1.7033E-01	1.6864E-01	1.6696E-01
1.8	1.6530E-01	1.6365E-01	1.6203E-01	1.6041E-01	1.5882E-01	1.5724E-01	1.5567E-01	1.5412E-01	1.5259E-01	1.5107E-01
1.9	1.4957E-01	1.4808E-01	1.4661E-01	1.4515E-01	1.4370E-01	1.4227E-01	1.4086E-01	1.3946E-01	1.3807E-01	1.3670E-01
2.0	1.3534E-01	1.3399E-01	1.3266E-01	1.3134E-01	1.3003E-01	1.2873E-01	1.2745E-01	1.2619E-01	1.2493E-01	1.2369E-01

Source: This table is abridged and modified from Table A2.2 in *Jelen's Cost and Optimization Engineering,* 3d ed. by Kenneth K. Humphreys, 1991 with the permission of the publisher.

TABLE A-3
TABLE OF FACTORS FOR CONTINUOUS COMPOUND INTEREST: VALUE OF $P/n\mathring{A} = \dfrac{1-e^{-ixn}}{i \times n}$

in → / in ↓	0.00	0.01	0.02	0.03	0.04	0.05	0.06	0.07	0.08	0.09
0	1.0000E 00	9.9502E-01	9.9007E-01	9.8515E-01	9.8026E-01	9.7541E-01	9.7059E-01	9.6580E-01	9.6105E-01	9.5632E-01
0.1	9.5163E-01	9.4696E-01	9.4233E-01	9.3773E-01	9.3316E-01	9.2861E-01	9.2410E-01	9.1962E-01	9.1517E-01	9.1074E-01
0.2	9.0635E-01	9.0198E-01	8.9764E-01	8.9333E-01	8.8905E-01	8.8480E-01	8.8057E-01	8.7637E-01	8.7220E-01	8.6806E-01
0.3	8.6394E-01	8.5985E-01	8.5578E-01	8.5175E-01	8.4773E-01	8.4375E-01	8.3979E-01	8.3585E-01	8.3194E-01	8.2806E-01
0.4	8.2420E-01	8.2037E-01	8.1656E-01	8.1277E-01	8.0901E-01	8.0527E-01	8.0156E-01	7.9787E-01	7.9420E-01	7.9056E-01
0.5	7.8694E-01	7.8334E-01	7.7977E-01	7.7622E-01	7.7269E-01	7.6918E-01	7.6570E-01	7.6224E-01	7.5880E-01	7.5538E-01
0.6	7.5198E-01	7.4861E-01	7.4525E-01	7.4192E-01	7.3861E-01	7.3531E-01	7.3204E-01	7.2879E-01	7.2556E-01	7.2235E-01
0.7	7.1916E-01	7.1599E-01	7.1284E-01	7.0971E-01	7.0660E-01	7.0351E-01	7.0044E-01	6.9739E-01	6.9435E-01	6.9134E-01
0.8	6.8834E-01	6.8536E-01	6.8240E-01	6.7946E-01	6.7654E-01	6.7363E-01	6.7074E-01	6.6787E-01	6.6502E-01	6.6218E-01
0.9	6.5937E-01	6.5657E-01	6.5378E-01	6.5102E-01	6.4827E-01	6.4554E-01	6.4282E-01	6.4012E-01	6.3744E-01	6.3477E-01
1.0	6.3212E-01	6.2949E-01	6.2687E-01	6.2427E-01	6.2168E-01	6.1911E-01	6.1655E-01	6.1401E-01	6.1149E-01	6.0898E-01
1.1	6.0648E-01	6.0400E-01	6.0154E-01	5.9909E-01	5.9665E-01	5.9423E-01	5.9182E-01	5.8943E-01	5.8705E-01	5.8469E-01
1.2	5.8234E-01	5.8000E-01	5.7768E-01	5.7537E-01	5.7308E-01	5.7080E-01	5.6853E-01	5.6627E-01	5.6403E-01	5.6181E-01
1.3	5.5959E-01	5.5739E-01	5.5520E-01	5.5302E-01	5.5086E-01	5.4871E-01	5.4657E-01	5.4445E-01	5.4233E-01	5.4023E-01
1.4	5.3815E-01	5.3607E-01	5.3400E-01	5.3195E-01	5.2991E-01	5.2788E-01	5.2587E-01	5.2386E-01	5.2187E-01	5.1988E-01
1.5	5.1791E-01	5.1595E-01	5.1401E-01	5.1207E-01	5.1014E-01	5.0823E-01	5.0632E-01	5.0443E-01	5.0255E-01	5.0068E-01
1.6	4.9881E-01	4.9696E-01	4.9512E-01	4.9329E-01	4.9148E-01	4.8967E-01	4.8787E-01	4.8608E-01	4.8430E-01	4.8253E-01
1.7	4.8077E-01	4.7903E-01	4.7729E-01	4.7556E-01	4.7384E-01	4.7213E-01	4.7043E-01	4.6874E-01	4.6706E-01	4.6539E-01
1.8	4.6372E-01	4.6207E-01	4.6043E-01	4.5879E-01	4.5716E-01	4.5555E-01	4.5394E-01	4.5234E-01	4.5075E-01	4.4917E-01
1.9	4.4760E-01	4.4603E-01	4.4448E-01	4.4293E-01	4.4139E-01	4.3986E-01	4.3834E-01	4.3682E-01	4.3532E-01	4.3382E-01
2.0	4.3233E-01	4.3085E-01	4.2938E-01	4.2791E-01	4.2646E-01	4.2501E-01	4.2357E-01	4.2213E-01	4.2071E-01	4.1929E-01

Source: This table is abridged and modified from Table A2.3 in *Jelen's Cost and Optimization Engineering*, 3d ed. by Kenneth K. Humphreys, 1991 with the permission of the publisher.

APPENDIX B

B-1 DEVELOPMENT OF $F(i,n)/A$, THE UNIFORM SERIES FUTURE WORTH FACTOR

Although a little lengthy, this development follows easily by using the future worth factor $F(i,k)/P(k)$. In Section 6-3 on annuities, with $P(k) = A$ for each kth year over a period of n years, we showed that the factor $[F(i,n)/A]$ on using k for a running index equals

$$\left[F(i,n)/A\right] = \sum_{k=n-1}^{0} F(i,k)/P(k) = \sum_{k=0}^{n-1} F(i,k)/P(k)$$

Recall that reversing the summation is possible because linear operations can be commuted. Now on substituting for the F/P's $= (1+i)^k$, with i in decimal form,

$$\left[F(i,n)/A\right] = \sum_{k=0}^{n-1} (1+i)^k$$

We use this equation below. First, let us consider an old mathematical manipulation. Divide $(1-x)$ into 1, by long division:

$$
\begin{array}{r}
1 + x + x^2 + \cdots \\
1-x\overline{)1} \\
\underline{1-x} \\
0 + x + \\
\underline{x - x^2} \\
x^2 \\
\underline{x^2 - x^3} \\
x^3
\end{array}
$$

We engineers recognize the pattern, that is,

$$\sum_{k=0}^{\infty} x^k = \frac{1}{1-x} \tag{6-6}$$

Let us now multiply both sides of Equation 6-6 by x^n, for $n \geq 0$. On expanding, we have

$$\frac{x^n}{1-x} = x^n \sum_{k=0}^{\infty} x^k = x^n + x^{n+1} + x^{n+2} + \cdots \tag{6-7}$$

Expanding Equation 6-6, we have

$$\frac{1}{1-x} = \sum_{k=0}^{\infty} x^k = 1 + x + x^2 + \cdots + x^{n-1} + x^n + x^{n+1} \cdots \tag{6-8a}$$

$$= \sum_{k=0}^{n-1} x^k + x^n + x^{n+1} + \cdots \tag{6-8b}$$

Note, the summation in Equation 6-8b includes the first "n" terms of Equation 6-8a.
Subtracting Equation 6-7 from 6-8b, gives

$$\frac{1 - x^n}{1 - x} = \sum_{k=0}^{n-1} x^k$$

On letting $x = 1 + i$, and using the expression developed above for $[F(i,n)/A]$,

$$\frac{1 - (1 + i)^n}{1 - (1 + i)} = \left[(1 + i)^n - 1\right]/i = \sum_{k=0}^{n-1}(1 + i)^k = \left[F(i, n) / A\right]$$

This is the expression used in Section 6-3 on annuities.

B-2 EXTENSION OF THE UNIFORM SERIES PRESENT WORTH FACTOR: PRESENT VALUE OF ANNUITIES

In Section 6-3, in the subsection on annuities, an expression for the present value of a series of annuities was developed. It is

$$P(i, n) / A = \frac{(1 + i)^n - 1}{i \times (1 + i)^n}$$

This equation could be used to evaluate i as the percent return that A gives on investment P.

Consider that A is the profit from an investment of P. Furthermore, consider that the stream of profit is over n years and that n is very long. Thus,

$$P(i, n) / A = \lim_{n \to \infty} \frac{(1 + i)^\infty}{i \times (1 + i)^\infty} = \frac{1}{i}$$

$(1 + i)^\infty \gg 1$.

Or, the rate of return $i \approx A/P$, the reciprocal of the payback time without interest.

The payback time can also be considered as the reciprocal of a stocks' P/E ratio presented in Section 6-1 and reviewed in more detail in Chapter 7.

Comstock

CHAPTER 7

FINANCIAL PLANNING
FOR ENGINEERS

Investing is the only way an individual can independently achieve financial security. Personal investing is a process of deferring present pleasures for future advantages. It requires strong self-discipline, for the advantages might never materialize. By applying the engineering approach, however, we can maximize the probability of achieving the advantages. Furthermore, responsible persons—to minimize impositions on family, friends, and society—have no choice. They must make plans based on the most probable outcomes and prepare for the unexpected. The word *personal* in personal investing is meant two ways. The first way refers to investing in our own personal growth, such as in education. The second way involves using the information and economic fundamentals presented in Chapter 6 to develop an investment strategy for our own financial security. The scope of this chapter is primarily limited to unchanging fundamental, defensive investing. Naturally, information from professionals and the financial literature will be needed in dealing with the changing financial environment. The prudent investor has two main objectives: preservation of capital (savings) and moderate real return in excess of inflation.

In the next two sections we review your present estate and how insurance can be used to establish an immediate, low-cost supplement. Insurance is suited only for the short term; for long-term goals and aspirations, conventional investments are best. Indirect investing through mutual funds is adequate and excellent for your long-term investment needs. Active investing provides our Child ego state the pleasures of playing the "Money Game." A passive strategy is effective, but an active strategy is more fun and has greater risk. Both strategies are presented in this chapter. Many concepts in these first sections are presented later in detail.

7-1 YOUR ESTATE NOW: SAVINGS, EDUCATION, AND SOCIAL SECURITY

Although probably inadequate, you already have an estate. It is the total of everything you own, which includes your capabilities. As an engineering student or young graduate engineer, you presently have valuable capabilities and will improve them in the normal course of your activities, in school or at work. You might have been fortunate in having already accumulated some monetary wealth; however, that is less important than your capabilities.

The first purpose for an estate is to reduce the effects of misfortunes. These range from fender bender–type auto accidents to your being severely disabled, or even your death. A review of your present estate and an analysis of possible misfortunes, to you or your family, are the first steps in determining the estate you will need. Social security is now (or soon will be) part of your estate. It is a base on which to build.

We and our employers pay for social security. Since it is involuntary, analyzing its worth as an investment would be futile; however, reviewing the possible benefits is definitely part of estate planning. For your present or soon anticipated state in life, search your personal characteristics and make a judgment regarding financial needs in the event of various misfortunes. For the lesser ones, a strong bank account will be adequate.

Social security will help by providing severe disability or survivor's benefits in the case of your death. The amount of these benefits depends on the number of years you and your employer have paid social security taxes; generally 5 of the last 10 yr are required. A person who has credit for 10 yr of work is sure of being fully insured for disability and survivors benefits. The disability benefits payments range from 50 to 80 percent of your salary, below $20,000. At higher levels, the benefits are greater but are a smaller percentage of your predisabled salary. Survivors benefits are 50 to 60 percent of predeath salaries for those fully insured. Social security also pays you in retirement.

A very important point is that social security does not provide complete disability or survivors' coverage. For these as well as for retirement, it is only a foundation. It must be supplemented by savings, insurance, and long-term investments. The pamphlet *Social Security, How It Works for You* gives more details and is free from any social security office [1].

7-2 ESTATE BUILDING

Financial security greatly influences a person's self-worth. It, in turn, is a very significant aspect of one's happiness, health, and competence. This security is achieved through building an estate. An adequate estate gives three wonderful benefits. The financial security indemnifies* us against the unexpected. By building an estate, we recognize that aspirations can be achieved and, finally, we are less inhibited (for example, in seeking new responsibilities). Through your engineering competence you will soon (or already) have professional employment, and through a prudent investment plan you can create the estate you need or desire. I included the word *desire* with hesitation, for I wish to stress that great wealth requires almost an obsession with work activities, with no time for anything else, and results in one's being deprived of the truly greater aspects of life. This could be the situation if a student strives to make all A's; actually,

---

Indemnify is Latin for "make (a situation) not damaging."

a transcript with more A's than B's and only a few C's is well appreciated by prospective employers and permits time for the "broader" education and pleasures in school and life in general.

Meeting Unexpected Needs

These needs are the first guides to the size of estate we should create. As stressed above, their costs primarily arise from the unexpected expenses that we might incur. If all continues normally, our present sources (parents, part-time jobs for students, or present salaries for the graduate engineers) would probably meet our needs. The first step to meet "rainy day" needs is a strong bank account. Insurance for health and accidents can be used to meet these greater misfortunes. Minimizing insurance costs is reviewed later.

A Strong Bank Account

The word *strong* is used to refer to the account's magnitude and its location in a bank which is financially strong, as indicated by two ratios presented below. One would be "insurance poor" if she or he tried to buy insurance against all possibilities. As reviewed below, the cheapest type is insuring yourself. This practice is wise for losses that one can afford to incur. The occasional loss can be very unpleasant, but the rewards from self-insurance are great and continuous. Naturally, the "unpleasantness" depends on the size of the loss as a fraction of your bank account.

As a first goal, build your bank account to an amount equal to at least 2 months' normal expenses. This process will be only slightly inconvenient if you deposit 10 to 15 percent of your take-home pay each pay period. Do so before any other outflows. After achieving your first "savings goal," relax for a short interval. After the break, extend the savings process to build to an amount equal to 3 months' normal expenses. Having significant "money in the bank" is truly a dominant factor in one's mental outlook. It gives a degree of independence and an opportunity of helping those less fortunate. A former student told me that when she first started working as an engineer, she felt restrained, but after saving $1000, she felt free and enjoyed interchanging with others at work. All of these advantages would be sadly lost if your bank "went under." Testing your bank by the indices presented below minimizes your dependency on the federal programs and the disadvantages noted.

A bank failure could at most wipe out your accumulations and at least cause you considerable inconvenience and expense. The federal programs, Federal Deposit Insurance Corporation (FDIC) for commercial banks and Federal Savings and Loan Insurance Corporation (FSLIC) for savings and loans, are grossly inadequate. The reserves can cover less than 5 cents of each insured dollar [2]. Of course, through the insurance "backed by the full faith of the U.S. government," in time, the government would cover your loss up to $100,000. In the short term, however, your funds would be unavailable for normal expenses such as food, rent, and mortgage or automobile payments. Furthermore, to get the moneys for payments, the government would have "to print" it. This is discussed in Appendix A-4 of this chapter.

Indices of Banks' Financial State

Before 1990, the federal insurance programs seem to have encouraged banks and the savings and loan institutions (S&L's) to take greater risks than were wise. The ratios for "liquidity" and "capital strength" indicate the states of these financial institutions [2]. Get a bank's financial statement to evaluate the following:

$$\text{Liquidity} = \frac{\text{total quick assets (current assets} - \text{inventories)}}{\text{total deposits}}$$

$$\geq 40\% \text{ for a financially strong bank}$$

$$\text{Capital strength} = \frac{\text{shareholder's equity}}{\text{total assets}}$$

$$\geq 5\% \text{ for financial strength}$$

As reviewed later in this chapter, a bank with capital strength greater than 8 percent is not using its assets fully.

Goals and Aspirations

After the basic needs, Maslow's hierarchy puts goals and aspirations next in importance. They are our next purpose in estate building. Well before marriage and the responsibilities of a family, it is wise to start building a financial estate. Even small, regular savings grow rapidly with the magic of compound interest. The discipline required to decline a "steak vs. hamburger" lifestyle becomes an asset. After graduation, along with a "good salary," this discipline enables you to buy a well-cared-for used car instead of obligating a major portion of your salary to pay for a flashy new one. Again, the substantial savings compound rapidly for later aspirations: a house, children's education, travel, vacations, and retirement.

The rule of 72* shows the advantage of starting early in saving for a goal. The rule is $k = 72/i$; i is the interest for the period considered, usually years, and k is the number of periods required to double the initial value. For example, an investment paying 6 percent per year will double in $\frac{72}{6}$ or 12 yr. For $i = 8$ percent, the investment would

---

*In Chapter 6, we developed an expression for the compound interest factor; by setting it equal to 2, we have

$$(1 + i)^k = 2$$

where k is the number of periods, at interest rate i, required to double the initial value. The value could be for money or for other things, such as population. On taking the natural logarithm of both sides, the number of periods, usually years, required for doubling is given by

$$k = \ln(2)/\ln(1 + i) \approx 72/i \text{ for } i \text{ in percent}$$

The approximation $\ln(1 + i) \approx i$ for i less than 0.2 is used.

double in 9 yr and double again in 9 more, thus quadrupling in 18 yr (just in time for college expenses).

Insurance

This is a method for coping with the unexpected. It is an arrangement by which one party (the insurer) agrees in the event of a misfortune to a second party (the insured) to lessen the damages to that party. The relief of damages is generally by the payment of money. Pure insurance is not an investment, but at a reasonable cost, it can be used to instantly build the estate required to meet the costs of casualties that might come to an individual or family. These include sickness, automobile accidents, fire, and death; insurance for each is usually purchased by separate policies. An exception is that insurance for automobiles, both collision and liability, is often combined with fire insurance for the home, as in a "home owner's" plan. It gives considerable savings over separate policies for each. Similarly, because of combined premiums and bookkeeping, employers can buy a group plan to cover health and hospital costs at great savings over individual plans. Also, the IEEE, ASME, and other engineering organizations make group plans available to their members. If, however, you do not have such a group plan available, perhaps even when in school, you have no prudent choice except to purchase protection from one of the many reputable insurance companies which offer health and hospital plans. Use A. M. Best, Standard and Poor's, or other guides to check the rating of the insurance company. These references are available in most libraries. Use a company, with a rating "A" or better, which offers the coverage (health, fire, or other) needed at the lowest cost.

It is always least expensive to insure yourself.* Insurance companies require about half your insurance premium to pay personnel, other expenses, and for profit. You can realize a significant savings in cost, first, by not purchasing unnecessary plans (discussed later in this chapter), and, second, by insuring yourself, that is, accepting the highest deductible amount your present estate can support. While not wanting to accept the risk of the possible total loss of your house by fire, you can have a policy with $500 or $1000 deductible. That is, for the latter, you might feel that while it would be unpleasant, your present estate could support a $1000 loss. On a $100,000 home owner's policy, $1000 deduction gives a savings in premium of about $100 per year. By "setting aside" $1000 in a savings account to cover such a loss, the $100 per year savings adds 10 percent (tax-free) to the taxable interest already being paid by the account. Even if the loss occurred, by time averaging over the long term, you would realize a profit just as an insurance company does, only more so, for you would not have the cost of personnel and other overhead. Evidently, the expense of handling many small claims is a major part of insurance companies' costs.

Many firms offer savings on the premiums for a home owner's policy: for example, (1) by having dead bolt locks on all outside doors, the premium is typically reduced by 3 percent; and (2) having smoke alarms and one or more fire extinguishers reduces the premium by 2 percent.

---

*Fundamentals of risks and insurance are reviewed in Chapter 2.

Deductibles on health insurance and automobile collision insurance also give significant savings. The savings act as additional interest on the amount in your savings account equal to that of the deductibles. Since the deductibles will be money invested on a long-term basis, as reviewed later in this chapter, higher rates are available. A safe driving record can reduce the premium on collision insurance, typically by 12 percent.

In an accident, whether in an automobile or at your home, you could be liable for damages. Regretfully, at present no insurance company offers deductibles on liability insurance, and no one with any estate can afford not to have this insurance.

Finally, but probably most significant, we must consider disability and life insurance. For disability, inadequate but significant funds are paid by social security; also there are other sources such as sick leave from your employer, worker's compensation, state funds, veteran's insurance, and group insurance. To supplement these, if you do not have a group plan at work, you can get private disability insurance. It is usually inexpensive because of the low probability of the occurrence.

In determining the probable situation of your family, in case of your disability or death, a careful analysis is important. It should include a close review of the family's regular living costs and costs of future plans such as college education for your children. Use the techniques of Chapter 6 to consider the time values of the future costs and incomes from social security etc., as noted above. Also include interest and dividends produced by your savings and stocks. By this analysis, you can determine the additional amounts that would be required monthly from disability sources or from a life insurance plan in case of your death. As you build your estate, the need of funds from insurance and other outside sources diminishes. Take advantage of this to reduce costs.

The Life Insurance Ploy

As in buying any service or merchandise, understanding of (and knowledge about) the "product" reduces the probability of one's being deceived. Here we review the understanding and knowledge of life insurance. Recall that insurance is an arrangement in which a party pays a fee, called the *premium,* to a second party for that party to assume a particular risk. The second party is generally an insurance firm which can profitably accept the risk by pooling the fees from many insureds. Using the statistical averages, fees are set such that their sum over an extended period will always exceed the total losses, plus other costs. Since the same actuarial data on deaths are available to all life insurance companies, they devise ploys to achieve individual advantages.

The written arrangement for the insurance, with all of the exemptions and other conditions, is called the *policy.* An amount of money, usually in multiples of $1000, is paid when the "insured against event" occurs. As is easily reasoned, the greater the risk is, the greater the premium per $1000 (of insurance). The small charge for insurance against accidental death indicates the very small probability of its occurrence for any particular person, especially for persons over 30 yr of age. Young people are far less likely to die of natural causes than older people; in turn, premiums for each $1000 to be paid on death are accordingly much smaller. As the saying goes, "Insurance is cheap when you are young." It cannot be a bargain, however, if there is no need. The objective of all insurance is an attempt to replace the value of the loss incurred on occurrence of the "insured against" event. Of course, the value of a life cannot be replaced, but its earning capability as needed to support and educate dependents can be replaced by life insurance. It is thus offered that one does not need life insurance until there are dependents to support or educate. An exception would be a case in which a student has borrowed money for educational expenses from family or a financial institution and wishes to insure against being unable to repay.

When need for insurance does occur, the following facts will help one meet it at the smallest cost. All life insurance is basically of two types, term and whole life. As might be inferred, term insurance is for a given period, usually 5 yr. The premium might be paid monthly, quarterly, semiannually, or annually; and for each the payment is the same over the entire period or term, e.g., 5 yr. The premium is charged on a per $1000 basis; e.g., for $5000 and $10,000, the premium would be, respectively, 5 and 10 times the per $1000 charge. If in that period the insured dies by any means (some companies exclude suicide, many exclude death by war), this total amount is paid to the beneficiary. On the other hand, if the insured doesn't die by the end of the period, the policy expires and neither party owes the other anything. Most firms do offer renewable terms to age 70 or greater. Since the risk increases with age, the premium increases with each succeeding term; in the last 40 yr, however, the increases have been small because of dividends and the reducing mortality rate.

The other type of life insurance is whole life; it will pay anytime the insured dies rather than within a specified period or term. The premium can be paid over one's entire life or over a specific period, e.g., 20 yr. Of course, the premiums for whole life are higher than those for term insurance. Even when paid over one's entire life, the portion of the premium used for risk is higher to meet the higher risk in later years. This type, also called *ordinary life,* has two parts: risk protection and investment. The risk part is equivalent to term insurance; the investment part is a forced savings account and forms the cash value of the ordinary life policy. With conscientious attention to many insurance salespeople in the last 25 yr, I have never found a company which paid interest on the cash value as high as the savings and loan associations. This could change in the 1990s. A policyholder, however, can borrow on the cash value at a rate only a little higher than that being paid by the insurance company.

Although it pays a lower rate of interest, ordinary life insurance does provide a means of forced saving for those who prefer not to exercise the necessary self-discipline. For them, it does establish an estate as will be needed for emergencies, education of dependents, and retirement. A serious disadvantage is that on death, ordinary life pays only the face value, regardless of the policy's cash value. So the real insurance is only the difference between the face value and the cash value. Of course this difference, and thus the real insurance, decreases every time a premium is paid.

In summary, although firms will have all sorts of catchy phrases by which they are called, and various involved ways by which premiums are paid, there are only two types of life insurance policies. The first is term; in it, one pays only for the risk. Its premium is commensurate with the risk during its term, and thus increases with age. It is worth noting, however, that when a person has accumulated a "nest egg" for the proverbial rainy day, paid for one's home, and educated the dependents, there is no need for life insurance; and it would not be prudent to purchase it even if the premiums were no higher than when younger. A term policy could, however, be used to pay estate taxes without having to quickly sell assets.

Finally, in all types of insurance, including life insurance, there is a charge for the risk. By writing an "ordinary life" policy, insurance companies add an investment plan. Furthermore, most companies are surreptitious about the fact that the real insurance, against the risk of death, is continuously being reduced as the cash value increases through the investment plan. The choice between term and any of the types of ordinary insurance must be based on the prudence of whether or not to purchase the added investment plan: (1) is it needed, and (2) does it offer the most advantageous investment plan for you? How much less return on investment are you willing to accept to force you to save?

7-3 TYPES OF INVESTMENTS

Investments can be made indirectly, as through the investment plans of insurance companies and mutual funds, or directly by buying equities through the brokers of financial firms. Indirect and direct investing are reviewed below.

While high risk does not necessarily mean high return, generally, high return (as a percentage of an investment of time and money) involves greater risk than lower return. It is possible that an adequately high return could justify its associated high risk, but only if the investor can afford the probable loss without significant damage. The investments we consider here are money market accounts, bonds, and stock, both preferred and common. Their risk and return increase in the order of their listing. Even with stocks, the strategies we consider are for the conservative, defensive investor. Later, however, we present higher-risk activities which could interest your Child ego state after you have established an estate adequate for your needs and responsibilities.

Size of Estate Controls Options and Expenses

Everyone who independently builds an estate starts small; this is true of an estate of knowledge as well as a financial estate. Your engineering knowledge already has provided (or soon will provide) many career options. From the basics presented here you can also develop financial sophistication and build an adequate money estate. Although often costly, the most effective teacher is experience. That is surely correct in financial matters; so keep losses small by starting with small amounts in those investments more risky than money markets and high grade bonds. The expense of investment options is significantly greater for small dollar magnitudes. In your growth stage, consider the greater expenses as tuition. Building on your "strong bank account," you can use part of it to provide the minimum deposit required for a higher-interest money market account. Check newspaper advertisements, and inquire among several banks for the best interest rate on insured deposits. Most important, for your larger accounts, check the indices of the bank's financial state, presented earlier.

Money Market*

This market is for short-term securities such as commercial paper and Treasury bills (often called *T-bills*). Maturities of these instruments are less than a year. Because of the large minimums ($10,000 to $100,000) required for purchase, individuals with lesser amounts can participate indirectly by owning shares in a money market fund. These funds often require only as little as $500. Money market checking accounts are available at many commercial banks and at savings and loan corporations. Minimums are usually $1000, with a limit of three check writings per statement period. Their interest is usually 2 to 3 percent above that of the regular checking accounts. These accounts are extremely safe: first, because of the quality and short maturities of the securities; and second, because those in banks are covered by the FDIC. Before presenting more risky investments, let us review the characteristics of their risk.

Investment Risks [3]

In the financial literature, risk and uncertainty are not used technically as defined in Chapter 2 on decision analysis. More informally, *risk* is defined as uncertainty in the anticipated return on investments. The total risk of an individual security j is measured by the standard deviation σ_j of its returns. Recall that the standard deviation is a measure of the average difference between a variable's actual value and its average or expected value. Here the difference is between the actual return on an investment and its expected return.

Total risk has two components: unsystematic risk, and systematic or market risk. The unsystematic risk is company- or industry-specific, such as the effect of VCR's on the theater chains. Systematic risks are inherent in the system and affect the entire market. Typical are summer droughts, oil embargoes, and political assassinations. Although systematic risks are totally uncontrollable, they can be reflected upon in a "what if" frame of mind. We can consider their effects and possibly recognize offsetting actions. In addition to market risk, systematic risks include interest rate risk and purchasing power risk.

Later in reviewing bonds, we will see how their valuations are affected by changes in interest rates. Purchasing power risk results from inflation. Consider an investment purchased 12 yr ago for $500 and sold today for $900. It would have a loss of $100 in purchasing power if the inflation rate was 6 percent. Recall the rule of 72: $k = \frac{72}{6}$ percent. It shows that the price of goods under an inflation rate of 6 percent would double in 12 yr. Thus, presently, $1000 would be required to purchase what $500 would purchase 12 yr ago. Accordingly, in an inflationary environment of d percent, any investment paying interest of i percent would approximately produce a net gain of only $(i - d)$ percent. In the late 1970s, this was frequently a negative value. It was about that time that investment companies started offering money market accounts. That

---

*As noted earlier, there is a glossary at the end of this chapter in Appendix A-8. It defines many financial terms. Hopefully, it will save your having to look back through the text for definitions. Terms not used here but used frequently in the financial literature are also included.

market requires interest rates in excess of inflation. We discuss inflation later in this chapter. The effect of systematic risk on the required interest rate is shown below by the *capital asset pricing model* (CAPM).

Company-specific (unsystematic) risks can be lowered by diversifying; thus, investors should not expect additional returns for this part of the total risk. Recall from your introductory probability and statistics courses [4] that the variance of a sample (such as stocks or bonds) can be reduced by increasing the number of samples N.

$$\sigma_s^2 = \sigma_{mkt}^2 / N$$

As noted, the risk of an investment's outcome can be measured by its standard deviation, the square root of its variance. Its relation to movements in the market is indicated by its beta. Beta can be calculated using the standard deviation of an investment's prices, the covariance between the asset and the market, and the variance of the market.* Beta indicates a particular investment's volatility (risk) relative to the average investment in that market. Thus, an average investment (in that market) has a beta of 1. A beta less than 1 indicates a defensive security, that is, less risky or volatile than the market. A beta greater than 1 indicates aggressive investments, for they tend to fluctuate more than the market.

The principle of risk compensation is reflected by the CAPM. It permits an investor to evaluate an asset's expected return relative to the asset's systematic risk as represented by its beta. This model shows that an investor should expect a premium from an asset in proportion to the asset's systematic risk. The corresponding equation is

$$E(R_j) = R_f + \beta_j \, [E(R_M) - R_f]$$

where $E(R_j)$ = the expected return of asset j

R_f = the nominal risk-free return; it is the sum of the real risk-free return plus a premium for inflation. The yield of Treasury bills is often used for R_f.

$E(R_M)$ = the expected return of the market

β_j = beta for the asset j

$[E(R_M) - R_f]$ is the risk premium for the market; it is multiplied by β_j to give RP_j, the risk premium for asset j. Beta values for many stocks are given in the *Value Line Investment Survey*, available in most larger city and university libraries.

Example 7-1

Consider stocks (or firm) A; it has a beta of 0.8. For T-bills the rate is 8 percent; the rate of the S&P 500 is 14 percent. What total return** should be expected for this stock?

---

*The market is generally defined as the Standard & Poor's index of 500 stocks.
**Total return is presented later.

Solution: By the CAPM

$$E(R) = 8 + 0.8(14\% - 8\%) = 12.8\%$$

Comparing this value with the actual rate can be used to indicate if a stock is over- or undervalued by the market.

The beta for a portfolio of stocks is the weighted average value for the group. That average is the sum of the group's individual betas, each multiplied by its respective dollar value; and the sum of these products is divided by the dollar value of the total group.

Example 7-2

Consider stocks A and B, with betas of 0.8 and 1.2, respectively. Determine the beta for a portfolio consisting of $3000 in A and $2000 in B.
Solution: The beta of this portfolio β_p is

$$\beta_p = \frac{\$3000 \times 0.8 + \$2,000 \times 1.2}{\$3000 + \$2000} = 0.96$$

This beta shows that the systematic risk of the portfolio is lower than B alone and greater than A alone.

The equation given earlier for the variance shows how the systematic risk of the sample variance (such as a group of stocks) is reduced by increasing the number of stocks in the group purchased in that particular market. Studies show that 10 to 15 different stocks will remove most of the unsystematic (or "diversifiable") risk. This is true only if the samples have a less than perfect correlation. The less the correlation (of the sample), the greater the systematic risk reduction effected by diversification. For instance, competition from a new supermarket chain would affect other supermarket chains, but motel chains would have only small or no correlation to supermarkets.

The asset allocation approach can be used to reduce the systematic risk of a particular market. As presented later, the reduction can be effected by diversifying among markets, such as foreign stock markets or precious metals, or both. We recognize the importance of diversifying, but brokers' commissions would make it unwise to try to spread even several thousand dollars over 5 to 10 different stocks and bonds. The problem is solved by buying shares in investment companies.

Investment Companies/Mutual Funds [3, 5, & 6]

Investment companies could well be a suitable course for the conservative investor. They offer the very important diversification stressed above and provide professional management at a small fee of $\frac{1}{2}$ to 1 percent of the net asset value (if the fund is chosen with care). Shares of the companies are sold to the public, and the proceeds are pooled and invested, usually in stocks or bonds, depending on the type of fund. The net asset value (NAV) is the total worth of all its investments at that time, divided by the total number of shares (of the company). Most companies operate under special provisions of the income tax law designed to relieve stockholders from double taxation on earnings. Accordingly, the funds must distribute all ordinary income less expenses. Furthermore, gains from sale of long-term investments can be paid out and considered long-term "capital gains" by the shareholders, as if the gains were on their own securities. Depending on the latest action of Congress, capital gains might be taxed at a reduced rate from ordinary income. In 1992, the income tax was the same for short- and long-term gains.

Open-End Funds   Two types of investment companies are known as investment funds: the open-end and closed-end funds. We review the open-end next.

Mutual funds are open-end; their shares can be purchased or redeemed on demand at net asset value. This can be a disadvantage; frequently the general public moves to excesses during times of "market crisis," as in October 1987. In such an event "everyone" wishes to redeem his or her shares, and the managers are forced to sell in a declining market. The "psychology of the market" often appears to cause investors to act irrationally.

Fund Expenses   Of course, for all funds, including the no-load ones, the total of all expenses must be paid on a pro rata basis by the shareholders. These costs are many, and a few seem "hidden." First is the sales commission. To promote growth, some *open-end* funds encourage the active selling of additional shares by "energetic" stockbrokers. A typical broker's commission is $8\frac{1}{2}$ percent of the fund's purchase price; that is, if you invest the normal minimum of $1000,* only $915 ($1000 − 0.085 × $1000) of your money purchases the fund's stock. Furthermore, the $85 is 9.23 percent of the net $915, but that could be part of your "tuition." This fee is called the front-end load, and funds which have them are called *load funds.*

Others, known as *no-load funds,* have no such charge; they are bought directly from the fund management. In those purchases, however, you must seek your own information and exercise your judgment as to which fund to purchase. In the no-load funds, management is content with the usual investment-counsel fees. Many studies have shown that there is no discernible difference between the performance of the two types of funds [7]. Thus, you can save the 9 percent "tuition" by an investment of time and effort to choose the best no-load fund. Guidelines and sources for the necessary information are reviewed below.

The broker-salesperson's commission, just discussed, is the most visible cost; however, many funds, even though advertised as no-load, make a charge when the shareholders redeem shares. Such funds are effectively back-loaded.

Another charge often made by no-load funds is a fee for distributing the fund's in-

*Many funds have smaller minimums for IRA accounts.

come less expenses. This "distribution fee' is usually called a 12b-1 fee, after the Security and Exchange Commission (SEC) rule that allows it. These fees are normally a fraction of a percent of the average fund's assets, but can run as high as $1\frac{1}{4}$ percent annually. This rule permits "hiding" market expenses charged to the fund's NAV.

The largest expense comes from trading costs [8]; more important than broker commissions, which are as little as 6 cents per share in a $500,000 block, is the impact of the block order. Even small block orders cause an appreciable increase in the spread between bid and ask prices (because of the larger than usual supply when selling, and demand when buying). Astute managers minimize this forced price spread by looking for offers already made by other buyers or sellers. The total trading costs per year increase proportionally to the fund's annual turnover of investments. Many have annual turnovers between 300 and 600 percent, but the better funds are under 100 percent, and these costs are accordingly less.

Expense Ratio Although the investment-counsel fee is normally only $\frac{1}{2}$ to 1 percent of the fund's average daily NAV, we recognize that it is not the total cost, as often implied in advertisements. Administrative overhead (for example, legal and transfer fees) and 12b-1 fees, if any, must be added to arrive at the annual total operating cost. This total is given in the fund's prospectus* as the annual expense ratio (to its NAV). The average stock fund has annual expenses of 1.6 percent of its assets' value. The most efficient funds have ratios only one-fifth of that; the most expensive, four times that value. Funds with high expense ratios are categorized as "dog funds" by *Forbes* [9]. Some load funds have smaller expenses than no-load funds; accordingly, on a long-term basis, the loaded fund would have lower total costs.

Pure No-Load Funds In these, you pay no sales commissions or fees when purchasing or redeeming shares. All of your money works for you; the other fees noted above come from the fund's return on its NAV. An exception could be a redemption fee if you require redemption shorter than a duration given in the prospectus. Often this fee is called a *contingent deferred sales charge.* Typically, it is a percentage of the total asset value of the shares redeemed, but there is no charge if shares of one fund are exchanged for those of another fund in the company's family. Exchanges are discussed later.

Total Return After deducting the expenses reviewed above, funds make distributions from two sources: (a) income from dividends and interest (their sum is often called the fund's *yield*) and (b) realized capital gains** less realized losses. The sum of the yield and net capital gains is the total return; it is usually expressed as a percentage of the fund's NAV. Naturally, when all distributions are reinvested, the NAV increases by the total return. It is important that you not let information from the fund

---

*Required by the SEC for all new issues, a prospectus gives all pertinent facts about the issue and issuer.

**Capital gains are those realized by the sale of capital assets such as stocks or bonds. The gains (or loss) equal the sale price minus the costs, including commissions and other transaction expenses.

mislead you into interpreting yield as total return. With a drop in interest rates, some income or balanced funds, defined later, continue paying high "yields" by cannibalizing, i.e., depleting the principal. The resulting depletion of the fund's capital is reflected by its per share NAV as reported in the newspapers (see below).

Closed-End Investment Companies   As with most corporations, the shares of their stock are traded only in the secondary markets,* such as the New York Stock Exchange, the American Stock Exchange, and others. Furthermore, their purchase involves no greater fee than that of any other stock purchased through a broker or discount broker, that is, about 2 to 3 percent for the former and about 1 percent for the latter. The closed-end funds do not continuously offer to buy and sell their shares as do the mutual (open-end) funds. Unlike other stock companies, however, they do not use their resources, financial and personnel, to sell goods and services. Their sole activity is the buying and selling of financial instruments, such as the bonds and stock, of "productive" firms. Because their shares are not being continuously bought back and more vigorously sold, the closed-end companies are generally smaller. More significant, however, is the fact that because they are not being aggressively pushed, in a few months or perhaps years after they have come public, they can be purchased at approximately a 15 percent discount to their NAV. Of course, if the fund continues to be priced at a discount, you are no better off than buying a pure no-load fund. Frequently, however, the discount reduces because of special situations (for example, they become an open-end fund). Such trading of closed-end funds is more nearly "trading" than investing and, thus, is not pursued further here. Like other stock companies, however, they can be a good investment as reviewed in Section 7-6 on the Money Game.

Funds' Investment Objectives   We will later review the importance of choosing those funds whose investment objectives match our own. Although they can be further categorized by methods and risk, funds' objectives can be broadly classified as income or growth.

Income Funds   These primarily involve fixed income investments such as money markets and bonds; the average maturity dates of each are closely correlated to the general risk-reward character of all investments. These factors are reviewed in detail below in the discussion on bonds, and sources for information on maturity dates are also given. Any stocks owned by income funds are probably preferred stocks, also reviewed below. Preservation of capital, that is, your money, should be stressed in the prospectus of this type of fund. In your own investing consider this type first. For income funds, the first page of the prospectus might state that "the primary objective is income through purchase of bonds and stocks that pay high interest and dividends, with capital appreciation a secondary objective."

Tax-Free Income Funds   These bond and money market funds differ from other income funds only in that their investments are restricted to municipal bonds and cer-

---

*The primary market is only for new issues. Secondary markets are the ones usually considered as the "market"; they are the stock exchanges by which financial instruments are bought and sold in auction fashion (by bids and offers in an open market).

tain short-term obligations. These investments are free of federal income tax and also free of state and local taxes if the investor lives in the state or locality of the issuer. Examples include bond funds for the states of California, New York, and New Jersey.

To determine if there is any tax advantage, the securities' equivalent taxable fund rates are easily calculated by dividing their rates by $(1 - t)$, where t is the sum of your top income tax rates (expressed as decimals) for local, state, and federal taxes. The first two, of course, are included only if you live in the locality and state which issue the financial instruments of the fund.

Growth Funds   These are equity funds that purchase stocks of companies which indicate the high probability of higher-than-average growth, in both earnings and market price. The prospectus will note that "income is secondary." With increased risk, these funds range from those which seek quality stocks of leading growth companies (with long records of excellent performance) to those which buy shares of new smaller companies that show high promise of spectacular rewards.

Balanced Funds   As noted in their prospectuses, these funds compromise between income and growth. They mix stocks with bonds or convertible bonds (discussed in Section 7-6). The Wiesenberger Report [10] shows that they have between 25 and 60 percent of their assets in preferred stocks and bonds, with the balance in common stocks. Graham [5] feels that it is better to stay only with income and growth funds. While developing adequate sophistication, little by little, the investor can develop guides for switching between these funds. The pure no-load funds do not charge for this switch within the family of funds of its company.

Specialized Funds   As in promoting aspirin equivalents, gasoline, or investments, firms are constantly trying to present their products as superior to the competition. In that vein, there are many specialized funds, some very risky. One type could be worthwhile: funds for foreign stocks. Using patterns presented later, you will be able to recognize times of a declining U.S. dollar and may wish to invest in a diversified portfolio of foreign stocks by way of a foreign stock fund. These are offered by many fund families. An alternative in such times is to open a savings account in a Swiss or other foreign bank.

Lord Kelvin, the nineteenth-century mathematician and physicist, said that you do not know much about a system until you can measure it. Fortunately, considerable data have been collected and are used to rate the different types of funds. We review them next.

Funds: Information Sources   There are three aspects about information on funds: first are their general characteristics as an investment vehicle; with limited scope, these are addressed in this section. The references cited give greater details about investment companies, mutual and closed. The other aspects are presented in the fund's prospectus.* The second concerns the fund's objectives. They are given on or near the first page of the fund's prospectus. Finally note the fund's performance, specifically its total return versus its total expense. The returns and expenses are re-

---

*Prospectuses are readily available from the funds' distributors; addresses and telephone numbers (many of them toll-free 800 numbers) are given in the sources noted in this section.

ported as ratios to their NAV. They are given as percentages, that is, as total return or cost per $100 of NAV. The prospectus also gives the credentials of the issuers. Since mutual funds are always promoting new concepts and selling new shares, their prospectuses change, and it is important to use the information from a current one. For determining the funds from which you might request prospectuses, it is convenient to have a source which presents the data of many funds. *Forbes* [11] reviews funds annually. Among other sources is *The Individual Guide to No-Load Mutual Funds* [6], published every June by the American Association of Individual Investors. These publications give pertinent data for analyzing and comparing all prominent mutual funds. Before and after purchase of one or more funds, it is convenient to check leading newspapers; they give the funds' latest quotations from the National Association of Securities Dealers Automated Quotations (NASDAQ).

The newspaper quotations are listed by families or groups, with specific funds listed under the family. The quotations are the prices at which the securities could have been sold (net asset value) that day or could have been bought, including any sales charge. To the right of each listing, sequentially, are the sell and buy prices, and the change from the preceding day. For a no-load fund, the buy and sell will be the same, or the letters *NL* replace the sell price. Most newspapers use several symbols: *r* indicates that the fund has a contingent deferred sales charge; *p* indicates that the fund has a 12b-1 plan, meaning that a percentage of the fund assets is used to pay broker commissions and other marketing expenses; and *t* is used by some papers for funds which have a redemption fee and a 12b-1 (distribution) fee.

Forbes [11] magazine annually rates all the leading mutual funds in the late August issue* (dated 2 weeks later—August 31, 1992). Using data from over three market cycles, separate ratings are made for both up and down markets. By making your best analysis, reviewed in Section 7-6, we can determine if the market is "going up" or "down"; then a corresponding rated fund can be chosen. Since your "marriage" to the fund will wisely be long term, the fund's expense ratio has a definite effect in net return. Expenses are continually paid in good times and bad, and all funds experience both.

Nearly the same information given in *Forbes* is also presented in a yearly publication of the American Association of Individual Investors: *Guide to No-Load Mutual Funds* [6]. The data are quite complete, including the stock funds' betas (defined earlier).

Information about the type of funds, their respective characteristics, and performance is of little value until we analyze our reasons for investing. We review them next.

7-4 INVESTING FOR OBJECTIVES, GOALS/ASPIRATIONS, AND RETIREMENT

Over one's career, investment purposes change from objectives, to goals/aspirations, to retirement. Respectively, these indicate investments for short-, medium-, and long-term purposes. The objectives of a mutual fund should match yours.

As presented above, your first need or objective is funds for emergencies; this can be met by a strong interest-paying bank account. As it becomes adequate through your

---

*This issue of *Forbes* also has many informative articles about mutual and closed-end funds.

regular saving, part of the funds can be moved to a higher-paying money market account; however, recall that these often limit check writings to usually three per statement cycle. The cycle is normally about a month, but often does not start on the first. Some investment companies offer short-term bond funds which pay about 1 to $1\frac{1}{2}$ percent higher interest rates than money market accounts. They also allow free check writing, but usually the amounts of the checks must be $100 or more.

Your medium-term goals would be funds for cash purchases of furniture, large appliances, or an automobile. We have already discussed the pros and cons of buying a low-mileage used car. Also in this medium-term group could be savings for a substantial down payment on a house. The pros and cons of owning a house can be analyzed by the technique of Chapter 6. The substantial down payment frequently avails you of a lower mortgage interest rate. Another plus is that the lending agency would be less concerned about your having to establish a strong credit rating and maintaining an escrow fund for taxes and insurance. No interest is paid you on these funds. It is pleasing to be able to carry out financial transactions from a strong base.

Perhaps of greatest importance in this group of financial aspirations are funds for children's education. A personal thought: Beyond moderate provisions for normal living needs and fashions, supporting our children's education is our most important responsibility. It seems that any major self-support by undergraduate students is a definite handicap for academic accomplishments, the main purpose of their being at the college or university. Doing well academically is probably one of the most important endeavors of their lives.

Investment accounts in your children's names can have tax advantages. The earned incomes of children any age and the unearned incomes of those 14 or older are taxed at the children's rate. For a child under 14, unearned income in excess of $1000 (or, if greater, $500 plus the child's itemized expenses attributable to the unearned income) is taxed at the parents' highest rate. By starting at an early age, children's earnings under $1000 each year can grow rapidly; the magic is compound interest. As described later, zero coupon bonds have the smallest amounts of interest in their early years. Special among these are the Series EE, U.S. Savings bonds. They are available in small amounts, and their interest is presently (1992) a minimum of 6 percent, although it is regularly adjusted to the T-bill rate. This interest is exempt from state and local taxes. It is (with conditions) also exempt from federal taxes if used by the owner or dependent for college education expenses.

Accumulations for these medium-term goals will typically be over a 10- to 20-yr period. The longer periods give diversification over several business cycles; in turn, they tend to average out the greater fluctuations of stock funds. Since 1926 stocks, as represented by the Standard & Poor's 500, have had an average total return (price gains plus reinvested dividends) of 10 percent per year [12]. Over that period, this return has exceeded all others, including precious metals.

Preparing for Retirement

Start now. This is your longest-term goal; a major part of the preparation is for financial security. The long period allows for time averaging over many market cycles; thus, growth stock funds are well suited for this goal. A great advantage for retirement investing is the available qualified (tax-deferred) plans; they permit you to accumulate returns on moneys which would otherwise be used to pay taxes. An individual retirement account (IRA) and 401(k) plans are the best known; however, for general information, we will also review others, but in less detail.

The IRA is a retirement plan available to those with earned income; an annual contribution to the plan may be made up to the lesser of the total earned income or $2000 ($2250 for someone with a spouse who has no earned income). The amount of the contribution deductible from taxable income is reduced if the individual or spouse participates in an employer-sponsored retirement plan or has an adjusted gross income (AGI) greater than $25,000 if filing a single return, or $40,000, for a joint return. The contribution is partially deductible with an AGI up to $35,000, if filing a single return, or $50,000, for joint returns. The reduction is 20 percent of the excesses for each case. Even with nondeductible contributions, the dividends, interest, or any returns from the investment accumulate tax-free. That is still a tremendous advantage for long-term retirement plans. Generally, only collectibles-type investments are not allowable.

It is important to keep accurate records by which to differentiate between the pretax contributions and the after-tax ones. Separate accounts are suggested.

Caution: The purpose of an IRA is for retirement; withdrawal of funds before age $59\frac{1}{2}$ will cause a penalty of 10 percent of the amount prematurely distributed. Furthermore, borrowing money using the IRA as security is considered a premature withdrawal equal to the amount used for securing the loan. If disabled or over age $59\frac{1}{2}$, you may elect to receive scheduled or nonscheduled payments. The entire interest in your IRA must be distributed to you (or begin to be distributed to you) no later than April 1 following the taxable year in which you attain age $70\frac{1}{2}$. By exercising an option under IRA, an amount can be made payable from it to any beneficiary at or after your death, and the amount is not subject to federal gift tax. The IRS Publication 590 gives more details on this.

A 401(k), so named after the IRS rule, is a tax-deferred profit sharing plan; both employer and employee may contribute to it. The limit of an employee's salary reduction contribution is subject to cost-of-living adjustments; the limit for 1992 was $8728. Rules for premature distributions are the same as those for IRA's. However, they are different from IRA's in that there is no penalty for borrowing from it. The loans must be repaid within 5 yr (unless applied to a primary residence), and the interest is not deductible from taxable income. Hardship withdrawals, however, are limited to the employee's contributions; no employer's matching contribution is eligible. If contributions are made by your employer, 401(k)'s are better than IRA's. Furthermore, all accumulations are yours and can be taken with you to a new employer, or rolled over into an IRA.

A Keogh plan is another profit sharing tax-deferred plan available for the self-employed; the plan must also cover all full-time employees with 3 yr or more of service. The contribution for the employees must be a "reasonable" amount relative to that of the employer. The Keogh Act allows the employer to make an annual contribution (15 percent of total income) deductible from taxable income up to $30,000. The contributions are nonforfeitable and belong to the employees.

Simplified employee pensions (SEP's) are salary reduction plans. They permit a business owner and employees to make contributions to SEP accounts up to $30,000 (15% of $200,000) each in 1992 (also subject to COLA's*). This plan is available to businesses with 25 or fewer employees. The SEP accounts may be IRA's, and the employees keep all of it on leaving the employer.

The IRS rule 403(b) allows a tax-advantaged plan for tax-exempt organizations; these include public school systems and most universities. Both employer and employee contributions are permitted up to $9500 each in 1992 (also subject to COLA's). An engineer might take advantage of this plan as a full-time or part-time instructor.

Income for Your Retirement Life Pattern

Naturally, the amount of income you will need in retirement must be the target for your long-term investment plan. In retirement, lifestyle at your preretirement level will require only about two-thirds to three-fourths of the amount needed before retirement. The major factor is that 25 percent of a working person's expenses are work-related (transportation, clothes, etc.). The report of the President's Commissions on Pension Policy shows that the required income for a preretirement lifestyle reduces from two-thirds for lower-income levels to one-half for incomes of $50,000 or more.

Life Cycle Investing

Table 7-1 shows the typical pattern of a professional's life cycle. To match this cycle, Smith Barney's (*Topics,* August 1989) suggests a slow linear change in asset allocation of 20 percent bonds and 80 percent stocks at age 30, to 80 percent bonds and 20 percent stocks at 70. While this plan serves as a reference, we engineers recognize this as an open-loop control system which does not consider, nor is suited for "changing times."

The essence of modern life is flexibility. Only by using a closed-loop feedback system can we adjust accordingly. In such a system, errors (due to changes in the system, environment, or objectives) are detected as differences between results and objectives; in turn, they are used to change actions so as to minimize the errors. The controller in

*Cost of living allowances.

TABLE 7-1
THE LIFE CYCLE: RISK TOLERANCE, HOLDING PERIOD, AND TAX EXPOSURE*

	Early career	Mid career	Late career	Retirement
Risk tolerance	High	High	Medium	Low
Holding period	Long	Long	Intermediate	Intermediate
Tax exposure	Low	High	High	Low

*Reprinted with permission from *The Individual Investor's Guide to No-Load Mutual Funds,* 10th ed. 1991, published by the American Association of Individual Investors, 625 N. Michigan Ave., Chicago, IL 60611; (312) 280-0170.

a closed-loop system directs the system's actions to minimize the errors. By using a dynamic portfolio of mutual funds, you can be that controller, adjusting distribution of your assets primarily between money market, bond, and stock funds.

The main intent of this chapter is to give you needed information. With it, you can detect economic changes and recognize the best actions. Furthermore, by using the ideas just presented, you can choose mutual funds for an investment portfolio that matches your goals and aspirations. The portfolio will also give financial security in retirement.

Guides for Choosing Mutual Funds

Generally, any significant decision and subsequent action has the potential for mental, physical, and economic loss. A guide in life, then, is to act little by little; do not be misled by the often used criticizing phrase "Too little too late." The better plan is this: Over time, use the results of small test actions to guide subsequent decisions. My experience has shown that this plan is the fastest and least costly method for achieving goals. We learn best from the experience of our own activities; when mistakes are small, they are probably the least painful. This concept is further developed in Section 7-6, when we discuss the Money Game.

The characteristics of mutual funds needed for prudent investing are their objectives, performance/risk patterns, costs, and services.

The first step in choosing a fund is ensuring that its objectives match yours at this time in your life cycle. The fund's objectives are stated explicitly on the title or first page of its prospectus. The objectives are presented implicitly in the fund's title (such as, money market funds and short-term bond funds for current income, bond funds for income, and growth funds for appreciation of invested capital). With higher attendant risk, aggressive growth funds seek to maximize capital gains; a 10- to 20-yr horizon greatly reduces its risk.

Since funds' performance information is available in the sources noted earlier, you are wise to select a fund which over the last 10 yr has had an average total return equal to or greater than that of the bench mark references given, such as the Standard & Poor's 500 average for stock funds (see no. 9, page 273). The choice must be compatible with your risk tolerance. This tolerance can be decided only by you, usually from experience. Your estimate is more accurate if the experience includes losses. Personally, I feel easier about a given assessment of risk after accumulating profits from earlier investments. The fund's beta rating given in *The Individual Guide to No-Load Mutual Funds* is a convenient quantitative risk indicator. The fund's annual average total return divided by the fund's beta can be used as a performance/risk reference for comparing stock funds. Recall that a greater beta indicates the greater risk (volatility or price fluctuation) some managers use to achieve higher returns (when lucky).

Forbes' rankings of "A" through "D" in both up and down markets is a general risk indicator [11]. Requiring a rank of "B" or better in a down market is suggested until you accumulate profits from income funds. All of the ratings are based on past performance. While they do not assure future results, they are our best guides. Even so, however, it is true wisdom to reflect on the present and possible future situations, as supplementary guides for the future. The indicators we have reviewed represent the consistency of a fund's performance pattern; another important quality is its efficiency. At what price was its performance bought?

The expense ratio as reviewed earlier is a significant part of the price of performance; however, as also noted, it does not include the brokerage costs incurred in each transaction. An index for these costs is the fund's annual turnover. It is expressed as a percentage, typically 90 percent. Higher rates suggest high brokerage costs, which reduce total returns. This rate expressed as a decimal can be added to the expense ratio for comparing funds. Also, high turnover rates indicate more realized net gains, which will be distributed as capital gains. The distributions could well be at a time not advantageous in view of your other financial activities, and thus more immediate taxes. A turnover rate of about 50 percent seems ideal.

The services offered by mutual funds are of two types: (1) those for tax-deferred plans, such as IRA's, Keoghs, and 401(k)'s; and (2) conveniences such as dividend reinvestments and no-charge, toll-free telephone switching within the family of funds (for example, from a bond or stock fund to a money market or short-term bond fund). These services will be clearly stated in prospectuses.

The time and effort used to closely study the funds' characteristics (as presented above) are a wise investment. The fund's prospectus and either the current year's *Forbes Mutual Fund Report* or the latest *Guide of the American Association of Individual Investors* will provide the information needed for your analysis.

Investing through mutual funds seems to be the method most efficient in time and effort for the defensive, prudent investor. Mutual funds offer diversification and professional management. Furthermore if one uses the guides reviewed above, these advantages are available at small costs. Next we will discuss a foreign bank account, which has many of the conveniences of a special mutual fund.

Foreign Bank Account [13]

After accumulating a substantial financial estate, options for further diversification and conveniences can be available through a foreign bank account. I feel that near zero inflation is the most important advantage of having an account in certain foreign currencies. The effective compound rate of return on an account in a stable country such as Switzerland is about 9 percent. This rate is determined by adding the $5\frac{1}{2}$ percent or greater U.S. inflation rate to their savings account rate of 3 to $3\frac{1}{2}$ percent. It is an excellent return on an account of only about one or two thousand dollars. For Swiss accounts, there is a 35 percent federal withholding tax on the interest, but by filing a Form R-82, all is returned except 5 percent. The form must be filed before December 31 of the third year after the withholdings. Also, for accounts in excess of $10,000 at any time during a year, U.S. residents must file a TD F 90-22.1 form with the Department of the Treasury by June 30 of the following year. Both of these forms are easy to execute. Finally, on transfer of amounts in excess of $5000, the United States requires that it be reported to Customs. It seems that an account in Switzerland, Austria, Germany, or the Netherlands is equally stable. History shows that they have had very low inflation in the last 50 to 75 yr. In 1985, in an effort to improve the U.S. balance of trade, the G7, an international group of seven western nations,* devaluated the U.S. dollar by approximately 50 percent. Of course, in only a few months, those with foreign accounts, such as in Swiss francs, nearly doubled the purchasing power of their original amounts, if converted back to U.S. currency. Internal Revenue tax on the capital gains would have to be paid. If the currency is spent in the corresponding country, however, there is no tax. Further, devaluations are politically the least painful correction if, as seems probable, the unbalance in foreign trade continues.

Regarding foreign accounts, there are other advantages as well: (1) Low commissions (only $1\frac{1}{4}$ to $1\frac{1}{2}$ percent) are charged by foreign banks for even small stock transactions. All that is required is an inexpensive telephone call. And all of these banks have competent English-speaking personnel. (2) No commission is charged other than the spread (between the buying and selling prices) for precious metals or coins. In addition to the spread, American firms charge 1 to $1\frac{1}{2}$ percent. (3) A Swiss "power of attorney" is different from that in the United States, which lapses upon the death of the principal. The Swiss version remains in effect after death. Thus when giving signature power to someone, such as your spouse, you are also designating your beneficiary. The advantage is that in the event of your death (or your spouse's), the funds in the account are available to meet any emergency expenses. They do not require probate. Through an inexpensive bank wire, the money is available in a few days. Interestingly, a bank wire from the United States to Europe requires a week; an air mail letter, only 1

*The G7 consists of the central banks of the United States, United Kingdom, West Germany, France, Japan, Canada, and Italy. Devaluations are effected by heavily selling $US on the foreign money market.

TABLE 7-2
A FEW SWISS BANKS SUITABLE FOR FOREIGN DEPOSITORS [13]

"Big three" banks	Cantonal (state) banks	Other banks
Union Bank of Switzerland Bahnhofstrasse 45 CH-8021 Zurich, Switzerland	Basler Kantonalbank Spiegelgasse 2 CH-4001 Basel, Switzerland	Ueberseebank Limmatqual 2-k CH-8024 Zurich, Switzerland
Swiss Credit Bank Paradeplatz 8 CH-8021 Zurich, Switzerland	Zurcher Kantonalbank Bahnhofstrasse 9 CH-8022 Zurich, Switzerland	Banque Indiana (Swiss) 50, avenue de la Gare CH-1001 Lausanne, Switzerland
Swiss Bank Corporation Paradelplatz 6 8021 Zurich, Switzerland	Deposito-Cassa der Stadt Bern Kochergasse 6 CH-3000 Bern 7, Switzerland	Cambio + Valoren-bank, Utoquai 55 CH-8021 Zurich, Switzerland
		Migros Bank Seidengasse 12 CH-8023 Zurich, Switzerland

or 2 days more. (4) For travel in Europe, there is a bank card available. It guarantees your check (to the equivalent of approximately $300) in any currency you wish to write it. Table 7-2 is a list of Swiss banks suitable for foreign depositors.

Opening an account is easy; you need only write a letter requesting application material and current interest rates. Definitely send it by air mail. In about 2 weeks you will receive a packet containing current information such as interest rates for the various types of accounts. Addresses for a few Swiss banks are given in Table 7-2. *Note:* Your signature on the application will have to be verified by a Swiss consulate, a notary, a bank official, or perhaps an existing client of the bank.

7-5 MUTUAL FUNDS TO THE MONEY GAME

Probably few pleasures are comparable to the enjoyment that comes from understanding an endeavor. The following sections are about understanding financial investments. Although extensive knowledge is not necessary to benefit from mutual funds, familiarity with the concepts and terms when reading the financial literature is a true joy. To play the Money Game, however, the fundamentals presented are required for the prudent individual. The Efficient Market Hypothesis (EMH) permits us to move from indirect investing in mutual funds to high-risk stocks in two stages: first, a passive strategy; and then, an active one. Both are reviewed below. Recall that the scope of this chapter does not include the greater risk of futures contracts and index options. These can be used to reduce risk but require trading competence beyond the concepts of this chapter.

The Efficient Market Hypothesis (EMH) is based on several concepts: (1) There are homogeneous expectations of risk and return for all investors. (2) There is a free flow of all pertinent information. (3) There are millions of investors, and they include at least 50,000 professionals [3]. Furthermore, these investors analyze the information and quickly act on it. Therefore, security prices fully reflect all available information. The degree to which this occurs varies. The weak form of the EMH states that stock prices are random; thus, future prices are completely uncorrelated with past prices. The semistrong form states that security prices reflect all publicly available information. The strong form suggests that all information, public and insider, is quickly reflected by security prices. Many studies [3, p. 359] have supported all forms of the hypothesis. These results indicate that no investment strategy, on a long-term basis, can outperform the market. The EMH suggests a passive investment strategy; it is the first stage in moving from mutual funds to the Money Game.

Passive Strategy

This strategy does not attempt to outperform the market, but rather gives our Child* ego state the pleasure of "being in the market." The EMH implies that all securities are correctly priced at all times. Accordingly, one only needs to buy and hold an adequately diversified portfolio. Since 1926 the stock market has outperformed all other invest-

*Obviously, this statement is based on my personal judgment.

ments. Its real return was 10 percent compounded annually. The passive strategy has several advantages. First, by always being invested, the portfolio benefits from the market growth due to rapidly developing technology and management ingenuity. Also, transaction costs are minimum, and the amount of time and effort required for analyzing information is small. Only record keeping and tax planning are needed. In an active strategy, significant time is required to carefully follow the 10 to 15 stocks needed for adequate diversification. Often the required time could infringe on other duties. This dilemma does not exist when using a passive strategy. It is generally recognized that the market is a highly efficient information processor, but inefficiencies do exist. Occasionally there is a time lag in the market's response to events. These lags make possible excess risk-adjusted returns ("beating the market") by using active stategies.

Active Strategy

This strategy is rational only in cases where an investor has advantage over others. Four possibilities are

1 Information advantage (an "insider")
2 Analytic advantage
3 Judgmental advantage
4 Peculiar restrictions of others (such as for insurance firms or trusts)

We immediately recognize the importance of an active investor being informed. Only then is there any chance of his operating on "a level playing field." The remainder of this chapter is addressed to providing the fundamentals required for the second stage: actively playing the Money Game.

7-6 THE MONEY GAME: THE ENGINEERING APPROACH

In review, indirect investing through mutual funds can meet the investment needs recognized by our Adult ego state, but it deprives our Child of the thrill of the decisions and the gains and losses associated with direct investing. I call this type of investing the Money Game. Berne [14] notes that games are different from other social transactions in two ways: their hidden quality and the payoffs.

In playing the Money Game, the (prudent) Adult analyzes the fundamentals stored in the Parent and limits the Child to moneys one can afford to lose. Without reducing the fun, use of the engineering approach enhances the probability of gains and reduces the risk (and losses) of the Game. It is still a game of probability, but use of this approach is definitely worth the effort. Through it we can truly be "intelligent investors." Recall that the approach consists of five steps.

Step One: Defining Objectives

The first step is defining the problem. For investing, it involves deciding on your objectives. It seems wise to approach them in two phases: preserving your wealth, and then enhancing its growth. Because of taxes and inflation, the first goal requires the latter. For example, a return or yield on investment of at least 9 percent is required to offset 5 to 6 percent for inflation. This is determined by dividing 6 percent by 1 minus the sum of your marginal state and federal tax rates, for example, $1 - 0.34 = 0.66$. The marginal state tax rate + the federal rate, respectively, might be approximately $0.06 + 0.28 = 0.34$. Of course, on a tax-free investment, only the 6 percent is needed to preserve the buying power of your savings. Also, reducing waste and prudently shopping for groceries and other daily needs can fight and significantly reduce the effects of inflation. For more than 20 yr, my family has enjoyed playing the Game and has been rewarded by it. Real growth, however, is surely to be expected when one defers the pleasures of overconsumption and develops the habit of saving. Nonetheless, we see that taxes and inflation must be considered. It is also true that investing at any nonzero rate of growth, in part, reduces the diminishing effects of inflation. For this first goal (preservation and moderate growth), the fundamentals presented below show that bonds and preferred stocks are effective.

An Overview: Steps Two Through Five of the Engineering Approach

Step two has two phases: (a) the diligent study of the relevant fundamentals for understanding and (b) analyses based on this understanding. The analyses are used in step three to select and evaluate alternatives. This requires an in-depth analysis and understanding of the system and its environment. We first present the fundamentals of the investment system and later review the environment. It is primarily controlled by the G7* central banks. Among them, our Federal Reserve is often dominant. This information guides us in the implementation of step four (buying or selling). Step five is an ongoing endeavor of the engineer investor: reviewing the results and constantly seeking information as a guide to improve the investment strategy. Reviewing the "Important Terms" below could be helpful before considering step two: learning the fundamentals. The terms are presented first for your convenience.

Important Terms

Other definitions are included in the glossary at this end of this chapter. Note that later terms use the definitions of preceding ones. As appropriate, the terms are evaluated for the AB Corp. using data for 1991 from Tables 7-3 and 7-4 (pages 262 and 264).

Basis points are used as measures of bond interest rates. One basis point equals one-hundredth of one percentage point.

Bear markets (opposite of bull markets) are those in which prices are falling. Primary bear markets are established by the long-term falling of prices. Shorter-term drops are considered technical corrections (see Appendix A-6 at the end of this chapter).

Beta is a measure of systematic risk. It shows the volatility of a security relative to the market, i.e., SP500.

Bull markets are those in which prices are rising. A primary bull market is associated with a long-term trend. Shorter terms, when prices rise for a few days, are considered technical corrections.

*Please see the footnote in our previous discussion of foreign bank accounts.

Capital surplus, frequent in financial reports, is the amount by which the prices paid by the original purchaser exceed the security's par value, also called *paid-in capital.*

Cash flow per share is the sum of the reported net profits and the depreciation minus preferred dividends, all divided by the number of outstanding common stocks. For AB Corp.

$$\text{Cash flow/share} = \frac{\text{net profits} + \text{depreciation} - \text{pref. div.}}{\text{number of outstanding common shares}}$$

$$= \frac{\$16,800 + \$20,000 - \$3,200}{10,000} = 3.36$$

Current ratio is the ratio of a firm's current assets to its current liabilities. A minimum of 2 is usually desired.

Cyclical businesses, also called the *cyclicals,* are the most sensitive to the overall economic activity. Their earnings and, in turn, their share prices rise rapidly with the economic upturns and drop quickly in down markets. Typically, these are the basic industries (paper goods, autos) and housing.

Defensive stocks hold their prices best in economic downturns (betas < 1). Foods, utilities, grocery retails, and pharmaceuticals are examples. A defensive stock should be purchased only after an analysis as presented above. In a bear market, it is probably best not to sell a normal stock and switch.

Dividend payout is the fraction of earnings (per share) after taxes, available to stockholders, which is paid in dividends.

For example, AB Corp.'s per share earnings after taxes (EAT) minus dividends to preferred stockholders were $1.36 (see p. 265); the total dividends were $8600, or $0.86 per share for the 10,000 shareholders. Using this data,

$$\text{Dividend payout} = \frac{\$0.86}{\$1.36} = 0.63, \text{ or } 63\%$$

Dividend reinvestment plan (DRIP) is a plan by which firms permit shareholders to use dividends to automatically purchase the firm's shares with no commission, and sometimes at a discount.

Dividend yield is the dividend amount divided by the per share price. For AB Corp. the present market price is $14.00; thus,

$$\text{Dividend payout} = \frac{\$0.86}{\$14.00} = 0.0614, \text{ or } 6.1\%$$

Dollar cost averaging is a strategy which is largely a discipline technique. An equal amount of money is invested on a regular basis, such as every month or quarter. In this way, one purchases a greater number of shares when prices are low and a smaller number when prices are high. The result gives a lower average cost than the average price per share. It is a convenient way to budget and carry out a long-term investment plan. An alternative is discussed in Section 7-5.

Dow Jones averages consist of the industrial, transportation, and utility averages. The industrial is the most popular; it is a price-weighted average of the 30 largest U.S. companies.

Federal funds is the name of the market among commercial banks. It is used for short-term loans for banks to meet regulatory reserves by borrowing from banks with an excess. The loans are often for overnight only and are for over $1 million. The interest of the loans is known as the **federal funds rate** and is set by the Federal Reserve. As discussed earlier, this rate is considered the most sensitive interest rate indicator.

Free cash flow/share. This has become of increasing interest to analysts.

$$\text{Free cash flow/share} = [\text{cash flow} - \text{plant costs}]/\text{shares} - \text{div.}$$

AB Corp. had no capital plant outlays, and common dividends were $0.86 per share; thus;

$$\text{Free cash flow/share} = \$1.56 - \$0.86 = \$0.70$$

Fully diluted earnings per share is the quotient on dividing the net income by the total of the number of outstanding common stocks plus all those which would be added if all convertible securities were converted to equity.

G7 consists of the central banks of the United States, United Kingdom, West Germany, France, Japan, Canada, and Italy.

Growth stocks are those which have increased their earnings per share in the past 5 to 10 yr and are expected to continue to do so in the future.

Junk bonds are those not of investment quality. They have a rating less than Baa by Moody's and less than BBB by Standard & Poor's (S & P).

Leverage is used by individuals and companies. It is the use of borrowed money at a fixed rate of interest to achieve a greater return on equity. As discussed later in this section, debt increases the financial risk.

Limit order is an order to buy or sell a stated amount of securities at a specific or better price. You might place a limit buy $1 or $2 below the present market price in the hope of catching the stock on any down swing. A sell limit order can be used to catch a rise of a few dollars. *Note:* A limit order never becomes a market order (see *stop order*).

Major stock indexes. The Dow Jones Industrial consists of the blue-chips; the OTC (over-the-counter) Composite firms are the most speculative. The S&P 500 and the *Value Line* indexes represent the in-between stocks.

Market order is an order to your broker to buy or sell stock at the "going" market price.

Money supply is measured by the Federal Reserve. It uses M factors. Reported weekly, M1 includes currency, checking account balances, and other demand deposits. It is the main measure of the "active" money. M2 is broader and perhaps more accurate, but is reported only monthly. It includes M1 plus money market and savings accounts. More inclusive are measures M3 and L, but these are less representative of the supply's rate of growth. M1 and M2 are given in *Business Week* magazine under Business Week Index, respectively, with the "Monetary Indicators" and the "Leading Indicators."

Moving average is an average of data, usually prices, over n past periods of time. An example might be a 10-day moving average. The moving average for n periods is the sum of the specific data for the last n periods divided by n, the number of periods. On each day, the earliest price is subtracted from the sum, and the price for that (the present) day is added before division by n. For the 10-day average, the last 10 prices, $62.625 (62 and $\frac{5}{8}$), 63.125, . . ., 64, 64.125, add to 635; on division by 10, the 10-day average is $63.50. Tomorrow, from the newspaper, we add the new price of $64.375 (64 and $\frac{3}{8}$) and drop (subtract) the earliest price of $62.625. The new sum is 636.75; on dividing by 10 (the number of periods), the new moving average is $63.675.

NASDAQ is an acronym for National Association of Security Dealers Automated Quotations. Owned by NASD, it is a computer network by which current over-the-counter (OTC) quotes are "immediately" available to brokers and dealers.

NASDAQ OTC composite is a market capitalization weighted average of approximately 5000 stocks traded over-the-counter.

Net current asset value is the net working capital minus the aggregate redemption value of the preferred stock. When this per share value exceeds the stock's market price, Graham suggests that the stock should be an automatic buy [15].

Net income (or profit) is the total income (from operating plus interest, dividends, etc.) minus all other costs such as interest and taxes.

For AB Corp.,

$$\text{Net income} = \$25,000 + \$3000 \text{ (from interest)}$$
$$- \$7000 \text{ (for interst paid)}$$
$$- \$4200 \text{ (for federal and state taxes)}$$
$$= \$16,800$$

This value is also represented by earnings after taxes (EAT). The terms *net income* and *EAT* will be used interchangeably as a reminder of their equivalence.

Net operating margin is defined by the expression

$$\text{Net operating margin} = \frac{\text{net profits (EAT)}}{\text{sales for AB Corp.}}$$
$$= \frac{\$16,800}{\$73,000} = 23\%$$

Net sales are the total of gross sales minus refunds. This equals $73,000 from Table 7-4 (no refunds).

"Net" working capital is often used as working capital less the long-term debt, including preferred equity. AB Corp. has a negative "net" working capital, which is not unusual for companies.

New York Stock Exchange composite is a capitalization weighted average of all the common stock traded on the NYSE.

Operating costs are the sum of the cost of goods sold (or materials used by manufacturing or processing firms), selling expenses, plus general and administrative expenses. For AB Corp., this cost of goods sold ($25,000) plus depreciation ($20,000) plus selling and administrative expenses ($3000) gives a total of $48,000.

Operating income (or profit) is net sales minus the operating costs. The earnings before interest and taxes (EBIT) is a frequently used acronym for operating profits (income). For AB Corp., EBIT = $73,000 − $48,000 = $25,000.

Operating margin (of profit). This ratio is determined by

$$\text{Operating margin} = \frac{\text{operating income}}{\text{sales}}$$

For AB Corp.

$$\text{Operating margin} = \frac{\$25,000}{\$73,000} = 0.342, \text{ or } 34.2\%$$

Price to dividend ratio (P/D) is generally considered more meaningful than the P/E. That is because it is possible to manipulate the apparent earnings by changing the depreciation schedule. Recall that the IRS permits accelerated depreciation, and although firms usually use the straight-line schedule to determine true earnings, they might not be reported accordingly. Also, non-reoccurring incomes require close attention to the footnotes of the financial report, for the investor to avoid being misled. The P/D for a stock is the reciprocal of its dividend yield. It is easily determined by merely dividing its price (as given in the financial section of most daily newspapers) by the share's dividend (also reported in the firm's listing). For comparison, the ratio for the Standard & Poor's 500 (many newspapers report it as the S&P composite) is readily evaluated by dividing the reported composite price by its current annual dividend. The S&P dividend is given in *Barron's,* near the back under "Indexes, P/E's & Yields." The implications of P/D values are discussed later in this chapter. To maintain their image, some firms pay dividends greater than their earnings, impossible in the long-term (see footnote on p. 272). For AB Corp.,

$$P/D = \$14/\$0.86 = 16.3$$

Price to earnings ratio (P/E) is simply the current price of a share of the stock divided usually by its earnings for the last 12 months. This is called the *lagging P/E ratio.* Often analysts will use an estimate of the earnings for the next 12 months; it is identified as the *leading P/E ratio.* The *Standard & Poor's Stock Guide* uses earnings reported by the firm for the past 6 months and the estimate by S&P's analysts for the next 6 months. This ratio is given in the current data for the company. For AB Corp., the lagging P/E is $14/$1.36 = 10.

Primary earnings is the quotient on dividing a firm's net income by the number of shares of common stock currently outstanding. No extraordinary (nonrecurring) earnings are included. It is wise to check the footnotes of a firm's financial report for the extraordinary earnings, especially when the earnings have a significant increase. Sources such as *Value Line Investment Survey* check to determine if possible conversions of long-term debt (including convertible preferred stock) would significantly change the reported earnings per share (see *fully diluted earnings* above).

Primary market is generally used to indicate a market for new issues. In it the proceeds of the sale, less commissions to a broker or underwriter, go to the issuer of the securities.

Primary stocks are those generally considered "large" or "prominent" because of a leading position in their industry. The 30 included in the Dow Jones Industrial average are primary stocks.

Return on equity (ROE) is evaluated by the following equation:

$$\text{Return on equity} = \frac{\text{net income} - \text{preferred stock dividends}}{\text{common stock equity*}}$$

For AB Corp., taking the value for common equity from Tables 7-3 and 7-4, the return on equity is

$$\text{ROE} = \frac{\$16,800 - \$3200 \text{ (to preferreds)}}{\$210,000} = 0.0648, \text{ or } 6.5\%$$

By dividing both the numerator and denominator of this equation by the number of outstanding common shares, we get earnings per share (EPS) in the numerator and the equity per share or book value in the denominator. Accordingly, ROE is also known as *return on book value*.

Return on invested capital (ROI) is

$$\text{ROI} = \frac{\text{earning after taxes} + \text{interest to long-term debt}}{\text{total capitalization (defined below)}}$$

Again for AB Corp., using data from Table 7-4,

$$\text{ROI} = \frac{\$16,800 + \$7000}{\$350,000} = 0.068, \text{ or } 6.8\%$$

Comparison of this factor with ROE shows how well management is using debt capital to improve earnings. The ROI is also correlated to leverage.

It is interesting to note that earnings after taxes, per asset (EAT/assets), equals the net profit margin (EAT/sales) times the firm's asset-turnover ratio (sales/assets). Accordingly, a firm such as a retail grocery chain might have an operating margin of only

---

*Common stock + capital surplus + ret. earnings.

2 percent or less; however, its ROI, return on invested assets, could be substantially larger because of a large turnover ratio. Conversely, for firms with small turnover ratios, adequate ROI's require a large margin of profit. Of course, you recognize that this explains the high volume of discount stores and the large profit margins for many funeral homes. Often firms try to achieve both high volume and large profit margins by using "marketing" techniques.

Round lots are orders to buy or sell stocks in multiples of 100's. Others are called *odd lots*. Since they are combined by an odd lot broker to form round lots, the commission is slightly greater (presently $\$\frac{1}{8}$ for less than \$40 per share, $\$\frac{1}{4}$ for over \$40). This extra charge now seems to be assessed less often.

Russell 1000, 2000, and 3000 are indexes: the 1000 is for big companies, the 2000 for small companies (market valuation from \$15 million to \$350 million), and the 3000 is for all.

Secondary market is a market in which previously issued securities are sold; examples are the stock markets with which we are familiar (see *primary markets*) .

Secondary stocks are the nongrowth common stock not considered primary (see above).

Standard & Poor's composite index of 500 stocks is a more comprehensive blue-chip indicator (than the Dow Jones Industrial average). It is a market-capitalization-weighted index of stocks on the New York Stock Exchange (NYSE), American Stock Exchange (AMEX), and the over-the-counter (OTC) markets. The 500 consist of 400 industrials, 60 transportation and utility companies, and 40 financial stocks. It represents 80 percent of the market value of all NYSE stocks and is often referred to as the S&P 500. The S&P 400 is only for industrials.

Stop order is an order to buy at a specific price above or sell at a specific price below the current market price. Stop sell orders are used to protect unrealized profit or to limit loss. A stop order becomes a market order when the stock sells at or beyond the specified price. The general guidelines in Section 7-7 review how they can be used to automatically implement your strategy.

Total capitalization is the aggregate of fixed debt (of maturity greater than 1 yr) plus the sum of the par value of preferred and common stocks, any capital surplus, and retained earnings. It also equals the company's total assets minus the sum of its current liabilities, intangible assets (recall our discussion of patents and copyrights), and (if any) income taxes due, as could have been deferred by accelerated depreciation. Since AB Corp. has no intangible assets or deferred taxes, its total capitalization = \$400,000 − \$50,000, or \$350,000.

Total return includes normal yield from an investment, such as interest from bonds and dividends from stocks preferred or common, plus the gain or loss in the per share price of the stock. As an example, suppose in the last 12 months the per share price of AB Corp. increased from \$14 to \$16; the increase is a capital gain.

$$\text{Capital gain} = \$16 - \$14 = \$2$$

or
$$\$2 \div \$14 = 0.143, \text{ or } 14.3\%$$

$$\text{Total return} = \text{dividend yield} + \text{capital gain (or loss)}$$
$$= 6.1\% + 14.3\% = 20.4\%$$

It is usual that capital appreciation is the larger portion. That expectation is the reason an investor might buy stock of dividend yield less than that available from a high-grade bond. Bonds, however, have capital gains and losses, as discussed later in this chapter.

Value Line composite index is an equally weighted price index of all the stocks covered by the *Value Line Investment Survey* (over 1700 in 1992). Most are growth equities; hence, the index is commonly used as a proxy for growth stock activity.

Working capital is closely related to the current ratio. It is obtained by subtracting the current liabilities from the current assets. Thus, it is the current assets available for operation of the business: paying for materials, salaries, etc. A current ratio of 2, of course, means that the working capital equals the current debt. Recently through the aid of computers, closer control of inventories and other costs has permitted firms' treasurers to reduce the required working capital. Nonetheless, adequate working capital is the first requirement for preserving good trade and bank credit. Of equal importance, however, is that adequate working capital can provide funds for emergencies and new growth projects. Conversely, working capital produces only small return (as in money market funds) or no return at all. Furthermore, an excess could encourage waste; perhaps, it would be better to distribute it to shareholders.

For AB Corp.,

$$\text{Working capital} = \text{current assets} - \text{current liabilities}$$
$$= \$100,000 - \$50,000 = \$50,000$$

Yield is a general term for the percentage return on a security investment. It is usually used with a qualifier, for example, a nominal (coupon) yield on a bond or a dividend yield on common stock, as presented above.

Yield curve is a plot of the interest on high-quality fixed-income securities, such as bonds, versus their term, usually in years. Normally, because of the indefiniteness of long terms, the rate increases with maturity. To slow the economy, sometimes the Federal Reserve increases short-term rates above those for long-term. This abnormal condition is called an *inverted yield curve.*

Step Two: Phase One—Investment Securities: Fundamentals

This section begins with the fundamentals needed to be stored in the Parent of the informed investor. We start with investments of least risk; as noted in our previous discussion of mutual funds, with increasing risk, these investments are bonds (regular and convertible) and stocks (preferred and common). In *The Intelligent Investor,* Graham offers that "an investment operation is one which, upon thorough analysis, promises safety of principal and an adequate return. Those not meeting these requirements are speculative" [16]. In accordance with our earlier stated objectives, the speculative areas of commodity futures, puts and calls, options, and index trading are not considered here.

The thread through Graham's book is the concept of "margin of safety." We engineers are familiar with these margins; they are used to allow for indeterminate factors.

To facilitate comparisons, we define the margin differently from Graham, but the idea is the same. Here, the "margin" is simply the excess by which the net asset value of the investment exceeds its price, divided by its price. Below we use data from a firm's financial report to determine the net asset values for bonds, preferred stocks, and common stocks. Later in this section, the book values of each are developed. These values are then used in our discussion of margins of safety, to present the safety margins in terms of the securities' prices and their ratios of book value to price.

Two Approaches in Direct Investing: Fundamental and Technical Analysis

The fundamental approach, also known as *value-oriented investing,* is a conservative method in determining whether or not to make an investment. For stocks, the evaluation methods are based on judgments about the economy and future corporation profits. The value of a stock (or any investment) is the net present value of all future incomes and expenses. This method is used in our forthcoming discussion on bonds and is also applicable for preferred stocks. In both cases the interest rate for near no-risk fixed investments can be estimated from long-term U.S. government securities (see Appendix A-1 in this chapter, source no. 34). For common stock, the interest rate used in the "present value" determination is based on judgment about the present and future economy. As an aid to your judgment, consider CAPM presented in this chapter's section on Investment Risks. In our discussion on market cycles we noted how central banks of the industrial nations influence interest, in an effort to improve the respective nation's economic welfare. Fundamental analysis, presented first, can be supplemented by the technical approach, presented later. The technical approach assumes that all fundamental knowledge about the company is reflected in its market price, and the technicians use charts of the series of stock prices to make analyses. As the last step in the engineering approach (strategy revisions to improve effectiveness), we review technical analysis as a guide in timing the buying and selling of shares that fundamental analyses show are "good investments."

Fundamental Analysis

This is the evaluation of financial securities using the fundamentals presented in this chapter. The market will remain efficient only as long as an adequate number of investors use fundamental analysis to evaluate all information about the securities on the market. The first requirement of this analysis is an understanding of the investment vehicles. Toward that understanding, we review the essentials of bonds and stocks. As prudent investors, we focus our attention only on those two investment instruments. Following step two of the engineering approach, we review the basic characteristics of these securities. Corporations issue them to raise funds for capital investments and operations. The total of these securities is the firm's total capitalization. The total of their market value is the firm's market capitalization; however, for important ratios presented below, we will use the values from the firms' financial reports. Their values are conservatively based on the lower of original costs or market values. Stocks, both preferred and common, make up the company's equity capital; bonds and other debts make up its debt capital. These securities are ranked in accordance with their claim against the assets of the firm.* First mortgage bonds rank above debentures (promissory notes secured only by the good faith and credit of the company), and these are senior to preferred stock. Preferreds are senior to common stock. We next review the essence of bonds and then stocks.

Bonds   These are mainly of two types: (a) regular, sometimes called *straight;* and (b) convertible. Both are debt instruments which represent an IOU or promissory note of a corporation (or a government unit—but for them, there are only straight bonds). Corporate bonds are usually secured by a mortgage, as on specific property. As noted above, debentures are promissory notes backed only by the general credit of the issuer. Bondholders are creditors of the firm or government unit. Bonds usually have a "face value" of $1000; that is their value at the time of sale and the value at which they will be redeemed at the expiration date. The time from issue to redemption is its term: intermediate-term bonds vary from 5 to 10 yr; long-term, from 20 to 30 yr.** Over the duration of this term the corporation (or government unit) pays interest to the creditor at a specified rate called the *coupon rate*. The interest is paid every 6 months.*** Today most bonds are "registered" on the books of the company in the name of the owner. They can be transferred only when endorsed by the registered owner. Bearer or coupon bonds do not have the owner's name registered by the issuer. The attached coupons are clipped as they come due and are presented by the holder for payment of interest. Usually they may be presented for payment at commercial banks.

Bond Valuation: Returns and Risk   The value of the interest payments, every 6 months over the term of a bond, is one-half the product of the interest rate and the bond's face value. Both are specified on the face of the bond. For example, on a $1000 bond paying 8 percent, you would receive $40 checks semiannually. Let us consider what could be called *the bond identity*. It is based on the sum of the present values of (a) all interest payments and (b) the redemption payment. Consider that the bond in this example has a term of 10 yr; thus, it provides 20 semiannual payments of $40 each. Denote its face value as FV. From Chapter 6, the present value is

$$PV = \text{int.} \times P/A\ (i,n) + FV \times P/F\ (i,n)$$

The first term is the present value of all the interest payments; the second, the present value of the face value received at maturity. Using the equations for the factors from Table 6-3, interest of 0.04 for the 20 semiannual payments, and a hand calculator,

---

*The terms *corporation, firm,* and *business* are used synonymously.

**U.S. government securities also include Treasury bills, with maturities of up to 1 yr, and Treasury notes, with maturities between 1 and 7 yr. For U.S. bonds, maturities are 7 to 30 yr.

***Zero coupon bonds are a type in which interest is not paid until maturity. Although credited to its owner at specific periods, such as every 6 months, the interest compounds and is paid with the original purchase price at the maturity date. The U.S. government's Series EE Savings bonds (p. 239) are of this type.

$$P/A(0.04,20) = \left[(1.04)^{20} - 1\right]/0.04 \times (1.04)^{20} = 13.59$$

$$P/F(0.04,20) = (1.04)^{-20} = 0.45639$$

Thus, substituting above, the present value of this bond is PV = $40 × 13.59 + $1000 × .45639 = $543.61 + $456.39 = $1000. The equality of the bond's present valuation and its face value is not coincidental, but fundamental. It is a result that might be called the bond identity. It states that at the *time of issue*—for any term, face value, and interest rate—the valuation (present value) of a bond, as determined from the present values of all of its future cash incomes, exactly equals its face value. As for other financial securities, the bond valuation, *at any time,* is the sum of the present values of all cash flows, discounted at the market interest rate at that time. As presented below, possible changes in market interest rates effect a change in the bond's valuation.

Effects of Changes in Interest   After issue, changes in the economy will cause interest rates to change. In turn, the valuation or market value of a bond will adjust to equal the present values of all future incomes, evaluated using the new interest rate. Generally, the market's estimates of inflation are the dominant influence on the change; lenders typically require an interest rate equal to the inflation rate plus approximately 3 percent. Let us consider the bond in our example for a change in the annual interest rate from 8 percent to 10 percent, 5 percent per half year. The present value factors using 5 percent and 20 semiannual, $40 payments as per the face of the bond are

$$P/A(0.05,20) = [(1.05)^{20} - 1]/0.05(1.05)^{20} = 12.46$$

and

$$P/F(0.05,20) = (1.05)^{-20} = 0.3769$$

Using these, the new market value of the bond is

$$PV = 40 \times 12.46 + 1000 \times .3769 = 498.40 + 376.90 = \$875.30$$

In most situations some of the 40 semiannual dividends would have already been paid. Accordingly, the bond's present value would be further reduced. As has been stated, inflation decimates* the bond market.

Conversely, if interest rates go down, the present (and market) value of the bond goes up. Consider the bond in our example for a decrease in interest rates to 6 percent (3 percent semiannually).

The new present value factors are

$$P/A(0.03,20) = [(1.03)^{20} - 1]/0.03(1.03)^{20} = 14.88$$

---

*Interestingly, *decimate* comes from Latin. As punishment, the Romans would kill every tenth (*deci*) soldier in the conquered army.

and

$$P/F\ (0.03,20) = (1.03)^{-20} = 0.5537$$

With these factors, we get the bond's new value as

$$PV = 40 \times 14.88 + 1000 \times 0.5537 = \$595.20 + 553.70 = \$1148.90$$

Sale of any security held for less than 1 yr produces a short-term capital loss if sold for less than the purchase price, and a short-term capital gain if sold for a greater amount. If the security is held for 1 yr or longer, the transaction produces a long-term loss or gain. For our example, sale of the bond when the interest rate went to 10 percent would cause a capital loss of $124.70, short-term or long-term according to its holding time as just reviewed. The loss would be reduced by any dividends already received. As you recognize, sale of the bond when the interest rate was 6 percent would cause a capital gain of $148.90, long- or short-term according to the time it was held. Again, dividends provide more gain.

Call Provisions   When interest rates decrease, a firm would like to redeem (i.e., call in) its higher interest rate bonds and issue new ones at the lower rate. So when purchasing a bond, it is wise to check its call provisions. In the past, firms have been able to call bonds within a short period by paying only a 5 percent premium or less. Thus, they denied the investor the possible capital gains when interest rates dropped. More recently long-term bonds are not callable for 10 yr more. It is still wise to check the call provisions, including the premium paid (to the holder) for redemption.* Government (U.S.) bonds give call protection for 20 yr or more. Other risks are reviewed later in this section. A special risk-reward type of bond (called a *convertible bond*) is offered by less established firms. This type, which is reviewed next, can be redeemed at any time for a specified number of shares of common stock.

Convertible Bonds   Recall that a bond is a long-term debt of the corporation. On purchase of a quality bond, the investor is ensured of the interest payments, even if the firm does not do well. Conversely, the investor's return does not increase if the firm prospers. A convertible bond is an arrangement by which the investor can benefit if the firm grows. The convertible bond can be redeemed for a specific number of shares of common stock. As an example, a convertible bond might be redeemable for 20 shares of stock. On purchasing this 6 percent bond for $1000, redeeming it for stock would be equivalent to buying the stock at $50 per share. This is called the conversion price (of the stock obtained by conversion, not the price of the bond). The worth of the bond determined from the worth of the stock for which it can be redeemed is the bond's conversion value. We review these terms below to reinforce the concepts.

There are two measures of convertible bonds: the bond's conversion value and its investment value [7].

*These data regarding call provisions are given in the *Standard & Poor's Bond Guide,* published monthly.

Conversion Value   This value of a bond equals the present price of the corporation's common shares multiplied by the number of shares to which the bond can be converted. Conversion price is used for the per share price of the stock when the bond is redeemed for stock. Above, the conversion *price* is $50/share ($1000/20 shares). If, however, the shares' current market price is $45, then the bond's conversion *value* is 20 × $45 = $900. Because of the dependable semiannual payments of $30 ($\frac{1}{2}$ of 6 percent of its $1000 face value), the bond could be selling for the original issue price of $1000. At this price, the $100 over its conversion value is called its conversion premium of $\frac{100}{900}$ = 0.1111, or 11.11 percent. This is a typical premium. As the share's price approaches the conversion price, the bond's price will rise. In our example, if the share price rises from $48 to $49, the conversion value of the bond will rise $20. This is true because the bond can be converted to 20 shares any time the bond's holder wishes. The resulting increase in the bond's value is recognized as a capital gain, if the new price exceeds the bond's price of purchase. Note that for bonds of firms whose share price is much less than the conversion price (of the stocks if the bond is redeemed), the bond's investment value is of greater importance.

Investment Value   When the bond to stock conversion puts the per share price well above the stock's current share price, the bond is priced more nearly like a "straight" bond. In this case, its market price reflects primarily the bond's investment value, similar to regular bonds, purchased mainly for income. In these cases, convertible bonds often sell at a considerable discount from their face value. In turn, the fixed interest payments account for a larger percentage of the bonds' value, and the bonds' discounted price might offer high yields. The straight bonds, however, are senior* securities to them, but they are senior to preferred stock.

Other things being equal, convertibles would be advantageous over the straight bonds; however, the conversion privilege betrays an absence of genuine investment quality. The revealed risk should be carefully studied. Aspects to be considered are reviewed later in this section.

Bond Quality   Moody's Investor Service and Standard & Poor's Corp. rate the safety and investment quality of bonds. Their ratings down to Baa/BBB level are shown in the table.

Moody's	S&P
Aaa	AAA
Aa	AA (±)
A1*	
A	A (±)
Baa1*	
Baa	BBB (±)

*Municipal bonds only.

*Senior debts have claim for funds over junior debts.

Lower ratings are not considered of investment quality and are referred to as *junk bonds*. Ratings directly affect bond prices and yields. Top-quality bonds, considered safer, can be sold at a slightly lower interest rate or yield than bonds of the next lower rating. Ratings go down to D. No investment grade bond (Baa/BBB or above) has ever defaulted, but ratings do change. S&P gives a weekly rating for bonds; however, it is quite expensive. The more moderately priced monthly ratings are available in many city and university libraries. The main criteria for bond ratings are their book value and, more significant, a firm's "times interest earned." The methods of these criteria are presented below for two reasons. One is for general understanding; the second is that in the "rapidly moving" environment of mergers and takeovers, perhaps there could be a need to make your own analyses. Since preferred stocks are similar to long-term debt, these same criteria are proper for them as well.

Preferred Stock   Next up the risk ladder are preferred stocks. They are similar to bonds in that there is a face value for their price (usually $100) and a fixed value for the rate of return (such as 7 percent). An important difference is that the bond is a debt instrument and its holder has no ownership of (or participation in) the firm. Owners of stock, preferred and common (presented next), own a share of the company and participate in the firm's management through voting at the general meetings (usually preferreds have voting rights only if dividends are in default). The shareholder can vote in person at the meetings or, as is most often, by proxy, (that is, by their agent; see Chapter 8).

For preferred stock with a 7 percent rate of return, a dividend of this rate times the face value of the stock is received each year, usually one-fourth every 3 months. For the rate noted, $0.07 \times \$100$ gives $7.00 per share, or $1.75 per quarter. Generally, the rate of return is lower than for bonds but higher than the dividends declared by income-type common stocks. Recall that the bond interest has to be paid unless the firm goes into bankruptcy. Dividends (declared for stockholders) are paid only when the company does well; preferred stockholders, however, have claim on the company's earnings before any dividends are paid to the holders of common stock. Furthermore, if dividends are skipped, for cumulative preferreds, dividends accumulate and are paid before any payments to holders of common stock. Also, if the company liquidates, although the claim of preferreds is junior to that of bondholders, they are senior to any claims by owners of common stocks. Finally, there is a participating preferred which entitles its owner to the stated dividend plus additional dividends on a specified basis when the company does especially well. The nonparticipating preferreds are considered perpetuities; that is, they provide forever a constant stream of the stated dividends, as above, $7.00 per year.

Convertible Preferred   The term *convertibles* is often used to refer to both convertible bonds and convertible preferred stocks. Both can be redeemed or converted into a fixed number of shares of the company's common stocks. The number is specified on the certificates. As with bonds, conversion and investment values are important in evaluating preferred stocks. The concept is the same as for bonds, but let us consider an example for preferreds. For a conversion number of 4 and a market value of the common stock of $20, obviously the conversion value is $80. If perhaps the current market price of the preferred stock is $100, then the $20 difference is the conversion premium, or 25 percent (of the conversion price). That would probably be too great unless independently its dividend justified an *investment value* near its market price. Again, as reviewed for bonds, the conversion privilege betrays a probable underlying weakness in the issue's investment quality. The conversion provision is primarily advantageous to the firms. The debt capital is acquired at a lower rate than by straight (nonconvertible) securities, and if converted, senior obligations are exchanged into equity. The resulting dilution of the common stock is obviously at the expense of other common stockholders, including the ones who do the conversion. Except possibly for corporations, it seems that preferred stocks are not the best investments. For corporations, 80 percent of the dividends from preferred and common stock is tax-free.

Call Provisions   As with bonds, when interest rates decline, firms like to redeem (call in) the outstanding preferreds and sell new ones at a lower dividend rate. The call provisions are stated on the face of the stock certificates. These data (premium and dates) are also given in the monthly *Standard & Poor's Stock Guide*.

Generally for convertibles (both bonds and preferreds), if the underlying stocks are not an intelligent purchase, neither are the convertibles. Perhaps it is wise to remember that financially, nothing is free.

Safety and Total Returns: Bonds and Preferreds

As quoted above, Benjamin Graham said that an investment should give safety of principal and adequate return. The yield or normal return of bonds and preferred stocks is fixed (thus, they are known as *fixed investments*). These rates, stated on the securities, are known and considered before purchase. The total return, however, includes capital gains (or losses) due to change in the interest rates (please see the preceding discussion on the effects of interest rates). The total return is more important to the investor. The safety of these securities, nevertheless, depends on the financial status of the issuing firm. The data for its determination are available from the firm's annual financial report. The factors, book values, and times interest (dividends) earned indicate the safety margins for bonds (preferred stocks). They will be defined and evaluated for the AB Corp. The required data for 1991 are given in Table 7-3.

Book Value of Securities

These values are developed for bonds, preferred stocks, and common stocks. Although the net assets per bonds and the net assets per share of preferred stocks are their book values, the term *book value* or just *book* is used mostly in reference to common stock. Book value for common stocks is reviewed later in this section. It is only one of various components in estimating their safety.

Net Assets for Bonds (NAB)   Bonds are the most senior securities, thus only current liabilities have a higher claim. For net assets, intangible values are subtracted from the total assets. On liquidation, the value of intangibles is questionable. Define NAB as the net assets for bonds. It equals the total capitalization and is the total assets

TABLE 7-3
Financial Statement for AB Corp.
December 31, 1991
Assets

Current assets:
Cash . . . . $ 20,000
Inventory   80,000
$100,000

Fixed assets:
Equipment   130,000
Buildings   170,000
$400,000

Liabilities

Current liabilities:
Accounts payable . . . . $ 15,000
Notes payable   35,000
$ 50,000

Long-term liabilities:
SBA note   30,000
Bonds (70 × $1000 ea.)   70,000

Equity

Capital:
Preferred stock, 400 shs $100 par value   40,000
Common stock, 10,000 shs $10 par value   100,000
Capital surplus   100,000
Accumulated retained earnings   10,000
$400,000

minus the sum of intangible values, deferred taxes, and current liabilities. Using the data from Table 7-3 for the AB Corp. (which has no intangibles or deferred taxes),

$$\text{NAB} = \text{total assets} - \text{intangibles} - \text{current liabilities}$$
$$= \$400,000 - \$0 - \$50,000 = \$350,000$$

To determine the net asset value per bond, you need only divide NAB by the number of bonds plus equivalents. For AB Corp., the SBA note is equivalent to 30 ($1000 per equivalent bond); adding the 30 to 70 bonds gives a total of 100. Now the net assets per bond (NAPB) is

$$\text{NAB/bond} = \frac{\text{NAB}}{\text{no. bonds}} = \frac{\$350,000}{100} = \$3500 \text{ per } \$1000 \text{ bond}$$

Net Assets for Preferred Stocks (NAP)   This amount equals the assets for bonds reduced by the value of the bonds and equivalents. Using the long-term debt data from Table 7-3,

$$\text{NAP} = \text{NAB} - \text{value of bonds} = \$350,000 - \$100,000$$
$$= \$250,000$$

$$\text{NAP/share} = \frac{\$250,000}{400} = \$625 \text{ per share of preferred}$$

These strong asset values for bonds and preferred stock ensure safety in purchasing the securities. Their strengths are due to AB Corp.'s high equity ratio, reviewed below. Safety and anticipated performance of common stocks are analyzed next; per share net asset value (or book value) is a component used.

Book Value, or Net Assets for Common Stock (NAC) Owners of common stock have the last claim for assets in case of bankruptcy; however, only they, and owners of common stock equivalents, can benefit directly from growth and improved earnings of the company. Continuing the progression,

$$\text{NAC} = \text{NAP} - 100 \times \text{the no. of shares of preferred stock}$$
$$= \$250,000 - (100 \times 400) = \$210,000$$

$$\text{NAC/share} = \frac{\$210,000}{10,000} = \$21$$

This net asset value per share is also known as the stock's *book value*. It is the amount per share that stockholders might get on liquidation of the firm, after all senior claims were met. The selling price (i.e., the market value) of common shares is usually greater than their book value.

Times Interest Earned More closely related to the quality of bonds and preferred stock is the firm's earnings available for payment of their interest and dividends for preferreds. Both are expressed as a ratio of the proper earnings, respectively, to the bond interest and the preferred's dividend. We will determine these for the AB Corp. The needed data are taken from its 1991 income statement, Table 7-4.

Bonds

The income available for interest payments is the total income. The "times interest earned" for bonds is that amount divided by the total bond and bond equivalent payments. For the AB Corp.,

$$\text{Times interest earned — bonds} = \frac{\text{total income}}{\text{long – term interest}}$$
$$= \frac{\$28,000}{\$7000} = 4$$

For an investment grade bond, this ratio should be 3 or 4.

TABLE 7-4*
Consolidated Income Statement for AB Corp.

1991 Net sales		$73,000
Cost of sales and operating expenses		
Cost of goods sold	$25,000	
Depreciation	20,000	
Selling and administrative expenses	3,000	
Total operating costs		48,000
Operating profits: sales − total costs		$25,000
Other income		
Dividends and interest		3,000
Total income		$28,000
Less interest on loans		7,000
Income before federal and state income taxes		$21,000
Provision for federal taxes**		4,200
Net profit for the year		$16,800
Accumulated Retained Earnings Statement		
Balance January 1		$70,000
Net profit for year		16,800
Total		$86,800
Less dividends:		
Preferred stock (8% on 400 shs.)		3,200
Common stock		8,600
Balance December 31		$75,000

*As of November 1987, the Financial Accounting Standards, No. 95, (issued by the FASB) requires companies, with their balance and income statement, to include a statement of cash flows. It is a summary of the income statement and accumulated retained earnings statement.

**A federal tax of 20 percent is arbitrarily chosen as representative of the individual and corporate rates for this magnitude of income.

Preferred Stocks

Since these are equity securities, income taxes must be paid before dividends. Thus, the income available for preferred dividends is the net or after-tax profits. The "times dividend earned" for preferred is that amount divided by the total of preferred dividends (see the Accumulated Retained Earnings Statement in Table 7-4). For the AB Corp.,

$$\text{Times dividends earned — preferreds} = \frac{\text{net income}}{\text{preferred dividends}}$$

$$= \frac{\$16,800}{\$3200} = 5.25$$

This value is influenced by the preferred to common stock ratio; here it is 1 to 25. The more frequent ratio of 1 to 100 would give a ratio in the hundreds. The greater magnitude would be preferable, but 5 or greater is investment quality.

As for safety criteria for bonds and preferred stocks, for the ratios developed above, analysts generally require a minimum of similar ratios: two ratios for bonds and two for preferreds.

For bonds, (1) the NAB/bond should be two to three times its face (redemption) value; (2) the "times interest earned" should be 3 to 4. Above, we see that AB Corp. meets both criteria: 3.5 for (1) and 4.0 for (2).

For preferred stocks, (1) the NAP/share (NAP divided by the stock's face value, usually $100) should be 5 or greater, and (2) the "times dividends earned" should be 4 to 5. Also, as developed above for AB Corp., both criteria for preferreds are met: 6.25 for (1) and 5.25 for (2).

Margin of Safety

Price-Earnings Ratio   The earnings available for common stock is the net profit less the dividends paid to preferred stockholders. This amount divided by the total number of shares (including the number of stock equivalents, discussed later) gives the earnings per share. Remember that our small company has only 10,000 shares; thus, using data from Table 7-4, the earnings per share is calculated below.

$$\text{Earnings per share} = (\$16{,}800 - \$3200) / 10{,}000$$
$$= \$13{,}600 / 10{,}000 = \$1.36$$

To determine the P/E ratio, the earnings per share is divided into a stock's current market price. Thus, if the AB Corp.'s stock was presently selling for $14, its P/E would be 10. More details on this ratio are given in "Important Terms" presented earlier.

Common Stocks   These, like preferred stock, represent ownership in the corporation. Each share represents a definite portion of the company's net assets; the value of that single portion is called the share's book value. It is determined above in our discussion of net assets for common stock (NAC). Typically the board of directors of a corporation will distribute 40 to 60 percent of a firm's earnings as dividends to stockholders. If the board thinks it unwise to distribute any dividends, it need not; however, bond interests *must* be paid. When distributions are made, preferred stockholders receive their cumulative, specified dividends first. Because of this accumulation, management often considers preferreds like debt and makes a special effort to pay the dividends at the specified times. Finally, common stockholders receive dividends. Common stock is different from all the other securities in that when a firm's earnings improve (from competent management or general economic improvements), common stockholders fully share the good fortune. Dividends might be increased, or greater retained earnings could increase the book value of the common shares. Both will probably increase the market value.

As implied, common stocks are at the top of the risk ladder for conventional investments.

A thought by A. P. Gouthey seems applicable: [17]

To get profit without risk,
experience without danger,
reward without work,
is as impossible as to live,
without being born.

Though they always exist, risk, danger, and work are minimized through an "engineering approach." The essence of this approach is positive actions after thorough analyses based on fundamentals and their understanding. When applied to investments, we will realize the "fundamental approach" is but a variation of the engineering approach. The task or "problem" is to get the greatest return to risk ratio, always keeping the risk well within the expendable funds of the investor.

Suggestions on Risk   A friend of mine enjoyed a weekly poker game. The stakes were adequate to justify care in decisions to "hold or fold," but well below the amount that could cause losses greater than what one might otherwise spend on an evening's entertainment. My friend, however, enjoyed "playing well the cards he held." He knew the odds on the various hands in the different types of games such as stud, draw, and the "wild card" games of baseball and seven-up. Early in the game, he played (held) only those hands which had a high probability of winning. After accumulating significant winnings, he took greater risks. When he lost $10 (in the 1940s, that was equivalent to about $100 in the 1990s), he quit for the night, for the month if with that night's losses the total losses exceeded his allotment of $25 per month. Occasionally his luck was hot and he did well; but often in taking the greater risks, while fun, they reduced his winnings to near zero or a small loss.

I suggest my friend's approach to investing in common stock. It is summarized as follows: (1) Risk only a small portion of your discretionary funds. (2) Keep your losses small, as by using "stop losses," reviewed below. (3) Let your winnings run.

The fundamentals can indicate with high probability a company's safety, risk, and future performance. This analysis is best when reinforced by judgment. An experienced broker observed that even a defensive investor needs the experience of two to three market cycles. Furthermore, as General William Westmoreland said, "Judgment comes from experience, and experience comes from (poor) judgment." Next we consider safety (preservation of principal) and then the risk (price fluctuations in expected performance).

Safety and Risk Analyses: Common Stocks   These terms are frequently used interchangeably; however, here we will follow the use in our financial references and develop safety as related to a firm's survival, especially in economic downturns. Risk is used as related to possible substantial fluctuations of the security's market price (here we are referring to common stock). These aspects are of prime consideration in a fundamental approach.

Step Two: Phase Two—Analytical Evaluation: Indices for Safety, Performance, and Growth

Indices of safety are presented first; then those related to the firm's volatility are reviewed. How does the fluctuation (in price) compare with the average market? This comparison is indicated by the firm's beta, discussed earlier in this chapter.

Safety   This is mostly dependent on the firm's debts, both short- and long-term [19]. Three indices, developed below, are helpful in assessing safety. They are (1) a firm's liquidity ratios; (2) its debt ratio, a capitalization ratio; and (3) the "paydown time" of the long-term debt from depreciation, the prepaid* cash flow. The terms of this third index are reviewed with its development. The indices require data from the balance sheet of the firm's financial report. These indices are developed next and evaluated for the AB Corp. The data are taken from Tables 7-3 and 7-4.

Liquidity Ratios   These are the first indication of probable financial troubles, such as impending bankruptcy. They are the current ratio and the quick-asset or acid-test ratio.

Current Ratio

$$\text{Current ratio} = \frac{\text{current assets}}{\text{current liabilities}}$$

For AB Corp.,

$$\text{Current ratio} = \frac{\$100,000}{\$50,000} = 2$$

This is usually considered a minimum value, especially for comparing firms in the same industry.

Working capital is closely related to the current ratio. It is obtained by subtracting the current liabilities from the current assets. Thus, it is the current assets available for operation of the business: paying for materials, salaries, etc. Adequate working capital is the first requirement for preserving good trade and bank credit. Of equal importance, however, is that adequate working capital can provide funds for emergencies and new growth projects. Conversely, working capital produces only small return (such as in money market funds) or no return at all. Furthermore, an excess could encourage waste; perhaps, it would be better to distribute it to shareholders.

For the AB Corp.,

$$\text{Working capital} = \text{current assets} - \text{current liabilities}$$
$$= \$100,000 - \$50,000 = \$50,000$$

---

*Remember that cash flow equals after-tax profits plus depreciation. I offer that depreciation cost recovered from before-tax profits is the prepaid part of cash flow.

"Net" working capital is often used as working capital less the long-term debt. This factor is positive only for firms especially strong financially. It is negative for the AB Corp.

For the AB Corp.,

$$\text{Net working capital} = \text{working capital} - \text{long-term liabilities}$$
$$= \$50,000 - \$100,000 = -\$50,000$$

Quick-Asset Ratio

The quick-asset ratio (also called the acid-test ratio) is similar to the current ratio. As a test for quicker liquidity, the current assets are reduced by the dollar value of the inventory.

For the AB Corp.,

$$\text{Quick-asset ratio} = \frac{\text{current assets} - \text{inventory}}{\text{current liabilities}}$$
$$= \frac{\$100,000 - \$80,000}{\$50,000} = 0.4$$

This is considerably below 1.00, the desired value of the ratio, when the inventory does not have a rapid turnover. The inventory-turnover ratio divided into 365 gives an estimation of the importance of the quick-asset test. This ratio requires data from the firm's income statement.

For AB Corp.

$$\text{Inventory-turnover ratio} = \frac{\text{annual sales}}{\text{ave. inventory}}$$
$$= \frac{\$73,000}{\$80,000} = 0.9125$$

This value shows that more than a year (1/0.91 = 1.113 mo) is the average time to normally liquidate inventory. Accordingly, for AB Corp., a current ratio of 2 is minimum. Long-term financial health is reflected by the capitalization ratios.

Capitalization Ratios [20]   These ratios show the proportion of each kind of security issued by a firm. In turn, they indicate the relative position for owners of each type of capital. We use the preceding financial report to determine AB Corp.'s bond ratio, preferred stock ratio, and common stock ratio. The ratio for each type is, respectively, the total dollar amount of that type issue divided by the firm's total capitalization. The total capitalization equals the company's total assets minus the sum of its current liabilities, intangible assets (such as patents and copyrights), and (if any) income taxes due, as could have been deferred by accelerated depreciation. From Table 7-3, the total assets are $400,000; the current liabilities are $50,000. Since in this simple example there are no intangibles or deferred taxes, the total capitalization equals $400,000 - $50,000, or $350,000.

Bond Ratio The "bond ratio" should consider all long-term debts; in the case of AB Corp,. other types of long-term liabilities are outstanding. Thus, from Table 7-3, we see that the total long-term debt is $30,000 + $70,000 = $100,000.

$$\text{The bond ratio} = \frac{\text{long-term liabilities}}{\text{total capitalization}} = \frac{\$100,000}{\$350,000} = 28.6\%$$

As discussed later, this is a proper value; less than 50 percent is desired. Often this ratio is called the (long-term) debt ratio.

Preferred Stock Ratio For preferred stock, the table gives $40,000; thus,

$$\text{The preferred stock ratio} = \frac{\text{preferred stock capital}}{\text{total capitalization}}$$

$$= \frac{\$40,000}{\$350,000} = 11.4\%$$

The sum of these senior securities is 40 percent.

Common Stock Ratio Since the total capitalization is 100 percent, the common stock ratio is about $(100- 40) = 60$ percent. Or, similar to the others, the ratio can be evaluated as follows:

$$\text{Common stock ratio} = \frac{\text{common stock capital *}}{\text{total capitalization}} = \frac{\$210,000}{\$350,000} = 60\%$$

Equity Ratio This ratio is the complement of the debt (also called the *bond*) ratio. It is the ratio of the sum of the dollar value of the preferred stock plus that of the common stock divided by the total capitalization. As above for the common ratio, the equity ratio can be determined by subtracting the debt ratio from 100 percent. Thus, the equity ratio is $(100 - 28.6) = 71.4$ percent, a favorable situation for stockholders. As noted above, 50 percent is considered minimum. This ratio can also be determined by the expression

$$\text{Equity ratio} = \frac{\text{dollar value of preferred plus common stock}}{\text{total capitalization **}}$$

$$= \frac{\$40,000 + \$210,000}{\$350,000} = \frac{\$250,000}{\$350,000} = 71.4\%$$

Paydown Time [19] This is the period of time required to pay the entire long-term debt, using only the "refund" part of the firm's cash flow. Although used loosely

---

*Common stock capital is the total stocks' par value plus the capital surplus (money paid for stock above par) plus the accumulated retained earnings.

**Often, for both the debt and equity ratios, total assets instead of total capitalization is used. Since the two methods differ by only a few percent, either is suitable. This is especially true for the differences cancel when the ratios are used to compare the relative status of firms.

in referring to any movement of money, cash flow is properly defined as the sum of a firm's net profit (earnings after interest and taxes), plus depreciation and amortization, and any increase in funds for deferred taxes. The "refund" portion is from depreciation plus amortization. Depreciation consists of expenses attributed to the current year's operations that were paid when the machines and other equipment were purchased. Recall that funds from amortization would be those payments on loans held by the firm; the payments include interest and also partial return of the principal, loaned by the firm.

$$\text{Paydown time} = \frac{\text{long-term debt}}{\text{prepaid cash flow}}$$

For AB Corp. whose nonprofit cash flow is only from depreciation,

$$\text{Paydown time} = \frac{\$100,000}{\$20,000} = 5 \text{ yr}$$

This is less than the 7 yr Fisher offers as the maximum [19].

The above ratios can be use to indicate the probable safety of investing in a company. Furthermore, they relieve us of having to consider a firm's liquidation worth in evaluating stocks considered for purchase. As mentioned earlier, the true value of any investment is its present valuation, that is, the discounted sum of all future earning. The interest rate used for discounting is called the *capitalization rate* and is discussed later when considering a stock's P/E ratio. Now, we consider factors used to anticipate the probable financial performance (earnings and growth) of a firm.

Performance Factors [21]   These are used to evaluate the level and stability of a firm's current performance. (Projected future improvements are reviewed later.) Each performance factor is reviewed and its significance discussed. The factors are classified into two groups: qualitative and quantitative. Of the latter, some are defined with their earlier presentations; others are defined in Important Terms in this section. A few are developed below. They are discussed in order of importance to the defensive investor. Sources of needed information are presented later.

Qualitative   Although these factors may have numerical indication, they involve subjective evaluations by the general investors. These qualities concern (1) management; (2) listing on an established stock exchange; (3) size and public acceptance; (4) earnings' stability; (5) dividend record; (6) the generally perceived prospects of the firm's industry, the firm's competition, and its share of the market; (7) whether it is a research leader; and (8) realistic accounting. After safety, these qualities are used to narrow potential stocks for purchase and to weigh the quantitative analyses presented later. Regretfully, there are no regular sources for some of this information. However, these qualities can be noted and recorded for later reference, when you read the financial literature. In addition to *Barron's, Business Week,* and the *Wall Street Journal,* other sources are listed in Appendix A-1. We next enlarge on these qualities.

1. *Management.* While this is surely one of the most important aspects in the success of an enterprise, as noted above, the information is not available in a single source. Naturally, the firm's annual report lists the corporate officers with minimum information on their backgrounds; however, the report at least gives their names for further investigation. It also gives other business connections; if the officers are or have been with other public companies, a quick check on their progress, sales, and profits will give further insight into their qualities. Their having had experience with a substantial company is significant; also, past successes and contacts are positive indications. The report also gives the firm's activities and management's stock ownership. It is advantageous for management's welfare to be strongly correlated to that of the stockholders. Again articles in the financial literature such as *Barrons, Business Week, Forbes,* and the *Wall Street Journal* are excellent sources about the firm's leaders. Finally, and probably most important, "by their fruits you will know them." These results are observable by the quantitative factors reviewed below. Of course, the validity of these indicators is greatly dependent on the duration a particular management has been active at a firm. Management change signals caution; study is required to determine if the switch will probably be favorable for the net performances.

2. *Listing.* Stocks listed on the New York Stock Exchange are generally of the largest and most financially secure corporations. This is because of the Exchange's registration requirements on total capitalization, annual sales, and operating profits. Stocks on the American Stock Exchange are of medium-sized firms and are considered more speculative.

3. *Size.* This measure screens firms not well known or well established. Normally, prices of the high-cap (capitalized) mature firms are less affected by the normal ups and downs of the industry or the nation's economy. Graham recommends industrial firms with annual sales of not less than $50 million and total assets not less than $50 million (for public utilities, we use this latter minimum as a general guide). Another measure offered by David Dreman [22] is to consider only those companies with over $1 billion in market value. This is often called *market capitalization* and equals the number of outstanding shares multiplied by the price of a share.

4. *Earnings' stability.* There should be some earnings for common stock in each of the past 10 yr.

5. *Dividend record.* A record of some continuous dividend over many years is one of the most respected indications of high quality. Graham [15] feels that a dividend payment for 20 yr or more is an important plus. "Indeed the defensive investor might be justified in limiting her purchases to those meeting this test."

6. *Prominence.* These firms have a strong advantage. They are in an industry in favor with Wall Street. *Prominence* implies that the firm is well established, is not having intense competition, and has a major share of that industry's product or service market. The nearer a firm's product line is to an indistinguishable commodity, the less control it has over its fate. For example, it may need to raise prices if uncontrollable costs rise. Also, if a strong competitor were to reduce the price on the firm's product or services, this would have a significant impact. In the case of an industrial parts distributor, for instance, the firm might offset a competitor's lower prices by offering quick, dependable delivery. Thus, the added utility can protect a firm from price re-

ductions, and the premium is reflected in the firm's profits. Finally, does the firm have a strong brand name, and is there a growing or established demand for the products of that industry?

7. *Research.* A good, practical research program indicates a forward planning management. Equally important, a research program opens opportunities for new products, more efficient processes, and new marketing methods. All of these can materially affect the firm's future.

8. *Realistic accounting.* An *Accounting Principles Board Opinion 30,* issued in 1973, encourages companies to report nonrecurring income and losses on their income statements as "extraordinary items." Extraordinary items are defined as events that were unrelated to the company's business and unlikely to recur. Other special events are to be reported as "unusual items." Nonetheless, Chakravarty reports that (as is often the case) a firm which in 1986 had a one-time $318 million sale reported it as operating earnings [23]. With diligence, these occurrences can be picked up in explanatory footnotes. They indicate misleading management practices. If the information about the firm needs to be closely scrutinized, that probably indicates the stock should be avoided (it requires unnecessary efforts and risk).

In determining the reported book value of a stock, the firm's total net assets are divided by the number of outstanding stocks. Responsible accounting considers the total number of outstanding common shares, and the possible shares that could be converted from preferred stocks and bonds.

Finally, conservative accounting uses LIFO (last-in, first-out) inventory evaluations. FIFO accounting (first-in, first-out) exaggerates profit. The most recently sold items, because of inflation, cost less to make (in the past) than their replacements. LIFO lowers earnings, but it cuts taxes and provides a more accurate evaluation of the company's profits. Furthermore, LIFO understates inventory value; it gives the company added financial strength.

Quantitative   To a large extent, these 13 factors measure the results of the firm's qualitative characteristics (reviewed above). They are defined and later evaluated for the GE Company.

1. *The price to earnings ratio (P/E).* This is probably the most popular ratio. For all New York Stock Exchange (NYSE) listings, it is given in the financial section of most daily newspapers. The value is usually calculated using the firm's current share price and its earnings, as given by the firm for the last 12 months. The P/E for the S&P 500 is an excellent reference for comparison. It is reported in *Barron's* (its location in that weekly is noted below, in our discussion of the ROE). Use of the P/E and other quantitative factors is reviewed later in this section.

2. *The price to dividend ratio (P/D).* As you realize, this is the ratio of the current price divided by the current annual dividend. Both are reported in most daily newspapers. In the discussion on strategies below, we note the usual* greater importance of the P/D over the P/E.

3. *The price to sales ratio (P/S or PSR).* This ratio could be more important than

---

*Some firms pay dividends which earnings cannot support. In 1991 Commonwealth Edison paid a $3.00 dividend; the earnings were $0.08.

the P/E, for it permits you to check if sales improve when the earnings increase. Non-reoccurring activities, often not clearly reported, can give a misleading value of earnings for the year of the extraordinary activity (see the discussion on realistic accounting above). You also note that the P/S indicates how much investors are willing to pay for a dollar's worth of sales. A typical value as 0.20 shows that the market is paying $0.20 for $1.00 in sales. In faster growing companies, a reasonable PSR might be 0.75 or less. Normal values are 1.5; 3.0 or more indicates overvalue.

4. *Price to book value (P/BV)*. Remember that the book value of a stock is the issuing firm's net asset value (total assets minus all liabilities) divided by the total number of common shares outstanding. In realistic accounting this total number of shares is adjusted to consider possible conversions from both preferred stocks and bonds. (For example, GE's book value in 1989 was $23.20; its price, $67. Dividing, the price to "book" ratio was 2.9.) Hardly any stocks sell at book value. The difference between the stock's price and its book value is the premium an investor has to pay for the intangibles. These include the value of an established firm, its brand names, patents, and general public perception.

5. *Cash flow per share*. This is the net profit plus the noncash (or prepaid) charges such as depreciation and amortization, less preferred dividends (if any), all divided by the number of outstanding common shares. Please see *fully diluted earnings* in the preceding Important Terms regarding the proper number for "outstanding shares." Related to working capital, this factor indicates a company's ability to pay running costs and finance new projects.

6. *Free cash flow per share*. This is the normal cash flow, defined above, minus all dividends paid and the per share running cost, that is, the regular plant outlays for capital goods. It is a stronger indicator of a firm's ability to develop new processes, projects, and other growth opportunities.

7. *The current ratio*. This ratio, defined earlier in the discussion on safety, is also important in performance. That is because it indicates the working capital available for taking advantage of new, profitable projects. The minimum recommended value of 2 indicates working capital equal to or greater than the amount of debt.

8. *Debt/equity ratio (D/E)*. This ratio is a measure of the leverage a firm's managers are using. Both safety and risk are highly sensitive to it. As noted earlier, safety is an important indication of a firm's ability to weather poor economic periods, without danger of being near bankruptcy. Risk corresponds to significant fluctuations in the stock's market price. For a healthy firm, the D/E should be less than 1; this approximately corresponds to an equity ratio greater than 50 percent. Approximating total assets $(TA) = D + E$, debt plus equity, $D/E = D/TA \div (1 - D/TA)$.

9. *Return on equity (ROE)*. This is indicative of a firm's profitability. For industrial firms, it is useful to compare their ROE's with that value for the S&P 500. The value can be determined using data from the Indexes' P/E & Yields column, usually the fifth page from the end (counting the back page) in *Barron's* under the heading "Market Laboratory." On July 27, 1992, this column gave earnings for the S&P 500 of $16.19 and a book value of $156.42. Dividing the former by the latter gives ROE = 10.4 percent; it was near 17 percent in 1990.

10. *Return on investment (ROI)*. This is a measure of management's use of total

capitalization. It should be considerably greater than (perhaps twice) the interest rate the firm is paying on its bonds and other long-term debt.

11. *Net operating margin.* As defined in the preceding Important Terms, this factor shows the fraction of total revenue converted into operating profits, after interest and taxes have been deducted. The before-tax fraction is simply the operating margin. Recall that the operating margin multiplied by the inventory turnover is the first term in the equation for ROI. A low margin suggests that operating costs might be too high.

12. *Beta.* This is another means of comparing a stock with the general market. Beta indicates the volatility in a company's market price, relative to the market. Remember that a value of 1 means that its market price, statistically, varies with the market (S&P 500). A value greater than 1 indicates greater price fluctuations; less than 1, less fluctuations. Diversification among the industrial stocks can be used to reduce the beta of the portfolio, even less than that of the market. This can be done by averaging in stocks with betas less than 1. The technique is presented below in the discussion on strategy.

13. *Working capital.* As noted earlier, the working capital is the current assets minus the current liabilities. It is also known as *net current assets.* This capital is necessary for sales growth; however, closer control (possibly through the use of computers) has reduced the amount previously required for safety. The importance of its growth with sales is reviewed below in the discussion on strategies.

Appendix A-1 lists sources of information for the qualitative factors and data for evaluation of the quantitative factors. The items are numbered for convenient identification. In the appendix, Table 7-5 lists these performance factors and gives the numbers corresponding to the sources (listed in Appendix A-1) which provide the necessary information. For step three, in the discussion on investing strategies, we review the techniques for using the factors as guides for investing.

Growth Although growth is not a significant goal for the defensive investor, preserving purchasing power is. This preservation is best achieved in selecting investments with prospects of moderate growth, as indicated by the historical trends in sales, earnings, and working capital.

The material just presented relates to the fundamentals of the system of investment for stocks and bonds. We will now review concepts regarding their selection and analysis.

Step Three: Selection and Evaluation of Investment Candidates

Selections Guided by our objectives, we select, from various information sources, possible investment candidates for evaluation. As engineers, we have a special source of information in our professional activities. Personally, in electrical engineering, I was exposed to Hewlett Packard in the early 1950s, then Texas Instruments and IBM in the 1960s, Intel in the 1970s, and Microsoft in the 1980s and 1990s. Also in the 1990s, manufacturers of modems (*modulator/dem*odulators) and ETHERNET cards should have a strong market. All home computers can use modems to communicate, via telephone lines, with information sources and stores for telemarketing. Regretfully,

Hayes, dominant in the modem area, is at present privately held. Prometheus, Everex, Racal-Vadic, and Rixon are a few small firms which also make modems. Large firms (IBM, DEC) and medium-sized Intel also market them. Coaxial cable connections and later fiber optics have capability of much greater transmission speeds (in bauds, or bits per second). The telephone systems will provide the fiber optics capabilities. The higher rates, via coaxial cables, are presently used for communications between computers. Only ETHERNET cards and proper software are required for this communication. They are marketed by firms such as IBM, DEC, and Western Digital. This communication is also used for electronic mail (E-mail).

In daily life we are exposed to guides for investing. It has been apparent for many years how progressive McDonald's has been; and from what I read in the financial literature, including that section in the newspapers, the company is continuing to do well. In this decade and the previous one, we have encountered other candidates from our engineering activities. Examples are IBM, Intel, and GE. Columnists in *Forbes, Business Week, Fortune,* etc., with partial analyses, have many recommendations. It is wise, however, to completely analyze them ourselves. Below (in step four) we analyze General Electric. Part of our data will be taken from a recent issue of *Value Line*. It ranks stocks for safety and timeliness (similar to the qualitative factors presented above); those stocks with ranks 1 or 2 in both safety and timeliness are possible candidates.

Evaluation   This analysis can be done using the indices for safety, performance, and growth. Although considering all of them might seem extensive, each will improve your understanding and knowledge of investing. Furthermore, the probability of a correct decision will be greatly improved by your greater understanding and knowledge regarding that particular stock. Recall from the decision theory in Chapter 2 that the expected or average value (reward) of your actions is the sum of the values of each possible outcome multiplied by the probability of that outcome. Consider a possible $2000 lost in the market. If an hour spent analyzing the investment improves by only 1 percent the probability of your making a correct decision, your reward (or avoided loss) would pay you at the rate of $20 per hour (tax-free and at your convenience!). Finally, the greater the investment is, the greater the reward will be. Also, the experience from each analysis will make future ones easier. In the next section, we note how knowledge gained in analyses improves our investment strategy. After this discussion, for an example, we will evaluate General Electric.

7-7 INVESTMENT STRATEGIES: PASSIVE AND ACTIVE

Like most activities in life, stock investing is an endeavor greatly influenced by probability. As reviewed in Chapter 2, the probability of favorable results from a decision made in complete ignorance is at most 50 percent, as in the flip of a coin. Of course, in a stock transaction there are two disagreeing parties (not considering emergency forced sales). The one with greater knowledge of the fundamentals about the economic environment and about the firm whose stocks are being traded has the greater probability of realizing favorable results.

Diversification

Earlier we showed how the randomness in the aggregate value of a portfolio is reduced as a function of the number of uncorrelated securities. Fortunately, diversification permits us to significantly reduce the results of our incomplete knowledge, and errors in general. Diversification is *the* first rule of investing. Nevertheless, the greater our ignorance is, the less diversification's effectiveness will be. We badly fool ourselves if we think that diversification can compensate for our lack of studying the investment's fundamentals. Let us review this statement in light of the "efficient market theory." Please accept repetitions of earlier material as re-emphasis.

Efficient Capital Market [3, 23]

As defined earlier, this is a market of capital and securities in which there is a free flow of all pertinent information. Furthermore, there are homogeneous expectations of risk and return of all investors. In turn, stocks and other securities are efficiently priced to reflect all available market information. The efficient price results because millions of investors, including at least 50,000 professionals, have access to the same information and quickly adjust prices in accordance with the relevant facts. Also, the market well anticipates changes in earnings, dividends, and similar factors. Random events and surprises (such as in earnings, etc.) are quickly considered (see the investment model in Appendix A-5).

A Passive Strategy An efficient market suggests a passive strategy. For it, one expects to do as well as the market and only needs to invest in a well-diversified portfolio. The risk is reduced as the number of uncorrelated investments is increased to about 10 or 15 stocks [3, p. 144]. Naturally, the stocks can be purchased on a regular basis over many years. In an active strategy, the desired close following of the 10 or more stocks would demand significant time and effort. But it seems we make time for what we want to do. Also, it might seems vain to think we can do better than professionals who devote their lives to the market. Nonetheless, investing only those amounts we can afford to lose, funds earmarked for entertainment, is surely rational.

Active Strategies Investing requires three decisions: selection, timing, and diversification. Only the second one is special for an active strategy: timing purchases and sales. For stocks, the fundamental approach consists of three steps: analyses of the economy, the industry, and the specific firm. We first consider the economy and its cycle.

Portfolio Allocation and the Economic Cycle

A paramount aspect of a prudent life is "preparation ahead of need." In investing, the "preparation" is evaluation of various investments before the best time for action. The defensive investor's portfolio should be divided among four types: (1) cash and demand checking accounts, also cash equivalents (CD's, T-bills, and "government-only" money market funds); (2) regular money market funds; (3) bonds; and (4) stocks (equities).

The best time for each type depends on the phase of the economic cycle. In Appendix A-5, the "forced" investment model shows how the economy responds to the environment. And the environment is strongly influenced by the world's central banks. We have noted that our central bank, the Federal Reserve (also known as the *Fed*), is often the dominant bank. However, the large amount of foreign borrowing needed to meet our national debt requires that the Fed keep our interest rates appealing to foreign investors.

In Appendix A-5 we show how the Fed's actions in phases one and three produce the reactions of phases two and four. In phase one, lower interest rates are caused by the open-market action of the Fed. The reduced rates effect an increase in the money supply (monitored mainly by M1), and in 6 to 9 months the economy starts to improve.* Phase two is the reaction to this stimulus; "to keep up with demand," businesses overexpand. These expansions, with increased consumer borrowing, further increase demands, and prices rise to inflationary levels. Now the Fed begins to reduce the money supply using one or more "tools" reviewed in Appendix A-4. This action starts phase three; the response to the reduced money supply is increased interest rates. As in all systems, the reaction is delayed, here by usually 12 to 24 months. Consumers are slow to give up "free spending." Eventually, the tight money brings reduced demands. Firms bring production in line with demand; in turn, workweeks are reduced. Later, with job layoffs, a recession begins. These reactions are phase four. Note that our substantial savings, recommended at the beginning of this chapter, cushion our vulnerability to these variations. Furthermore, we can time our investments to the best advantage. As reviewed next in our discussion of investment timing, money market and short-term bond funds can be used to earn high interest. They have the highest low-risk interest rates. It is convenient to use them while waiting for phase one or two to purchase stock.

Investment Timing [24]   This is analogous to doing your Christmas shopping early in the year, completing it before Thanksgiving. Retailers report that a major portion of their annual business is done between Thanksgiving and Christmas. It seems probable that you do not get pleasure from this "rat race." By avoiding the competition, Christmas shopping can be pleasant and much less costly. Similarly, a sort of "contrarian" investing, out of phase with the group, is advantageous. Let us now consider each phase of the business cycle and the best investments for it. For completeness, commodities are included, but not recommended. Perhaps after you have built a truly substantial estate you might permit your Child more excitement, but at considerably greater risk. Real estate is also included, but not recommended because of the great amount of time and emotional energy it seems to require. An exception is probably the purchase of vacant lots when developments first open.

---

*The reverse action by the Fed can be used to achieve reverse reactions.

Phase one Recession: interest rates lowered	Best: common stocks. Good: bonds, industrial commodities (e.g., copper and palladium). Fair: real estate. Poor: gold and collectibles (also called *tangibles*).
Phase two Rising economy	Best: commodities, including gold,* real estate. Good: stocks, collectibles, money market funds.** Fair: bonds.
Phase three Higher interest, slowing economy	Best: gold* and collectibles. Good: money market funds.** Poor: stocks and bonds.
Phase four Slow economy	Best: money market funds.** Good: (after the economy has definitely begun to slow) stocks and bonds. Poor: gold* and other commodities.

*After shopping among several dealers, I suggest that gold be purchased in bullion-type gold coins. They are easily stored and do not require assaying when sold.

**For diversification, money market funds are suggested as opposed to one or several money market accounts.

Recall that the *actions* of changing the interest rate, down in phase one and up in phase three, effect the delayed *reactions* of phases two and four. Let the anticipation of these reactions guide your investments. Next we consider signals which indicate the most probable, current phase of the economy.

Identification of the Economic Phase

Phase One   Unemployment is increasing; retail sales, auto sales, sales of other durable goods, and housing starts are down. Other signs of slow economic activity are apparent on visits to shopping malls and when traveling. Well into this phase, the Fed will reduce the interest rates. A leading indicator of this action is a change in the federal funds rate. It is the rate banks charge each other for overnight loans of $1 million or more to meet their reserve requirements. This usually triggers changes in the U.S. prime rate and the home mortgage rate. These indicators are given in the financial section of daily newspapers. They are also available regularly in *Business Week* (see the feature "Business Week Index" immediately after the table of contents) and in *Forbes* (see the "Forbes Index").

Phase Two   In this expansive period employment, production, prices, wages, interest rates, and profits are all rising. These data are available in the "Business Week Index." An equally weighted average of relevant components is given in *Forbes* (the "Forbes Index"). On the same page, the prime rate and the change in the GNP are also indicated.

Phases Three and Four   These are the reverse of phases one and two, respectively. Accordingly, the indicators for phases three and four are the opposite of the indicators for phases one and two.

Preparing Ahead of Need, Valuations of Stocks
In an example above, we chose GE as a candidate for investing. To be ready when phase two of the economic cycle occurs, we will now make an evaluation of GE as an investment. We first test for safe-

ty and then use the 8 qualitative and 13 quantitative performance factors presented earlier. After our performance analysis we will check for growth.

Safety   This status is normally indicated by three factors presented earlier: current ratio, debt ratio, and the paydown time. We will evaluate each for GE, using 1990 data from *Value Line* (source no. 24, Appendix A-1).

$$\text{Current ratio} = \frac{\text{current assets}}{\text{current liabilities}} = \frac{\$15.136\_B}{\$14.170\_B} = 1.07$$

Since this is much less than the normally desirable value of 2, we will determine GE's inventory turnover ratio.

$$\text{Inventory turnover ratio} = \frac{\text{annual sales}}{\text{inventory}}$$
$$= \frac{\$416 \text{ billion}}{\$7.3 \text{ billion}} = 57$$

This indicates a turnover of less than 1 week; thus, it suggests no difficulty with the current ratio near unity. Although it does indicate a weakness, the average of all American industries in that same year was only 0.4. Also, as presented next, GE's debt ratio is extremely small. It is only 17 percent. This value is considerably less than the suggested maximum of 50 percent [19].

Paydown Time

$$\text{Paydown time} = \frac{\text{long-term debt}}{\text{prepaid cash flow *}} = \frac{\$6.9 \text{ billion}}{\$1.85 \text{ billion}}$$
$$= 3.73 \text{ years}$$

This value is considerably less than the 7 yr suggested as the upper limit [19].

By these three factors, we see that GE well meets the first screening for safety. We will now check its performance for the past 15 yr.

Performance

Qualitative

1. *Management.* For a well-established firm such as GE, management qualities are reflected by past performance as revealed by the quantitative evaluations. As you realize, this is not the case for a new growth company. Nevertheless, from source no. 7 in Appendix A-1, we learned that GE's CEO (chief executive officer) and chairman of the board is John (Jack) Francis Welch. He was born in 1915. From source no. 8, we learned that he became vice president in 1971, senior vice president in 1977, vice chairman of the board in 1979, and its chairman in 1981. He received a bachelor's de-

*Please see footnote, page 267.

gree from the University of Massachusetts in 1957 and a master's degree from the University of Illinois in 1960. This indicates that he has the capabilities of advanced formal education and has worked his way up in the company, with at least 17 to 20 yr of experience.

From source no. 12, we learned that management seems to be responsible in cleaning up hazardous waste. The expense in 1988 was $48 million, and was estimated at $60 to $100 million over the next 2 yr. Also, in 1987, GE acquired the French electronics firm Thomsen, S.A. It is well established in medical equipment. This acquisition should prepare GE to participate in the economic growth expected of the European Economic Community formed in 1992.

These sources also give the other corporate officers with their education and promotions over the past 15 to 20 yr. Many libraries offer computer cross-referencing between a firm's name and a particular individual (such as a CEO). It indicates articles in *Business Week, Forbes,* and other financial magazines which give current information on the person's activities. For example, checking on Bob Wright, president of the GE division at NBC, a Forbes article (p. 124, April 17, 1989) gave his formal education and details of his career history. It also told of his success with GE's financial service (operating profits doubled to $400 million). As president of NBC he has had similar successes raising revenues and increasing operating profits. This improvement was made while ABC and CBS had almost no operating margin.

McDonald's was also an investment candidate. A computer cross-check of Michael R. Quinlan and McDonald's provided an article in *Business Week* (p. 67, May 8, 1989). It told of their enterprise in using NASA photos to study traffic patterns, and their testing of spicy fried chicken to capture a greater market share. Further information on McDonald's management is available in *Forbes* (p. 189, March 20, 1989, and p. 24, April 1990). For our evaluation example, GE was chosen over McDonald's because of GE's better overall situation. It is probable that both will justify your analysis. I believe they will do well into the twenty-first century. Time will tell where wisdom lies.

Continuing our qualitative analysis, the following data on GE were obtained from source no. 24, the April 1990 issue of *Value Line.*

2. *Listing.* GE is listed on the NYSE.

3. *Size.* The magnitude of sales is a popular measure of size. Sales per share in 1989 was $46.30; and the number of common stock, 9 trillion. The product of these indicates sales over $416 billion. Another indicator of size is a firm's total worth. GE's book value is $23.20; this number multiplied by the number of shares is $208 billion.

4. *Stability.* Sales and earnings have grown steadily over the last 15 yr; earnings rose from $0.84 to $4.36 in 1989.

5. *Dividends.* They have steadily increased over that period from $0.40 to $1.70.

6. *Prominence.* GE is one of the world's largest and most diversified companies. Its activities include three broad categories: technology (aircraft engines, medical, and plastics), services, and core manufacturing.

7. *Research.* This expenditure has been a proper 3 percent of sales. In the 1990 newspaper report of companies granted the most U.S. patents (over 300), IBM and GE were the only U.S. firms.

8. *Realistic accounting.* GE uses a conservative LIFO method for inventories and does not include extraordinary income in operating revenues. Also, it has no pension liability. Regarding book value, *Value Line* (our source of data) adjusts as proper, for possible conversion to common stock from preferred and bonds.

Quantitative

Again, the factors or data for their evaluation were taken from *Value Line,* source no. 24 (April 24, 1990).

Factors 1 through 3 are P/E = 15, P/D = 34.5, and P/S = 0.92.

Factor 4. Price to book value. Price (3/30/90) $64.88, BV $23.25, or P/BV= 2.8.

Factor 5. Cash flow/share = $6.10 or total cash flow = $6.10 × 9 billion shares = $54.9 billion.

Factor 6. The free cash flow/share is $2.02. Recall that the free cash flow is the regular cash flow minus the sum of the dividend payments and plant outlays. GE has no preferred stocks, so the sum of the common dividend $1.88 and capital outlay per share of $2.20 (in 1989) subtracted from the regular cash flow/share gives the free cash flow/share as $2.02, or a total amount of $18 billion.

Factor 7. Current ratio = $15.1 B/$14.1 B = 1.07.

Factor 8. Debt/equity 17 percent/83 percent = 0.116.

Factor 9. ROE (% earned net worth) = 18.5 percent.

Factor 10. ROI (% earned total capital) = 16.0 percent.

Factor 11. Operating margin = 14.9 percent.

Factor 12. Beta = 1.1, closely follows S&P 500.

Factor 13. Working capital, $1.0 billion.

Growth   Again, this is not a major objective of the defensive investor. Preservation of capital even with a low inflation environment, however, requires some growth to offset it. Graham [21] recommends seeking firms whose earnings increased by one-third in the past 10 yr. This is determined by comparing the first 3-yr average earnings with the average for the last 3. For GE, the average per share earning for 1980 through 1982 is $28.79. The average for 1987 through 1989 is $44.29, a 54 percent increase.

Weighing the Tests and Factors   GE did well in the test for safety. From an information search and weighing of its qualitative factors, my judgment perceives them as strong. In turn, the quantitative factors were evaluated. They are considered below.

Over the years, my judgment has been formed by experience and heavily by literature research. I use that judgment below to review the quantitative factors. As in dead-reckoning navigation (Appendix A-2), with experience, you will decide on corrections (improvements) which more nearly suit your objectives and risk characteristics.

First recognize that the first four factors, the P/X ratios, reflect the market's present evaluations of the firm's X's: earnings, dividends, sales, and net asset value per share. It is interesting how three of these are related.

P/BV = ROE × P/E, where ROE = E/BV is in decimal form. And P/S = net operating margin, E/S × P/E; the net operating margin = E/S, is also in decimal form. Now, consider the other factors for determining the price at which the stock would be an excellent investment.

Two other factors, ROI and ROE, appear to be most important and are used in the development below. We will also show how they are related, dependent only on the debt ratio. The remaining seven factors are like leading indicators. They give indications of the company's smooth operation and continued growth. We could form a weighted average of them and compare it to a standard. Such a standard could be the same average evaluated for the S&P 500 or *Value Line's* annual industrial composite. Feeling that such an average could have only little value, I recommend reviewing the six factors, comparing them with other firms, and using your best judgment. (If in your best judgment you feel that the average would be useful, great!) In the conclusion, I encourage your independent approaches, but always risking only a small part of your discretionary funds. Surely, experience and mistakes are the most effective learning methods. Try to keep their consequences small.

ROE, the Paramount Factor (or the Bottom Line Factor)

The following development seems exciting; thus, even though it is rather involved, it is not "hidden under a bushel" in the appendix. Its focus is the firm's ROE, but it also uses its P/E and P/BV. All amounts are in dollars; naturally, the ratios are nondimensional. *Eq* is used to distinguish equity from *E*, for earning. Consider an amount of money; its magnitude is designated by *M*. Thus, $M \div$ P/BV equals the amount of *Eq* (the firm's equity) that can be purchased with *M* dollars. Remember from above that P/BV = ROE $\times$ P/E. Accordingly, *Eq*(0), the amount of equity purchased at time zero, can also be expressed as

$$Eq(0) = M(0) \div \text{P/BV}(0) = M(0) \div [\text{ROE}(0) \times \text{P/E}(0)]$$

This equivalence is used because ROE and P/E are often more available than P/BV. *Eq*(0) indicates the equity you purchased at time zero. The same method is used to indicate time zero for the other terms that might change with time.

Having obtained the equity, we can use the techniques of Chapter 6 to determine its value, *Eq*(*n*), *n* years in the future, compounding at rate ROE.* We first consider that ROE and P/E do not change. Again, ROE is in decimal form.

$$Eq(n) = Eq(0) \times [1 + \text{ROE}]^n$$

Now, let us determine *M*(*n*), the amount expected from selling this equity. The equation for *M*(*n*) is developed from the one above for *Eq*(0):

$$M(n) = Eq(n) \times \text{P/BV}(n)$$

---

*For a firm's equity to compound at ROE, all the after-tax earnings would have to be reinvested in its equity. That is nearly the case except for preferred and common dividends. Since preferred stocks are usually a small portion of total capitalization (zero for GE), reinvesting your dividends effects a near total "earnings reinvestment" for you. Most firms have a dividend reinvestment plan (DRIP) with no commission, and frequently the plan offers shares at a small discount.

Then using the above expression for $Eq(n)$,

$$M(n) = Eq(0) \times [1 + \text{ROE}]^n \times \text{P/BV}(n)$$

Again using P/BV = P/E $\times$ ROE gives

$$M(n) = Eq(0) \times [1 + \text{ROE}]^n \times \text{P/E}(n) \times \text{ROE}$$

Finally, using the expression developed above for $Eq(0)$, we have

$$M(n) = \frac{M(0)}{\text{ROE} \times \text{P/E}(0)} \times [1 + \text{ROE}]^n \times \text{P/E}(n) \times \text{ROE}$$

If ROE and the P/E did not change, they would cancel in this equation. Thus, the final value $M(n)$ would be the original investment $M(0)$ compounded over n years at interest rate ROE, that is,

$$M(n) = M(0) [1 + \text{ROE}]^n \qquad (7\text{-}1)$$

In well-established noncyclical firms, stability of ROE and P/E would be a close approximation. In a well-established growth-oriented firm like GE, the final values of both P/E and ROE would probably be greater. This, of course, would further increase $M(n)$.

The bottom line then is that for outstanding investments, seek firms that (1) are safe, (2) are stable, which includes those with steady growth, and (3) have a high ROE.

Example 7-3

Let us consider investing the amount of $2775 in GE using the development above and data from *Value Line* [26] (Spring 1992).

$$M(0) = \$2775$$
$$\text{ROE}(0) = 0.185 \ (18.5\%) \qquad \text{and} \qquad \text{P/E}(0) = 15$$

Thus,

$$\text{P/BV}(0) = \text{ROE}(0) \times \text{P/E}(0) = 2.775$$

Determine $M(10)$, the value of the investment after 10 yr.
Solution: From the relation above,

$$Eq(0) = M(0) \div [\text{ROE}(0) \times \text{P/E}(0)]$$
$$= \$2775 \div 2.775 = \$1000$$

This $1000 in equity now grows for $n = 10$ yr at the interest rate ROE = 0.185. Therefore,

$$Eq(10) = \$1000[1 + .185]^{10} = \$5459.88$$

Upon sale, this equity sells for a greater amount because of the price to equity ratio, P/BV = ROE × P/E. Accordingly,

$$M(10) = Eq(10) \times ROE(10) \times P/E(10)$$
$$M(10) = \$5459.88 \times 0.185 \times 15 = \$15,151.17$$

if ROE(10) = ROE(0) and P/E(10) = PE(0).

When ROE(10) and the P/E(10) equal their initial values or more, the stock which represents this amount of equity sells for $M(10)$ or more, as evaluated by the abridged Equation 7-1 (in which the ROE's and P/E's at times 0 and n cancel). Thus,

$$M(10) \geq M(0) \times [1 + 0.185]^{10}$$
$$\geq \$2775 \times 5.4599 = \$15,151.22$$

As expected, within the three significant figures used in our calculations, the minimum value for $M(10)$ is the same as above.

Recall that near the beginning of Section 7-6 on the Money Game we recognized that in order not to lose purchasing power, inflation and taxes required an investment to be compounded at 9 percent annually. In that earlier calculation, we assumed (as typical) a sum of state and federal taxes of about 34 percent. On yearly income such as dividends etc., this tax would have to be paid each year. In a long-term stock investment, paying small or no dividends, however, this tax (such as 34 percent) would not compound and would not have to be paid until the year of its sale. Thus, the investment, somewhat like an IRA, would compound at the rate of ROE tax-free until the year of sale. In turn, except for that year, only inflation need be considered to determine the investment's gain to $M(0,10)$, its purchasing power after 10 yr, in terms of year zero dollars. Specifically, the effective rate (over inflation) during the n years of the investment would be

$$ROE(eff) = [(1 + ROE)/(1 + d)] - 1$$

where d is the decimal value of inflation, a devaluator of the dollar. For GE with ROE of 18.5 percent, in a period of 5 to $5\frac{1}{2}$ percent inflation, for the worse case of $5\frac{1}{2}$ percent, the above equation gives

$$ROE(eff) = [(1.185)/(1.055)] - 1 = 0.1232, \text{ or } 12.32\%$$

As you realize, this rate of return considers inflation only. Let us now use it for a 10-yr investment. Then, we will consider the true net, or its after-tax growth.

Again, $M(10,0)$ is the magnitude, after 10 yr in terms of year zero dollars. For $M(0) = \$2775$ we have

$$M(10,0) = M(0) \times [1 + 0.1232]^{10} = 2775 \times 3.196 = \$8868.17$$

Upon sale and after taxes, and using ROE (eff),

$$\text{Net } M(10,0) = M(0) \times [1 + 0.1232]^{10} \times (1 - t)$$

On sale, a typical combination of state and federal taxes such as 34 percent would be paid on the long-term capital gain. The long-term tax will probably be reduced to less than the short-term in the near future. But, as of this writing, they are the same. The regular income taxes, however, are to be adjusted for inflation. Because of the adjustment, we can get the equivalent tax by using 34 percent on the gain determined in terms of the original dollars. Thus, the real gain is determined by subtracting the original investment of $2775 from the inflation adjusted $M(10,0)$. Thus,

$$\$8869.96 - \$2775 = \$6094.96$$

and the tax in time zero dollars is

$$\$6094.96 \times 0.34 = \$2072.29$$

This subtracted from $M(10,0)$ leaves $6797.67, the after-tax value of the investment, in terms of year zero dollars.

The net effective interest rate after taxes and above inflation can be determined as below.

$$M(10,0)/M(0,0) = \$6797.67/\$2775 = 2.45 = [1 + g]^{10}$$

where g is the growth rate after taxes and above inflation (for the values assumed). By use of logarithms,

$$\log 2.45 = 10 \times \log [1 + g] = 0.389166$$

or

$$\log [1 + g] = 0.0389166$$

and in turn

$$1 + g = 10^{0.0389166} = 1.09373$$

Thus, $g = 9.37$ percent, in excess of inflation and after taxes. The investment's nominal dollar value (at time of sale) is its year zero dollars multiplied by the inflation rate, compounded over the investment period (here 10 yr). This gives

$$M(10,10) = M(10,0) \times [1 + 0.055]^{10}$$
$$= \$6797.67 \times 1.708 = \$11,610.42$$

Determining this amount might seem vain; Keynes called these inflated amounts the "money illusion." There are two reasons for presenting it. One is that sources such as *Value Line* [26] (see "At a Glance, Inflation in Selections and Opinions," part 2 in the weekly reports) show that some items are less inflated than others; thus, by seeking them, you can more nearly realize the nominal, higher level from the investment. Presently auto sales and housing are down. Electronic goods, TV's, etc., are down 24 percent in the last 20 yr, a benefit of engineering and management. Also, reduced demand always suppresses inflation, as in phases one and four of an economic cycle. The second reason is that by shopping, one can always obtain less-inflated prices; seek the best value per dollar. It seems that all businesses constantly strive for the highest price for a given item or service. We next review the relation between ROE and ROI.

ROE and ROI

These returns differ only because of debt. They would be equal if there was no debt. Judicious management uses a small amount of long-term debt financing, preferably no more than 40 percent of the total capital. GE's long-term debt (1990) was only 17 percent of total capitalization. Many firms through long-term borrowing (the junk bond technique) employ leverage to dangerous levels. As defined in Important Terms in Section 7-6, leverage is the practice of borrowing money, obtained at a certain interest rate, and seeking to use it in enterprises expected to give much higher rates of return (ROI's). For instance, GE borrows money at a rate near or at the Aaa grade bond rate, perhaps 9 to 10 percent, and through its enterprises earns 16 percent (their ROI). As you realize, the difference increases the return on equity, their ROE. We will discuss leverage later when we examine the use of margin to purchase stock. You have probably already sensed that ROE and ROI are related by the bond or long-term debt ratio presented earlier. We will now develop this relationship. It uses the equality TA (total assets) equals D (debt) plus Eq (equity). As previously, E is used for earnings. Thus,

$$\text{ROE} = \frac{E}{Eq} = \frac{E}{TA - D} = \frac{E/TA}{1 - D/TA} = \frac{\text{ROI}}{1 - D/TA \text{ (debt ratio)}}$$

$$\text{ROI} = \frac{E}{Eq + D} = \frac{E/Eq}{1 + D/Eq} = \frac{\text{ROE}}{1 + D/Eq \text{ (debt - equity ratio)}}$$

or

$$\text{ROE} = \text{ROI} \, (1 + D/Eq)$$

and, as homework problem 19 shows,

$$D/Eq = \frac{\text{debt ratio}}{1 - \text{debt ratio}} \qquad \text{also,} \qquad D \text{ ratio} = \frac{D/Eq}{1 + D/Eq}$$

Mathematical Models   Models are fun. In our youth we enjoyed model airplanes, cars, and dolls. These "look like the real thing" types are known as *iconic* (from the Greek for "image") models. They do not have the versatility of a mathematical model, where we can set up solutions on a computer and test various "what if" situations. As you know, this arrangement is known as a *simulation*. It can be used to optimize the system being simulated. I hope the ROI-based model that we developed above was enjoyable to follow and will be fun to use in the future when evaluating investment possibilities. *A word of caution:* You know that all models are approximations. Complexity always prohibits considering many aspects. Other aspects are unknown. Furthermore, in a socioeconomic system such as a firm, random industrial, national, and world crises make their consideration impossible. Nonetheless, use of models is a wise first approach for a rational analysis of most situations.

Step Four: Action

Getting Started   We are now ready to implement the decisions to buy or sell one or more of the selections analyzed in step three. Unless you have inherited or been given stocks or bonds, your first activities will be buying. Nonetheless, after any investment, decisions for selling should be reviewed periodically, maybe the first weekend of every month. A stop-loss order placed at the time of purchase limits the possible loss. Stop orders, considered later, should be set 10 to 15 percent under the purchase price and any later increased price. Seldom will a buy and hold strategy be as rewarding as one of ongoing evaluation of investments. We now review the activities to start investing.

Open an account with a full-service broker. Probably it will require only a telephone call. They will be pleased to mail you proper forms for completion and signing. After setting up an account, you have 5 business days after a transaction to send them the money (or stock certificate, if selling). I suggest that we never have too much relevant information. The broker can give you current data not normally available. The broker's firm has a research department which has special contacts with firms and with the market in general. As you realize, we pay for the information through higher commissions (as opposed to using a discount broker). Information is truly of great value. The higher commission to your broker is a good investment. Later, you might use a discount broker when you do not need additional information. It is difficult to know when that is the situation. Sometimes, however, by using information from an investor's newsletter and perhaps from a service such as *Value Line* [26], you will have already made your analysis and decision.

It is important to note that we have presented brokers and their firms as sources of information, not as the *decision makers*. You and you alone will gain or lose from the investment decisions.* No one, however considerate, can be as interested in your well-being as you. Thus, discuss your decisions of step three with your broker as an additional information source, but know that the decisions are yours. Also, many (if not most) brokers do not make their own analyses. Rather, they depend on the analysts from their firm and, thus, promote the firm's recommendations as a source of commissions and, in turn, their livelihood.

Regarding further information, an excellent source is a trial subscription to *Value Line Investment Survey.* With it you receive a subscriber's guide. The guide is a key to accessing the data needed for the quantitative factors and, in part, for the qualitative factors. The weekly publication has three parts: (1) "Summary and Index" gives the latest median P/E and dividend yield for the 1700 equities included in the *Value Line Index.* This part also ranks 95 industrials, and includes 20 screens as the best performing stocks in the last 13 weeks and the best cash flow generators. (2) "Selections and Opinions" presents current investment articles, insider transactions, and (under "Fixed-Income View") interest rates and money supply rates of change. Of particular importance is the "Quarterly Economic Review," helpful in determining the present phase in the business cycle. (3) "Ratings and Reports," on a rotating basis of 13 weeks, reviews each of the 1700 equities. These are the equities in the *Value Line Composite Index* . This part gives the data needed for the quantitative factors. Also, it includes a 400-word report on recent company developments and prospects, helpful for the qualitative factors.

The trial subscription is for 10 weeks and costs about $50 to $60. In the 10 weeks' time you will be able to get adequately familiar with the reports and then be able to quickly access needed data in short visits to the public or university library.

The next step in prudent investing is determining the phase of the economy's cycle. We search for clues in the financial section of the daily paper and other financial publications (as referenced often in this chapter). Having used our best judgment, we decide on the present phase and make the investments presented above in the review of business cycles. Trying to anticipate the changes is helpful in making the proper investments early in the cycle. We have discussed many of the signals earlier. The strongest economic influence is the money supply and its effect on interest. Observe these, and the economy will follow in 18 to 24 months. Then you will have already taken the appropriate action.

Because of the greater risks of investing in stocks, let us consider guidelines to minimize that risk and possible loss.

General Guidelines   Regard as discretionary funds those which are not allocated for emergencies, major goals, or retirement. That is, they are truly available for some risks. Guidelines (3) and (5) concern "timing" and involve the technical approach. The details are presented in Appendix A-6.

1. *What stocks to buy?* Consider only those selected and analyzed by techniques similar to those presented earlier. Unless you wish to gamble at substantial risk, do *not*

---

*Here I am not considering the lesser broker fees which they gain from all transactions, on your buying or selling.

follow suggestions of friends, or even brokers and financial columnists, without the proper evaluations.

2. *When to buy, how much?* For your first stock purchases, use no more than 25 percent of your discretionary funds. Your first purpose should be to develop competence and knowledge about stock investing, not to make a great profit. After having analyzed the economy as reviewed earlier in this chapter and having decided to invest in a stock you have evaluated, simulate the process for a month. In the simulation you would consider that you made the purchase on a given day. Over the next month watch its price performance. A rising price would give you a degree of confidence in your selection and evaluation. To judge whether the price increase was primarily due to the stock or a bull market (which tends to "lift all boats"), use the relative strength indicator (RSI) presented in Appendix A-6. Recognize that if you can identify a good firm and a rising economy, you will achieve a greater gain. A firm whose RSI has risen 10 percent is definitely a good investment. Personally, I believe that as your discretionary funds permit, it is wise to purchase more stock each time its price or RSI rises 10 percent. Each gain gives a margin of safety for price declines. As a part-time investor, it is probably best never to purchase stock using borrowed money or to sell short. A method of borrowing money from the broker to pay for a portion of your cost is called *buying on margin.* Although it is strongly discouraged here, for general knowledge the method is explained in Appendix A-7. Short selling is outside the scope of our purpose.

3. *When to sell.* Above, I suggest buying more shares when a previously evaluated stock's price or its RSI rises 10 percent. Of much greater importance is "cutting your losses." Sell on a 10 percent drop from an earlier price, especially in your first years with little experience. Remember the experienced stock analyst who said that we all need the benefit of investing through several market cycles. If you feel you have special information, you might consider accepting a 15 percent drop from an earlier price. Particularly important is the rapidly increasing "one over x" relation concerning the percentage gain required to offset a loss of percent x . The relation for gain, G in percent, required to recapture loss of percent L is

$$G = 100 \times [1/(1 - L/100)] - 100$$

For $L = 10$ percent, G required is 11.1 percent; $L = 15$ percent, $G = 17.6$ percent. However, for $L = 25$ or 50 percent, G must be, respectively, 33.3 percent and 100 percent. From this equation we see the wisdom of cutting losses early. In the 30 yr since I learned this axiom, I have always been wrong by not following it, usually at considerable loss.

There is a well-known Wall Street saying: "Cut your losses, but let your profits run." Again, over the same 30 yr, I have erred when I sold a stock before its RSI went down 10 percent or before it dropped 10 percent from a previous price. This strategy is a technical approach known as the "relative strength" rule. Fair but less effective and requiring less effort is the *filter rule.* By it, a stock is bought only after it increases x percent and is sold on a decrease of y percent. As alluded to in other sections of this text, I have found $x = y = 10$ percent efficacious (the power to produce the desired results) values.

Finally, regarding the selling of any security, you need to periodically review your holdings with the same scrutiny you used before buying them, preferably monthly, but definitely within 3 months.

4. *Safety.* We have noted the importance of diversifying, for both firms and industries. A general rule is to limit your investments in any industry to 25 percent of your funds. Naturally, in your first investments, you can only diversify step by step. Accordingly, your first five purchases should be in five different industries. Recall that *Value Line*'s "Summary & Index" ranks 95 industries each week and gives the page (in part 3, "Ratings & Reports") of the most recent report on them. For the initial investments, consider staying with industries ranked in the top 20. Ideally, wide diversification is best; however, giving adequate attention to firms' activities and situations is wise. The required time and effort limit most individuals to keeping informed on 10 to 15 companies.

5. *Timing.* The durations of the economic phases vary. Give close attention to changes in interest rates and in the money supply measure M1. An increase in M1 greater than 4 percent, on an annual basis, heralds the end of phase one in 6 to 9 months. Low unemployment, such as around 5 percent, is great for the politicians, but along with rising interest rates it indicates the end of phase three. This end is confirmed when interest rates on the short-term Treasury bills go above the Federal Reserve discount rate. Near the ends of these phases (one and three) are the times to change investments.

If you feel that a phase is not clear, remember a well-accepted "truth" for long-term investors. Companies with low P/E's tend to outperform the market; and furthermore, firms with established brand names that pay good, secure dividends tend to hold value in down markets.

6. *Balance.* This is similar to diversification in discreet investing. Balance is broader, however, in that in addition to diversifying over industries and stocks, it indicates proper limits on your Money Game. It is advantageous to regularly readjust the balance of your investments among three types: (1) cash and equivalents (primarily checking and money market accounts), (2) bonds, and (3) quality growth stocks. Review your position in life as a guide to the best distribution.

Step Five: Review and Improvements

This is definitely the lifetime activity of every prudent individual, and all successful engineers and investors. The real challenge is determining when further efforts are not justified. It seems that the only guide, as in all control systems, is comparing the results with the reference, namely, your objectives. Furthermore, as indicated earlier, these objectives need to be reviewed regularly in accordance with your state in life.

7-8 SUMMARY

The competent engineer often needs creativity. This and general satisfaction in life are impossible without financial security. The material in this chapter, while more than adequate for that quest, introduces only the fundamentals of the Money Game. For that game, you are encouraged to constantly study, develop your strategy, and be cautious. Do not be fooled by good luck. Financial risk is directly proportional to debt. Again, I suggest the practice of a friend of mine. He never bought anything on credit (except for less than 70 percent of his house), has no credit cards, and has had no problems cashing checks, nationwide.

CONCLUSIONS

This information has been directed to the prudent, defensive investor. The presentation has focused on two aspects: (1) As offered by Graham [27], the rate of return on one's investments is proportional to the time and effort the investor is willing to give to the task. Dividing the task with a friend or spouse reduces the stress and can be very pleasant if not thrilling, especially when the economy is in phase two. A substantial estate can be readily built by starting early and using the magic of compounding. (2) Building an estate of excessive wealth requires significantly greater risks, unacceptable for prudent persons unless they dedicate their total life to the endeavor. Such a narrow life seems inconsistent with being prudent. As the country preacher said, "What you own, owns you."

DISCUSSION QUESTIONS AND PROBLEMS

7-1 YOUR ESTATE NOW: SAVINGS, EDUCATION, AND SOCIAL SECURITY

1 Give your thoughts about the needs for and value of financial security.
2 Relative to financial security, review the contributions of (a) a strong bank account, (b) social security, and (c) insurance.
3 Discuss the types of life insurance relative to needs and costs.

7-2 ESTATE BUILDING

4 Review the aspects of fixed income and equity investments.
5 Define *direct* and *indirect* investing.
6 Why is it usually best for the small investor to use mutual or closed-end funds for investing?
7 Define (a) *load* and *no-load* mutual funds, (b) the *expense ratio,* and (c) *NAV* (net asset value).
8 Explain the difference between primary and secondary security markets.
9 a Explain debt and equity investments.
  b Discuss stocks and bonds relative to risks and rewards.
10 How does Benjamin Graham define an investment operation?
11 Define *total return.*
12 a Give a written review of systematic and unsystematic risks.
  b Give a written review of the concepts of diversification and uncorrelated investments.
13 Give your thoughts on life cycle investing.
14 In writing briefly review the two traditional approaches to investing: (a) fundamental analysis and (b) technical analysis.
15 In writing review the efficient market hypothesis and its relevance to technical analysis.
16 A 10-yr corporate bond (face value $1000) has a coupon rate of 8 percent. If the bond is exactly 3 years old and the current market interest rate is 6 percent, what is the valuation of the bond? Remember that dividends are paid every 6 months.

17 A blue-chip (well-established) stock pays a quarterly dividend of $1.25 (expectedly "forever"). The present market interest rate (for the S&P 500) is 4.5 percent. What is the stock's valuation? *Hint:* It could be evaluated similarly to a bond.

18 An industrial firm earned $20 per share last year; a share's market price is $60.
   a What is its P/E ratio?
   b If the S&P 400 has a ratio of 14, what type of stock does it seem to be?

19 In a given year an industrial firm has a net income of $308 million; its total assets equal $8733 million, with stockholders equity of $2373 million. Determine (a) its ROI, (b) its ROE, (c) its debt ratio, (d) and show how the D/Eq and the debt ratio are algebraically related.

20 Consider yourself 3 to 4 yr after graduation. You have accumulated $10,000, perhaps in a "strong bank" account, money market funds, savings bonds, and a few stocks. Your take-home pay (after taxes, etc.) is about $2000/month, and your normal expenses run $1000 to $1200 per month. Review your financial status, and decide a proper asset allocation. For the bonds and stocks allocation, over several months (really too short for a true market study), simulate an investment portfolio. Using real-time data from various sources, buy and sell stocks and bonds following a combination of fundamental and technical analyses. Carefully consider the economy (phase of the cycle), the industries, and the firms, based on indices of safety, performance, and growth. For a realistic simulation include transaction commissions (each way), with $45 minimum, or 2 percent of the total transaction for stocks and $5 per bond.

REFERENCES

1 *Social security, how it works for you.* Washington: Social Security Administration, SSA Publication no. 05-10006, 1988.

2 Band, R. How safe is your bank? *Transcripts, personal finance seminar.* Arlington, VA: KCI Communications, 1985, p. 323.

3 Moses, E. A., & Cheney, J. M. *Investments analysis, selection, & management.* New York: West, 1989.

4 McClave, J. T., & Benson, P. G. *A first course in business statistics* (4th ed.). San Francisco: Dellen, 1989.

5 Graham B. *The intelligent investor* (4th ed.). New York: Harper & Row, 1984, p. 116.

6 *The individual investor's guide to no-load mutual funds* (8th ed.). Chicago: International Publishing, 1989.

7 Quinn, J. B. *Everybody's money book.* New York: Delacorte, 1979, p. 656.

8 Perritt, G. Hidden costs. *Forbes,* Sept. 4, 1989, p. 182.

9 Addis, R. Dog funds. *Forbes,* Sept. 4, 1989, p. 176.

10 *Investment companies.* New York: Arthur Wiesenberger, 1989.

11 Annual fund ratings. *Forbes,* Sept. 4, 1989.

12 Frank, A. Is it prudent to speculate some more? *Personal Finance.* Arlington, VA: KCI, Nov. 8, 1989, p. 170.

13 Kelder, J. A Swiss bank primer. *Personal Finance,* May 28, 1980, p. 161.

14 Berne, E. *Games people play.* New York: Grove, 1967, p. 12.

15 Graham. *Intelligent,* p. 157.

16 Graham. *Intelligent,* p. 1.

17 White, R. B. *Great business quotations.* Secaucus, NJ: Lyle Stuart, Inc., 1986, p. 62.

18 Jurgunson, E. B. *The inevitable general.* Boston: Little Brown & Co. 1968.

19 Fisher, K. L. Three easy steps. *Forbes,* Oct. 30, 1989, p. 253.

20 *How to read a financial report* (5th ed.). New York: Merrill Lynch et al., 1984, p. 17.

21 Graham. *Intelligent,* p. 184.

22 Dreman, D. The glories of low-P/E investing. *Forbes,* June 27, 1988, p. 173.

23 Chakravarty, S. N. Unreal accounting. *Forbes,* Nov. 16, 1987, p. 74.

24 Curran, W. S. *Principles of corporate finance.* New York: Harcourt Brace Jovanovich, 1988, p. 90.

25 Band, R. E. Successful investing in one lesson. *Personal Finance,* Aug. 18, 1982, p. 193.

26 *Value line investment survey,* weekly, New York, 10017.

27 Graham. *Intelligent,* p. 40.

28 Quinn. *Everybody's,* p. 604.

29 Shaw, A. R. *Technical analysis.* Homewood, IL: Dow Jones Irwin, 1975.

BIBLIOGRAPHY

Jenks, J. C., & Jenks, R.W. *Stock selection.* New York: Wiley, 1984.

Pessin, A. H., & Ross, J. A. *Words of Wall Street.* Homewood, IL: Dow Jones Irwin, 1983.

APPENDIX

A-1 SOURCES OF DATA FOR FACTORS EVALUATIONS*

These sources are grouped according to two types of factors: (1) qualitative and (2) quantitative. The sources are numbered sequentially, regardless of the groups. These numbers correlate with the entries in Table 7-5.

1. *Qualitative.* There is some overlap with sources mentioned in the review of these factors in the text.

Management Information

Again, information on management is of greatest importance. While the quantitative factors of the firm are the best indicators of its competence, you can get a "sneak preview" of a new chief executive officer (CEO) by checking the performance of a significant firm from which she came. Perhaps the best approach might be to keep an indexed notebook. In it, keep relevant data from your readings on the personnel of firms that have passed your earlier screening criteria. Also, as in part noted above, suggested financial literature is listed below:

Newspapers Many local dailies carry the listings of the New York Stock Exchange, the American Stock Exchange, and the OTC (over-the-counter) market. The last is listed under NASDAQ, which is an acronym for National Association of Securities Dealers Automated Quotation. This computerized system is used by brokers to carry out transactions for OTC bid-ask offers in the OTC securities. The *Wall Street Journal* and the *New York Times* are two dailies

*With permission, this list was largely taken from Clemson University Reference Guide No. 3, prepared by Freddie B. Siler, revised Spring 1991 by Ellen M. Krupar.

TABLE 7-5
INFORMATION SOURCES FOR EVALUATING STOCK FACTORS

Factors	Sources as numbered in Appendix A–1
Qualitative	
Management	6, 7, 8, 9, 13 (insiders)
Listing Exchange	14 through 24
Size	15, 16
Earnings stability	14, 15, 24
Dividend record	14, 15, 18, 21, 24
Prominence	16
Research activity	10. Newspapers, annually, "Companies Granted the Most U.S. Patents"
Realistic accounting	24, each firm's annual report
Quantitative	
P/E, price to earnings	14, 15, 19, 20, 21, 23, 24
P/D, price to dividends	14, 15, 18, 19, 21, 23, 24
P/S, price to sales	14, 15, 19, 21, 22, 23, 24
P/BV, price to book value	14, 15, 20, 21, 22, 23, 24
Cash flow per share	14, 22, 23, 24
Free cash flow per share	24
Current ratio	14, 19, 21, 23, 24
D/E, debt to equity	14, 17, 19, 21, 23, 24, 32
ROE, return on equity	21, 22, 23, 24, 32
ROI, return on investment	21 through 24, 32
Operating margins	14, 21, 22, 23, 24
Beta	24
Working capital	14, 19, 21, 23, 24

among many which have excellent sections. *Barron's,* a weekly, is an excellent source of financial information.

Magazines   *Business Week, Fortune, Money,* and *Forbes.* All of these have regular articles about the effects of management on the respective company.

Guides to the Literature*

1 *Where to Find Business Information.* HD30.35.B76. Focuses on current and up-to-date sources presently published in the area of investment and financial information.
2 *How to Find Information about Companies: The Corporate Intelligence Source Book.* HD2785.H68. A guide to key facts and figures about public and private companies, domestic or foreign. Also lists annotated sources such as directories and investment services.
3 *Guide to U.S. Government Publications.* Z1223.Z7A574. An annotated guide to the important series and periodicals currently being published by government agencies, especially by the Department of Commerce.

*The Library of Congress call numbers follow each reference.

Dictionaries

4 *The Investor's Dictionary.* HG4513.L91. Short definitions of investment and financial terms.

5 *Dictionary of Business, Finance, and Investment.* HF1001.M76. Brief definitions of words and phrases particular to business, the financial community, and the stock market. Charts are used to illustrate terms.

Directories: Biographical Data

6 *Standard and Poor's Register of Corporations, Directors and Executives* (vol. 2). HG4057.P82. Arranged alphabetically, it presents biographies of 79,000 individuals serving as officers, directors, trustees, partners, etc.

7 *Dun and Bradstreet's Reference Book of Corporate Management* (4 vols). HD2745.D85. Contains biographical information on 200,000 principal officers and directors of over 13,000 leading U.S. companies.

8 *The Corporate 1000: A Directory of Who Runs the Top 1000 U.S. Corporations.* HG4057.A15646. This is a guide to the persons who direct and manage America's largest corporations.

9 *Standard and Poor's Register of Corporations, Directors and Executives* (3 vols.). HG4057.P82. Reviews over 37,000 corporations and over 405,000 corporate executives in all areas of U.S. business and industry. Vol. 2 contains biographical data.

Directories: Businesses and Corporations

10 *Directory of American Research and Technology.* T176.I65. Has an alphabetical listing of public and private firms in the United States performing research and development.

Financial Information: Corporate Annual/10K Reports

11 *Career Resources Information (CRI).* Usually in the reference area of the library. This source provides profiles on microfiche of approximately 2500 publicly owned corporations, private companies, government agencies, and educational institutions. It contains the annual reports, 10K and 10Q (when available), of companies filed alphabetically.

12 *Directory of Companies Required to File Annual Reports with the Securities and Exchange Commission.* Gov. Doc. Sel. 27:986. Lists companies required to file annual reports under the Securities Exchange Act of 1934. Includes companies with securities listed on national exchanges and others traded over-the-counter.

13 *U.S. Securities and Exchange Commissions. Official Summary of Security Transactions and Holdings.* Gov. Doc. Sel. 9. A monthly summary of security transactions and holdings reported by officers, directors, and other "insiders."

2. Quantitative Investment Information

14 *Moody's Industrial Manual.* HG4961.M8. Gives key financial data on industrial firms listed on the New York, American, and regional exchanges, as well as many leading international companies. Includes business and products, capital structure, financial records, subsidiaries, financial and operating ratios, officers, and directors. Updated weekly.

15 *Moody's Handbook of Common Stocks.* HG4905.M8h. Provides statistics and background data on 900 companies. For each, it includes background, recent developments, prospects, financial statistics for the past 10 yr, top officers, number of stockholders, date of annual meeting, and stock exchange symbol.

16 *Moody's Investor's Industry Review.* HG4501.M665. Provides a ranked list of approximately 4000 leading companies in over 140 industry groups.

17 *Moody's Industrial News Report.* HG4961.M723. Gives the latest news on (and size of) industrial contracts.

17a *Moody's Bond Record.* HG4905.M78. Gives statistical information on municipal, corporate, and government bonds, as well as corporate convertible bonds and preferred stocks. Data include interest rates, yields, ratings, face value outstanding, interest coverage, total assets, and call price.

18 *Moody's Dividend Record.* HG4937.M63. Gives the most up-to-date dividend information on 11,000 issues, common and preferred dividend-paying stocks, nonpaying issues, income bonds, mutual funds, and foreign securities.

19 *Standard and Poor's Security Owner's Stock Guide.* HG4915.S67. Presents pertinent financial data monthly on more than 5100 common and preferred stocks. (It does not include book value or total capitalization. For that information, see no. 20, listed below.) This manual also covers 300 mutual funds.

20 *Standard and Poor's Bond Guide.* HG4905.S435. On a monthly basis, it presents descriptive and statistical data on 4800 corporate bonds and their issuing firms. It also gives information on 600 convertible bonds and 200 foreign bonds (including the firm's total capitalization).

21 *Standard and Poor's Corporate Records.* HG4501.S7663. Six loose-leaf volumes contain the company's history, current operations, capitalization, earnings, and finances.

22 *Standard and Poor's Stock Reports.* HG4501.S76672. Four volumes give all significant financial data on firms for their most recent 10-yr history.

23 *Standard and Poor's Analysts' Handbook.* HG4519.S772. Annually reports corporate per share data from 1946 to present for over 90 industries. Statistics and percentages (13 components) include sales, operating profits, depreciation, earnings, and dividends.

24 *Value Line Investment Survey.* HG4501.V26. This weekly report analyzes 1700 stocks in 98 industries. The present and 10-yr history includes data on 23 key investment factors plus estimates of future values for the next 3 to 5 yr. A weekly letter, "Selection & Opinion," gives views on business, economic outlook, investment advice, and *Value Line*'s stock price averages. Ratings and reports of the 1700 firms are updated every 13 weeks on a rotating basis.

Newspapers and Newspaper Indexes

25 *Wall Street Journal.* (The latest issue is usually at the reserve desk of the library.) It includes business and financial news; relevant, interesting articles; and stock market prices with P/E's and sales for the NYSE, AMEX, OTC, and others. It also contains the Dow Jones, *Value Line,* and Standard and Poor's averages.

26 *New York Times.* (The latest issue is usually at the reserve desk of the library.) The financial section contains news, price quotations, and P/E's for listings on the NYSE, AMEX, and OTC.

27 *Barron's National Business and Financial Weekly.* HG.B2. Excellent articles on industries, companies, and other financial topics. It quotes Dow Jones averages, S&P 500 average, P/E, and book value.

Indexes

28 *Wall Street Journal Index.* HG1.W179. A monthly index, cumulated yearly, listing all articles. They are organized by subject and listed chronologically. For articles after 1982, use InfoTrac (listed next).

29 *InfoTrac.* 1984– Computerized data base available at many libraries. It provides general interest topics, business information, and technical data. These and current data are given for the *Wall Street Journal* and the *New York Times.*

30 *New York Times Index.* A121.N53. A biweekly index, cumulated yearly. It lists all articles published in the *Times,* arranged by subject and chronologically.

Magazines

31 *Business Week,* weekly. HF5001.B97. Source of information on business outlook and economic developments, with information on specific companies and industries. Special features include "Corporate Scoreboard" (quarterly, third issues in March, May, August, and November); "Bank Scoreboard" (annual, mid-April); "International Scoreboard" (annual, mid-July); "Investment Outlook Scoreboard" (annual, last issue in December).

32 *Forbes,* biweekly. HF5001.F69. Has short informative articles related to business and finance, including prospects of industries and individual companies; the *Forbes* index of the general economy; and the *Forbes* after January 1992, Wall Street Review, which presents market indices such as the Barra All-US Index, the Dow Jones Industrial, the S&P 500, and others. Special features include "Annual Report on American Industry," an excellent source of data for analysis (first issue in January) which includes 1150 companies with return on equity, growth, sales and net income, and averages over the last 10-yr period; "Report on International Business" (first issue, July); earnings forecast for the *Forbes* 500 (last issue in November); and a mutual funds report (last issue in August).

33 *Fortune,* biweekly. AP2.F7. Source of information on U.S. and international companies, economic and financial trends, industries, and government regulations. Special features include "First Largest 500 Industrial Corporations" (first issue in May); "Second Largest 500 Industrial Corporations" (second issue in June); "The 50 Largest Commercial Banks, Life Insurance Companies, Diversified-Financial, Retailing, Transportation, and Utility Companies" (second issue in July).

34 *Money,* monthly. HG179.M59. Contains popular articles on personal and family finances, investments, consumer behavior, taxes, insurance, careers, and the nation's economic status. Annually in the February issue, it ranks the top NYSE stocks of the previous year; in the May issue, it ranks mutual funds, but includes "Fund Watch" in every issue. Every issue also includes key economic data in the section titled "Money Scorecard."

A-2 DEAD RECKONING: NAVIGATING THE FINANCIAL SEAS

Perhaps this is one of the earlier uses of the engineering approach. Before World War II, dead reckoning was the main method for navigating ships and aircraft. By this method, the fundamentals of physics are used to calculate the position of the ship or aircraft after a given period of time. For example, consider an aircraft traveling 200 mi/h (miles per hour) over point A at time zero, in a direction $30°$ from north. In 15 min we would expect its location to be 50 mi from A, in the same direction, $30°$ from north. So it would be, if there was no wind. Suppose there was a wind of 40 mi/h from the east ($90°$ off north). This can be resolved into perpendicular components: a head wind of 20 mi/h (that is, from $30°$) and a cross wind of 34.64 mi/h (from $120°$). We recognize that this wind would slow the aircraft by 20 mi/h to 180 mi/h. Accordingly, in the 15 min, the movement in the $30°$ direction would be 45 mi. The cross wind in that time would blow the craft 8.67 mi off course, in a perpendicular direction. As a result, the aircraft would actually be 45.8 mi in a direction $19.1°$. Wind velocities (that is, the magnitudes and directions) and sea currents are not known precisely. In dead reckoning only the best estimates of aircraft performance and wind velocities can be used. After an interval of time, the expected position is determined using the estimates and the fundamentals. The results are then checked against observed data, and new estimates are made for midcourse corrections and the next projection. Dead reckoning is used by NASA in space navigation; the noted Kalman filter, employing fundamentals of celestial mechanics, periodically projects the future positions of the spacecraft. Observations by theodolites are used for corrections. For ships, ocean currents have the same effect as the wind on aircraft. In the financial seas, presently, the U.S. government (specifically the Federal Reserve) produces the greatest currents. To navigate your investments, I suggest you use the fundamental approach for "dead reckoning" and technical indicators (Appendix A-6) to make observations for midcourse corrections.

A-3 ECONOMIC MULTIPLIERS

When only a fraction of a total is absorbed or consumed and the balance is passed on, a special multiplying effect results. As an example, consider the situation in a society in which the average family has a marginal propensity to consume (MPC) of 0.8. If its income is increased by $100, it spends $80 more, and the receiver of the $80 spends $64 ($0.8 \times 80$); in turn, the receiver of that $64 spends 80 percent of it, and so on for each receiver. The total increase of the money supply in that society is determined by the series

$$\text{Total increase} = \$100(1 + 0.8 + 0.8^2 + \cdots) = \$500$$

By using long division, we see that if $r < 1$ (see Appendix B in Chapter 6)

$$\frac{1}{1-r} = 1 + r + r^2 + \cdots \qquad \text{or} \qquad \frac{1}{1-0.8} = \frac{1}{0.2} = \frac{1}{\text{MPS}} = 5$$

We recognize that for the *Marginal Propensity to Save* (MPS): MPS + MPC = 1.

You probably recall in your study of electronics that in a transistor, only a small fraction of the emitter current is absorbed in the base. Thus, if this fraction is 0.02, then 0.98 (MPC for the transistor) passes into the collector, and a change in the base current is multiplied by 50. This produces a 50 times change in the collector current, or a current amplification of 50. Next, we will note how the multiplier effect can be used by the Federal Reserve System in influencing our economy.

A-4 TOOLS OF THE FEDERAL RESERVE

There are 12 regional Federal Reserve banks; they are governed by a seven-member Board of Governors, usually known as the Federal Reserve Board. The governors are appointed by the president of the United States for a 14-yr term. However, the Federal Reserve, popularly known as the Fed, is directly responsible to Congress. The Federal Reserve banks constitute the "central bank" for the United States. Every modern country has a central bank, such as the Bank of England, the Bank of France, and the Deutsche Bundesbank of Germany. Through the central bank, the government handles its transactions, coordinates and controls the commercial banks, and (most important) helps regulate the nation's money supply and credit conditions. There are three primary tools of the Fed: (1) change in the required reserve ratios that the member commercial banks must hold as a percentage of their demand and time deposits; (2) open-market operations, that is, purchase or sale of government securities (on the open market); and (3) control of the discount rate, that is, the interest rate at which the Federal Reserve lends money to commercial banks. These loans are called *discounts*. The way in which the Fed uses these tools is called its *monetary policy* (as opposed to fiscal policy, presented later). Let us review how these monetary tools work.

Tool 1

Demand deposits, as implied, are available to the depositors on demand, for example, checking accounts. Time deposits are different. Money is available, without penalty, only after the agreed-upon time. The reserve ratio controls the fraction of demand deposits which the commercial banks must hold for unexpected withdrawals. The reserve is held either internally or in the regional Federal Reserve bank. For example, the reserve requirement in the early 1990s is 12 percent. This means that for every dollar deposited in a demand account, the bank can loan $0.88. In turn, by the multiplier discussed above, every $100 deposited with the bank can increase the money in circulation by the multiplier factor to $833.33. That is,

$$\text{Multiplier} = \frac{1}{1 - 0.88} = \frac{1}{0.12} = 8.33$$

Only a moderate increase in the reserve requirement, such as to 20 percent, significantly reduces the multiplier to 5.

Tool 2

The Federal Reserve can increase the reserve of a commercial bank by purchasing U.S. securities in the open market. If the seller is a commercial bank, then its reserve is immediately increased. In turn, the bank can increase its loans by the increase in its reserve (the amount of the government securities) multiplied by the multiplier. It is 8.33 for the presently required fraction of 0.12. If the seller is a corporation, it is expected that the corporation would deposit its check in a commercial bank, giving the same multiplier effect.

Tool 3

The Federal Reserve acts as the bankers' bank. As commercial banks make loans to the public, the Fed makes loans, also known as *discounts,* to the commercial banks. The interest rate for these loans is naturally called the *discount rate.* While the banks have to initiate these loans, the Fed can control their appeal by adjusting the discount rate. The borrowing, however, is intended only as a temporary way for banks to reestablish the required reserves. This tool permits the Fed to offset, as necessary, tighter policies implemented by the other two tools.

A Minor Tool

Stocks and bonds can be purchased in part by borrowed money. This method, presented below in Appendix A-7, permits the purchaser to pay a certain fraction, called the *margin,* of the security's price and borrow the balance from the broker firm. The interest rates are competitive with regular bank rates. The fraction or margin allowed to be borrowed is controlled by the Fed. Changes in the margin affect the speculator, but the main effects of these changes seem to be mostly psychological. In the early 1990s, the margin is 50 percent. In 1929, the margin was 10 percent, quite a leverage.

A-5 THE INVESTMENT SYSTEM: A MODEL

A system and its environment constitute the universe. There seems to be a universal law; its relevance to physical and social systems is only slightly different. For physical systems, the law indicates that when stable, these systems tend toward equilibrium, i.e., a balance of forces or influencers, both those internal (to the system) and those external (from its environment). In social systems the tendency is toward equality; if external restraint is too great, instability results, and revolutions eventually flare up to establish a status closer to equality.

The System

We realize that the investment system is both social and physical. In all systems, delay between stimuli and responses can and often does cause varying or cyclic behavior. Also, as you recall from your mathematical and engineering courses, all systems have two types of behavior: (1) natural, which because of friction is transient (dies out); and (2) forced, also called the *steady-state response.* As the name implies, the system response is forced into conformity with the external, forcing stimulus. We consider the natural or unforced behavior first. Wesley McCain [28], a pension fund manager, observed four phases of this behavior in the business or economic cycle.

Natural/Unforced Behavior [A Self-Correcting System]

Phase One   This is arbitrarily started in the period of a recession.* Business is poor, and unemployment is high. Because of this state, no loans are sought for increased productivity or expansion. In turn, demand for money is small and short-term interest rates fall. This short-term rate is indicated by the interest offered on Treasury bills. In this state, the market hits bottom and turns up; however, few recognize that there is now an excess of money. We realize that

---

*Technically, a recession exists when the gross national product (GNP) declines for two sequential quarters. Statistics are delayed. It is best to monitor sources such as the *Forbes* index. It is reported bimonthly and equally weighs components such as personal income, retail sales, and other significant indicators.

prices cannot or will not go up unless buyers have or can borrow money to make purchases at a higher price. As discussed later, the interest rate available on no-risk Treasury securities tends to put a limit on the P/E ratio that the prudent investor will accept in purchasing a stock.

Phase Two   The lower interest rates resulting from reduced loan demand in phase one re-tard and finally halt the economic decline. Recovery begins. The change is indicated by modest increases in business and consumer loan demands. With the increased demand, the interest nudges up slightly, but its relative level is still low. Accordingly, part of the "easy money" moves into investments and the prices of stocks and bonds rise. Many investors are still afraid to buy and miss the best opportunities. Nonetheless, as discussed in Section 7-7, small, cautious movement into the market is wise.

Phase Three   Businesses are recovering well, and loan demands by businesses and con-sumers increase strongly. Thus, the interest rate also starts to increase, but not enough to pull money out of stocks. The market continues to rise, with most of the money going into firms with expanding profit margins. Others only hold their prices. Because of the psychological effects of the recession (phase one), many are slow to invest in stocks. Thus, this phase is usually longer than phase two.

Phase Four   This phase begins when the interest rate gets high enough to attract money from stocks, and their prices start to fall. Investors may be misled by the delay in the economic downturn. It does not come until the interest rate gets adequately high to choke the business loans, consumer loans, and home mortgages. The cycle now begins again with phase one.

The Investment Environment

State variables (as in control systems) correlate with the energy and vitality of a system. From above, we quickly recognize that the interest rate is the most important state variable in the nat-ural, self-correcting behavior of the investment system. In 1936, John Keynes, a prominent British economist, published a very influential book, *The General Theory of Employment, Inter-est, and Money.* Keynes advanced an idea for government influence. In times of high unemploy-ment the government should increase spending to stimulate the economy. Under such action, the system shifts to a forced behavior which can diminish the cyclic behavior. If, however, the stim-ulus is incorrectly timed, it could drive the system to greater instability. A delay in the action (such as an increase in the money supply) could cause the response (reduced interest rates) to come when the natural behavior has already produced adequate "easy money."

Forced Behavior, Response to Environment

In the United States, our central bank, the Federal Reserve, acts for our government to influ-ence the economic system. We now consider the four phases of its "forced" response [25].

Phase One   Again we start the cycle at the bottom of a recession. The government feels that it must "get the economy moving again." The Fed uses tool 2 as presented above in Ap-pendix A-4. It aggressively purchases Treasury securities in the open market. This action gives the commercial banks excess reserves, and of course they act to use it for increased profits.

They do this by actively making loans in the industrial and commercial sectors. Their inducement for firms to seek loans is reduced interest rates; the banks get no interest on their reserve in their federal bank. Money (through loans) is "created out of thin air." By the multiplier effect (Appendix A-3), the money supply increases rapidly as the bank customers spend the borrowed money. The economy improves, and prices stop falling. As reviewed in Important Terms, (Section 7-6), the money supply is measured by the Federal Reserve. M1 includes currency, checking account balances, and other demand deposits. It is the main measure of "active" money. M2, M3, and L are defined in Appendix A-8. Naturally, the increase in money supply devaluates it. Labor economists anticipate the devaluation. They take advantage of the increased consumer demand to gain higher wages and benefits. Although labor costs are about 80 percent of the cost of all goods, management feels that there will be no difficulty in passing the greater labor costs to the consumer by increasing prices. The price spiral starts and persists through phases two and three. Foreign competition is not considered in this model.

Phase Two　The economy accelerates. Businesses and farmers begin making unwise investments, and the general public makes unwise purchases. Interest rates are still low; management expands capital spending "to keep up with the demand." The increase bids up load demand and interest rates. With the increased demand, prices start to rise rapidly. The increases spread from machinery and equipment to consumer goods.

Phase Three　Prices become "exorbitant" because of the inflated money supply. Citizens are unhappy and want the government to "do something." The public could, however, recognize that the increased prices signal inadequate supply at the original prices. (This is normal for citrus fruits after a drought or freeze.) The only proper solution is to buy and consume less. With the public not yet so educated, the Fed moves to correct for the public's error and raises the interest rate. This is a natural result of the Fed's reducing the money supply by one of the tools reviewed in Appendix A-4.

The reduced money supply reduces consumers' borrowing limit, and they are forced to slow their spending. A recession begins, exposing the mistakes of businesses, farmers, and consumers. They had not reflected and realized that the boom would end.

Phase Four　The recession progresses. Consumers pay down their debts. Workers are laid off, and businesses deplete their inventories to bring production in line with demand. Prices of raw materials and commodities drop, and after a lag, consumer prices decline. When public clamor becomes too great, the Fed restarts phase one.

In Section 7-7 on investment strategies, we note that different types of investments are best in each phase.

Caution: There is a real danger of judgmental error by the Fed. Even when headed by giants such as Paul Volcker in the past and now Alan Greenspan, the Federal Reserve is composed of human beings and they could err. Too tight a money supply could cause multiple bank failures and, in turn, an extended recession. The opposite, too free a money supply, could allow the inflation to accelerate beyond control. Our close coupling with the world economy is a stabilizing factor. A foreign bank account could be protection for the individual.

Random Events　Physical and social systems are constantly experiencing random excitations, such as hurricanes, droughts, oil spills, and deaths of prominent corporate officials and politicians. Since these events, relatively speaking, are of short duration, they have little long-term effects on investments. This is true because of the normal discounting over time by social systems and friction in physical systems. These occasions offer excellent buying opportunities.

A-6 TECHNICAL ANALYSIS [29]

This Appendix reviews only those concepts needed for the "prudent fundamental investor." Technical analysis evaluates stocks entirely in terms of historical market prices. The theory is that all fundamental knowledge is included in the stock's present market price. This approach is similar to the state variable approach in automatic control engineering. The price is the state variable of the stock. These prices are plotted on charts, and their patterns are interpreted by the technician. A good technician is first a good fundamentalist. There are three assumptions required to support the approach: (a) Events are usually discounted (i.e., reflected in prices) in advance by the action (buying or selling) of informed buyers. (b) As informed buyers "accumulate" shares of the stock, its price will not rise until the supply "distributed" at that price is depleted. This often occurs within a neutral trading range and is called *consolidation*. Obviously an uptrend in prices denotes on-balance buying; a downtrend is indicative of greater selling. Thus, assumption (b) is that there is always a period of accumulation before an uptrend in a stock's price. The skilled technician seeks to identify it in the neutral range, to anticipate rather than react. (c) This assumption is an extension of the first two. It states that a short period of consolidation, really a phase of "backing and filling," will be followed by a short-term movement, up or down, in the stock price. Conversely, a long consolidation period will be followed by a greater potential for price move.

In technical analysis, three basic types of charts are used; they are the line, the bar, and the point and figure. Only the line is reviewed. Depending on the investment horizon, the prices are plotted daily, monthly, or quarterly. At each interval, a vertical line is drawn between the high and low prices, and a short horizontal line is drawn through the vertical one at the closing price. The range of the vertical lines in a consolidation phase before an uptrend is called the *support area,* in cases of later declines. The consolidation range before a downtrend is the *resistance level,* in cases of later rises.

Trend Analysis

Many practitioners of technical analysis use this as a caution signal. The price trends give this warning. A modified line chart is useful, with only the closing prices indicated. As in Figure 7-1, the trend is shown by a straight line through a sample firm's (GE) closing prices in the period considered. Also plotted is a 25-day moving average.

Moving Averages

These act as low-pass filters that smooth out most of the short-term random events (see Important Terms in Section 7-6). Time periods of 10 days, 25 days, and 200 days are used, respectively, for short-, intermediate-, and longer-term trend analysis. At a given time, the moving average for an n-day period is the sum of the prices (we use the closing price) for the last n days divided by the number n. The average for the next day is the original sum plus the latest price minus the past $n + 1$ day's price, again divided by n. In turn, after obtaining the original sum, later averages are easy to determine. To start calculations, you might start with a 10-day average and each day extend it until reaching 25.

Relative Strength

For comparison, a standard is needed. For determining the moving average, many use an alternative to the stock's closing price. Instead of using the stock's "raw" price, the method uses the stock's closing price divided by that day's S&P 500 closing price (some newspapers call it the S&P composite). Multiplying the ratio by 100 puts it in percent, and the larger value seems more convenient. An increase or decrease indicates respectively if the stock is doing better or worse than the market. For longer-term investing, this is a very significant indicator. Its importance depends on the belief that over the long term, the market will rise to reflect the increase in wealth of our overall system.

Since our Child is in the Money Game, we might desire to get in or out of a stock depending on how it is doing, rather than the market as a whole. For this purpose we develop a relative strength indicator for the stock relative to its past performance. For such an RSI (relative strength indicator), I suggest using the ratio of the stock's closing price to its 25-day moving average, then forming a 5-day moving average of this ratio multiplied by 100. As an engineer you might feel some other reference more valuable. In the text, we discuss using this RSI as a signal to buy or sell. This RSI for GE is also plotted in Figure 7-1. Note a characteristic of moving averages; they lag the raw data.

A-7 LEVERAGE: BUYING ON MARGIN

This is an arrangement of borrowing money to purchase stock. Presently, the Federal Reserve has set the margin at 50 percent (the percentage of the stocks' cost at the time of purchase that must be paid in cash). We use an example to show possible gains and losses. An individual has

FIGURE 7-1
Trend analysis.

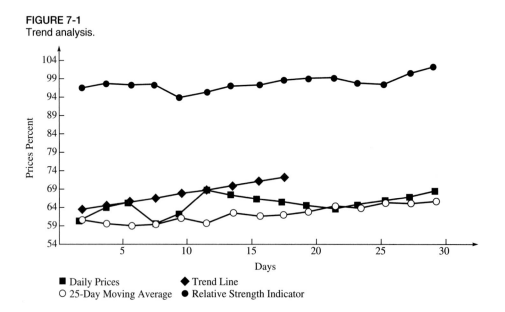

$2000 with which to buy stock. For simplicity, we neglect the nonnegligible broker's commission. The balance sheet equation is applicable. Consider the following cases:

A Cash Transaction: $2000 Invested

1. Price per share of $20:
   Number of shares   Assets = Equity + Liability
         100                $2000 = $2000 + $000.00
2. Price per share goes to $25:
         100                $2500 = $2500 + $000.00
   Gain, unrealized until the stock is sold, is $500.
3. Price per share drops to $15:
         100                $1500 = $1500 + $000.00
   Loss, realized when the stock is sold, is $500.

A Margin Transaction: $2000 Invested, $2000 Borrowed

1. Price per share of $20:
   Number of shares   Assets = Equity + Liability
         200                $4000 = $2000 + $2000
2. Price per share goes to $25, and a margin transaction:
         200                $5000 = $3000 + $2000
   Gain, unrealized unless the stock is sold, is $1000.
3. Price per share drops to $15, transaction, 50 percent margin:
         200                $3000 = $1000 + $2000
   Loss, realized when the stock is sold, is $1000.

We immediately recognize that in a margin transaction, there are greater potential gains and losses. The risks are increased by a factor equal to the reciprocal of the margin, expressed as a decimal. In 1929, the margin was only 10 percent. With the corresponding factor of 10, the market was extremely unstable. It was like combining power steering with the "high factor" of an emotional driver.

A-8 GLOSSARY*

Not all these terms appear in the text; others are included because they occur frequently in the financial literature.

accounts payable   Debts to suppliers of goods or services.

accounts receivable   Amounts due from customers.

bond   A written commitment to pay a scheduled series of interest payments plus the face value (principal) at a specified maturity date.

certificate of deposit (CD)   A marketable time deposit.

commercial bank   A privately owned, profit-seeking institution that accepts demand and savings deposits, makes loans, and acquires other earning assets, particularly bonds and shorter-term debt instruments. In the United States, a commercial bank may receive its charter from the federal government (in which case it is a national bank) or from a state government (a state bank).

---

*Material on pp. 305–308 is reprinted from *Economics,* First Edition, by Wonnacott and Wonacott, published by McGraw-Hill Publishing Company, 1979.

consumer price index (CPI)   A weighted average of the prices of goods and services commonly purchased by families in urban areas, as calculated by the U.S. Bureau of Labor Statistics. Over the past 20 yr, this index has tracked the price of gold.

corporation   An association of stockholders with a government charter which grants certain legal powers, privileges, and liabilities separate from those of the individual stockholder-owners. The major advantages of the corporate form of business organization are limited liability for the owners, continuity, and the relative ease of raising capital for expansion.

current liabilities   Debts that are due for payment within a year.

depreciation   (1) The loss in value of physical capital due to wear and obsolescence. (2) The estimate of such loss in business or economic accounts, that is, a capital consumption allowance.

discounting   (1) The process by which the present value of one or more future payments is calculated, using an interest rate. (2) In central banking, lending by the central bank to a commercial bank or banks.

discount rate   (1) In central banking, the rate of interest charged by the central bank on loans to commercial banks. (2) Less commonly, the interest rate used in discounting.

Eurodollars   Deposits in European banks, denominated in U.S. dollars.

exchange rate   The price of one national currency in terms of another.

externality   An adverse (or beneficial) side effect of production or consumption for which no payment is made. Also known as a spillover or third-party effect.

FASB   Financial Accounting Standards Board.

fiat money   Paper money that is neither backed by nor convertible into precious metals but is nevertheless legal tender.

financial intermediary   Institution that issues financial obligations (such as demand deposits) in order to acquire funds from the public. The institution then pools these funds and provides them in larger amounts to businesses, governments, or individuals. Examples are commercial banks, and savings and loan associations.

financial market   A market in which financial instruments (stocks, bonds, etc.) are bought and sold.

fiscal policy   The adjustment of tax rates or government spending in order to affect aggregate demand.

floating (or flexible) exchange rate   An exchange rate that is not pegged by monetary authorities but is allowed to change in response to changing demand or supply conditions. If governments and central banks withdraw completely from the exchange markets, the float is clean.

fractional-reserve banking   A banking system in which banks keep reserves (in the form of currency or deposits in the central bank) equal to only a fraction of their deposit liabilities.

Gresham's law   Crudely, this says that "bad money drives out good."

incremental cost   The term which business executives frequently use instead of marginal cost.

investment bank   A firm that merchandises common stocks, bonds, and other securities.

investment good   A capital good, such as plant, equipment, or inventory.

leading indicator   A time series that reaches a turning point (peak or trough) before the economy as a whole.

leverage   The ratio of debit to net worth.

limited liability   The amount an owner-shareholder of a corporation can lose in the event of bankruptcy. This is limited to the amount paid to purchase shares of the firm.

line of credit   Commitment by a bank or other lender to stand ready to lend up to a specified amount to a customer on request.

M1 The narrowly defined money stock; currency (paper money plus coins) plus demand deposits held by the public (that is, excluding holdings by the federal government, the Federal Reserve, and commercial banks).

M2 The more broadly defined money stock; M1 plus savings and time deposits in commercial banks (excluding large certificates of deposit).

M3 An even more broadly defined money stock; M2 plus deposits in mutual savings banks, savings and loan shares (deposits), and credit union shares (deposits).

macroeconomics The study of the overall aggregates of the economy (such as total employment, the unemployment rate, gross national product, and the rate of inflation).

margin call The requirement by a lender who holds stocks (or bonds) as security that more money be put up or the stocks (or bonds) will be sold. A margin call may be issued when the price of the stocks (or bonds) declines, making the stocks (or bonds) less adequate as security for the loan.

margin requirement The minimum percentage which purchasers of stocks or bonds must put up in their own money. For example, if the margin requirement on stock is 60 percent, the buyer must put up at least 60 percent of the price in his or her own money and can borrow no more than 40 percent from a bank or stockbroker.

market share Percentage of an industry's sales accounted for by a single firm.

microeconomics The study of individual units within the economy (such as households, firms, and industries) and their interrelationships. The study of the allocation of resources and the distribution of income.

mixed economy An economy in which the private market and the government share the decisions as to what shall be produced, how, and for whom.

monetarism A body of thought which has its roots in classical economics and which rejects much of the teaching of Keynes' general theory. According to monetarists, the most important determinant of aggregate demand is the quantity of money; the economy is basically stable if monetary growth is stable.

monetary policy Central bank policies aimed at changing the quantity of money or credit conditions, for example, open-market operations or changes in required reserve ratios.

money illusion Strictly defined, people have money illusion if their behavior changes in the event of a proportional change in prices, money incomes, and assets and liabilities measured in nominal (not real) money terms.

money market The market for short-term debt instruments (such as Treasury bills, municipals, and bonds issued by local governments).

near-money A highly liquid asset that can be quickly and easily converted into money. Examples are savings deposits or Treasury bills.

negotiable order of withdrawal (NOW) A checklike order to pay funds from an interest-bearing savings deposit.

nominal Measured in money terms (current dollar as contrasted to constant dollar or real).

normative statement A statement about what should be. Contrast with a positive statement.

NOW account A savings account against which a negotiable order of withdrawal may be written.

open-market operation The purchase (or sale) of government (or other) securities by the central bank on the open market.

opportunity cost The amount that an input could earn in its best alternative use. The alternative that must be forgone when something is produced.

Pareto improvement Making one person better off without making anyone else worse off (named after Vilfredo Pareto, 1848–1923).

perpetuity (or "perp") A bond with no maturity date that pays interest forever.

preferred stock   A stock that is given preference over common stock when dividends are paid; that is, specified dividends must be paid on preferred stock if any dividend is paid on common stock.

price-earnings ratio (P/E)   The ratio of the price of a stock to the annual (after-tax) earnings per share of stock.

prime rate of interest   The interest rate charged by banks on loans to their most creditworthy customers.

profit-and-loss statement   An accounting statement that summarizes a firm's revenues, costs, and income taxes over a given period of time (usually a year). An income statement.

prospectus   A statement of the financial condition and prospects of a corporation, presented when new securities are about to be issued.

rate of return   The annual revenue from the sale of goods or services produced by plant or equipment, less operating costs (labor, materials, etc.) and depreciation, as a percentage of the depreciated value of the plant or equipment.

real capital   Buildings, equipment, and other materials used in production, which have themselves been produced in the past: plant, equipment, and inventories.

recession   A cyclical downward movement in the economy, as identified by the National Bureau of Economic Research. A downward movement in the economy involving at least two consecutive quarters of decline in seasonally adjusted real GNP.

risk premium   The additional interest or yield needed to compensate the holder of bonds (or other securities) for risk.

rule of 72   A rule that tells approximately how many years it will take for something to double in size if it is growing at a compound rate. For example, a deposit earning 2 percent interest approximately doubles in $72 \div 2 = 36$ yr. In general, a deposit earning x percent interest will double in about $72 \div x$ years.

secular trend   The trend is relevant to economic activity over an extended period of many years.

short sale   (1) The sale of a borrowed asset (such as a stock) to be repurchased and returned to the lender at a later date. (The purpose is to make a profit if the price falls prior to the repurchase date.) (2) A contract to sell something at a later date for a price specified now.

speculation   The purchase (or short sale) of an asset in the hope of making a quick profit from a rise (fall) in its price.

state bank   A commercial bank chartered by a state government.

Treasury bill   A short-term (less than a year, often for 3 months) debt obligation of the U.S. Treasury. It carries no explicit interest payment; a purchaser gains by buying a bill for less than its face value.

undistributed corporate profits   After-tax corporate profits less dividends paid.

yield   Of a bond, the discount rate at which the present value of coupon payments plus the repayment of principal equals the current price of the bond.

LEGAL ASPECTS OF ENGINEERING

CHAPTER OUTLINE

The legal and ethical aspects of engineering serve as guides for behavior.* The former are based on laws of human origin and are enforceable at law. The ethical aspects appeal to our higher being and are addressed in Chapter 9. In this chapter we present the legal fundamentals and how they govern the relationships between people. An engineer in professional and personal life will have many legal demands, primarily concerning contracts and property liability. From the material presented, you can gain a degree of legal competence. It will be adequate to anticipate the effects of carelessness in dealing with others, to recognize trouble, and to know when an attorney is needed. Again by "preparing ahead of need," you will be able to prevent costly lawsuits for your company and for yourself. Contracts, presented here in detail, outline the rights and responsibilities of parties. Their close study is also a preventive measure. After contracts, property (real and personal) and the essence of criminal and civil law are reviewed. Finally, the presently very important elements of product liability are presented.

8-1 LEGAL ASPECTS OF ENGINEERING [1]**

Laws began as social customs. Tribal chiefs, and later their priests, developed the idea that the laws were of divine origin. Two codes of divine law influenced Western civilization: the code of Hammurabi (about 2000 B.C.) and the Ten Commandments.

History

Our main legal systems are English common law and Roman civil law. The former is used in most English-speaking countries; the latter, in the rest of the Western nations. Most of the early settlers in the United States brought the system of common law with

---

*Material on pp. 312–320 has been adapted from *Legal Aspects of Engineering,* 4th Edition by Richard C. Vaughn. Copyright 1983 by Kendall/Hunt Publishing Company. By permission.

**A collegiate dictionary such as *Webster's Ninth* defines many law terms.

them; states settled by the French and Spanish still have remnants of Roman civil law. As just presented, the term *civil law* originally referred to law based on the Roman code. It now refers to our system of private law, as opposed to criminal law. Hence, its use in the rest of this chapter means the "system of private law."

Types of Law

We have four basic types of law: constitutional, statute, common law, and equity.

Constitutional Law   This sets the operation of government, that is, the governmental powers and limitations as well as the fundamental principles in the relation between citizens and state. These include rights which may not be infringed.

Statute Law   This is based on statutes. They state the express declaration of the will of the legislature on the subject of the statute. Ambiguities in statutes are frequently resolved in the courts by reference to perceived legislative intent. Hence, judges wield great power and play a major role in application of statutory law. Judges who give relatively broad scope of interpretation to statutes are sometimes called *activists,* while more conservative statutory interpretation is said to be *strict construction* of a statute.

Common Law   This was established in England after the Norman Conquest. Before the conquest there was only local law, peculiar to each town and shire. Common law developed from the practice of judges writing their opinions and giving the reasoning followed in arriving at the decisions. When the situations were similar, other judges followed the past opinions. In legal terminology, this practice is known as *stare decisis* (standing decision). The large body of decisions and judgments of past cases is the major source of the common law system. Thus, it is nonstatutory in nature and is known as *unwritten law.* Although there are statutes in the system, they mean only what the courts interpret them to mean. Hence, the hierarchy of the sources in common law is constitutions, statutes, and cases.

Each state is thought of as having its own body of common law, though there are typically considerable similarities. There is generally no body of "federal" common law.

There are only three remedies available by common law: money damages, return of real property, and return of personal property. For example, if a power company was about to cut a tree on your property, without your prior permission, common law has no remedy. In a court of equity, however, you could quickly get an injunction restraining the power company from cutting the tree.

Equity   This is the fourth type of law; it evolved to meet the limitations of common law. Common law cannot:

1 Prevent a wrong from occurring
2 Order persons to perform their obligations
3 Correct mistakes

Remedies in courts of equity are directed to a person. Common law acts upon things. The most significant equity remedies are the injunction and specific performance.

Injunctions   These can be temporary, preliminary, or permanent. A temporary injunction (a restraining order) can be quickly obtained for cause. An attorney, as a member of the bar in the court's jurisdiction, is also an officer of the court. Accordingly, she can request a temporary injunction from the court, even without notice to the "other side." This is known as ex parte. After it is signed by the judge, anyone who knowingly does not obey the injunction is in contempt of court. The injunction holds the status quo (for a limited time) until a hearing can be held. Preliminary and permanent injunctions are normally issued after a hearing with both (or more) parties before the judge. When the injunction is permanent, the order holds as long as the cause exists. If all parties agree to an injunction, sometimes the judge's order is termed a *consent order,* also known as a *consent decree.*

Specific performance   This remedy is used in contracts involved with unique things, such as land, an antique, or a rare painting. Consider a contract for the sale of a specific parcel of land. If the seller refuses to deed the land to the buyer, the seller may be forced to do so by an order of specific performance from a court of equity.

Characteristics of Equity   There are three: choice, response, and privacy.

Choice   Generally, the plaintiff cannot choose between courts of equity or common law. Two requirements determine in which court the case will be heard. For courts of equity both must be met: (1) Remedy at law must be shown to be inadequate, and (2) a property right must be involved.

Response   As noted above, a temporary injunction can be quickly obtained. Also, since there is usually no jury in equity cases, the extensive time for selection of and deliberation by a jury is eliminated. However, a judge can take a decision under advisement with virtually no time limit. Rendering a decision in the courtroom is called *ruling from the bench.* A judge can choose to have a jury if he feels the need for one.

Privacy   The probability of privacy is greater without a jury.

A court of equity can invent whatever remedy it feels is required for proper equity. It will not, however, give a remedy which is contrary to common law, nor will it act when adequate remedy exists under statute or under common law.

8-2 CONTRACTS [2]

The modern world is one of contracts. They permeate our daily lives and are especially important in our professional interchange. All agreements, morally at least, are contracts. Formally, a contract is an agreement enforceable at law.

Contract Classifications

An *executory* contract is one in which the agreed upon promises or actions have not yet been done. An *executed* contract is one in which the promises or actions have been accomplished.

Expressed and Implied Contract

An *expressed* contract is one in which the terms have been explicitly stated orally or in writing. Nothing is inferred. An *implied* contract is one in which the terms have been inferred from general implications or deduced from the general language.

The Five Elements of a Contract

There are five elements of a valid contract. Each must exist, according to law.

1 Agreement
   a Offer
   b Acceptance
2 Competent parties
3 Consideration
4 Lawful purpose
5 Form

Agreement   As implied in the list of contract elements above, an agreement consists of an offer by the offeror and an acceptance by the offeree.

Offer   Normally, it makes two statements, explicitly or implicitly: (1) what is sought (by the offeror) and (2) what the offeror is willing to do in return. When the intent is clear, no formal statement is required. Solicitations, however, are a class not viewed by the court as offers. This class includes advertisements.

Acceptance   It is binding at the time given. It must, however, comply in every detail with the offer. The "time" for an oral offer is the instant of oral acceptance. Silence is not ordinarily considered an acceptance. For offers made by mail or telegraph, the "time" is set when the letter is delivered to the mail system or the telegram is given to the telegraph agent.

Termination of Offer   It may be terminated in the following ways: acceptance, revocation or withdrawal, rejection, death of a party, or expiration of a time period if one was stated in the offer. Those other than acceptance (reviewed above) are reviewed in the subsequent paragraphs.

Revocation

The offer may typically be withdrawn by the offeror any time before its acceptance. The withdrawal is not effective until the time it is made known to the offeree. The "time" is the same as that given above for the acceptance.

Rejection

This could be explicit, or it could be effected if the offeree makes a qualified acceptance or request for modification. Inquiry about a change, however, does not act as a rejection.

Death and Expiration

Death of either party terminates the contract, for of course there could be no "meeting of minds" as implied by the condition of agreement. The expiration of an option supported by consideration, however, ends only on expiration of the time period for which it was open.

Competent Parties   Two or more minds must meet in agreement. A person cannot contract with himself. As an example, an executor of an estate cannot contract for his own services for the estate.

Agreement implies competence. Legal competence requires that all parties be sane adults. In the past, courts declared contracts void if one party was insane. The minds (of the parties) could not meet in agreement. Today, court decisions hold that such contracts are voidable. Generally a voidable contract is one in which one or both parties may avoid the contract if desired. In this case, the option of voiding the contract belongs only to the insane, after sanity is recovered. For normal contracts, courts do not examine the value of the consideration. They do, however, in those involving an insane. If it seems that one party took advantage of the insane, the court allows the insane to avoid the contract.

Intoxicated persons are generally held responsible by the courts. They hold that intoxication is a voluntary state and that persons should have foresight to avoid getting drunk.

Legal adults must be 21 years of age. According to common law a person is an infant until 0:00 hour on the day before his twenty-first birthday. There is no legal adolescence. Interestingly, different from partnership with another adult, an adult in partnership with a minor (legally an infant), on the occasions of bankruptcy, can lose only whatever values were contributed.

Contracts of more than two parties can be joint, several, or joint and several. In joint contracts, persons merge their interest and form a party to a contract; together (jointly) they agree to do a certain act according to the terms of an agreement. The liability is similar to a partnership. In the several contract, the persons who make up a party agree severally to pay a consideration (restricting themselves to a certain amount each). The liability of each is limited to the amount stated. The agreement of cooperating individuals can be jointly and severally in a contract. Accordingly, such a contract is treated as a joint and several contract. If necessary, action can be taken jointly or severally, but not both.

Consideration   This is the act or forbearance one party gives in exchange for the act or promise of the other. There are four essentials of a valid consideration: It must (1) have value, (2) be legal, (3) be possible, and (4) be in the present or future.

Lawful Purpose   The agreement must be to do or not to do a lawful act. Contracts that are executory will not be enforced if they are (1) in conflict with the written law, (2) opposed to common law, or (3) contrary to public policy. If the contracts are executed, the courts will not consider their cases.

Form It seems wise to put in writing any nontrivial agreement, even with yourself. The process clarifies the agreement to you and to a second party as in a contract. In most circumstances an oral contract can be as effective as a written one; however, by law, certain kinds (reviewed below) are required to be written. The requirement results from the English Statute of Frauds. In 1677, it was passed as "an Act for Prevention of Frauds and Perjuries." The purpose was to relieve the courts of the necessity of considering certain types of contracts unless their terms were set forth in writing and signed. Only words adequate to establish the material provisions of the agreement are necessary. Also, only the signature of the defendant is required, and it could be initials, rubber stamp, or in ink or pencil, anything intended by the party to establish identification and assent. It may be anywhere on the document.

In the English Statute of Frauds, only sections 4 and 17 affect contracts. Except for minor changes, they have become law in all of the states in the United States.

Under section 4, no action of enforcement can be brought for certain types of agreements unless they are in writing and signed by the defendant or her agent. These types involve promises; only two are probably of interest to the engineer: (1) those involving real property and (2) those which cannot be performed within a year.

Section 17 puts constraints on contracts for sale of goods or merchandise which involve considerations in excess of certain amounts. The amount was originally 10£ (10 pounds sterling). Today, each state has set its upper limit for oral contracts. These limits vary from $50 to $2500 (Ohio); most are between $50 and $500. For an agreement of consideration over the limit to be enforceable, one of the following must be done to secure it:

1 Part of the goods, wares, or merchandise must be accepted by the buyer.
2 The buyer must pay something in earnest to bind the bargain.
3 Some note of the agreement must be written and signed.

In reviewing forms, the fifth contract element, we learn that most contracts can be and are oral. Although written agreements "make best friends," usually only those with considerations of significant amounts, and all those for sale of real property, are required to be written.

We next consider agents, the important extension of busy engineers and managers.

Delegation

Extensive delegation of duties and responsibilities is of greatest importance to large corporations. Delegation may involve one of three relationships: (a) employer-employee, (b) owner–independent contractor, or (c) principal-agent.

Employer-employee relation exists when an engineer's activities are mainly in the plant. She might do research or design, working with other company personnel. The personnel would be peers and management. As management's focus changes from specific influence to seeking specific results, the relation changes to that of an independent contractor.

Owner–independent contractor relation is used to accomplish a specific job. The degree of liability of the employer (owner) is determined by the relationship created.

The degree of control exercised by the employer is the criterion in legally determining that relationship.

The agent acts for his employer (the principal) in dealing with extracompany persons. The agent is still an employee, but in dealing with others, he has special duties and responsibilities. The engineer is often in this position when making a proposal, working on a customer's project, and negotiating financially with the customer.

As implied above, an agency involves three people: (1) the principal who is represented by the agent and is the source of the agent's authority, (2) the agent who represents and acts in place of the principal, and (3) the party or parties with whom the agent deals as a substitute for the principal.

Establishing Agency The agency relation may be established in four ways: by agreement, ratification, estoppel, or necessity.

Agreement Both parties must intend to establish an agency relation. The way in which they express the intent is unimportant. The agency contract is like any other; it may be expressed or implied. For certain purposes, however, a written or sealed instrument may be required. If the instrument to which the principal is to be bound requires a seal, the agent's authority must be in writing and sealed. The frequently needed power of attorney is of this type. It establishes the agent as an attorney-in-fact. Another form is the corporate proxy. By it, the shareholder appoints a certain person to act for her.

Ratification A person without authority may act for another, or an agent in fact may exceed his authority. If in either case the principal assents to the act, an agency is established by ratification. To be binding, however, the principal must have full knowledge of all material facts, without misunderstanding or misrepresentation. Furthermore, the principal must accept all or nothing.

Estoppel This arises in the situation where a person appears to have authority to act for another. Furthermore, despite the lack of authority, the person does act for another. If the principal acts in such a way that the person appears to be his agent, thus deceiving the third party, he is estopped from denying the person as his agent.

Necessity An emergency can create an agency by necessity. It is rare, but to save an employer from a disaster, an employee must deal with others without opportunity to obtain authority.

Agent's Authority This authority, expressed or implied, comes from the principal. When implied, the authority is based on previous activities between the parties or local customs. If the agent exceeds his authority, the principal is not bound. This burden on the third party advises caution on her part. The safest approach is to get evidence of the agent's authority. An agent is chosen because of his particular capabilities and fidelity.

Agent's Duties The principal's trust and the agent's loyalty are implied in this fiduciary (Latin for "trust") relationship. The law enforces them; breach is cause for action. An agent owes his principal a duty of strict obedience in ordinary circumstances. It is not the agent's function to question or judge the wisdom of the principal's orders. Rather, it is the agent's function to do all possible to carry them out. Exercise of the necessary care and skill is tested by that expected of the reasonable, prudent person. Finally, an agent has the duty to act for and accept compensation from only one principal. For instance, he cannot represent both buyer and seller in a sales contract.

Principal's Duties to His Agent   There are three: (a) The principal must compensate the agent in accordance with the contract. If there is no contract, the principal must pay a reasonable amount for the services. If, after the agent has preformed his duties, the principal and the third party do not complete the transaction, compensation is still due the agent. (b) The principal must reimburse the agent for expenses properly and reasonably incurred, for and on behalf of the principal. (c) If in following the principal's orders the agent injures someone and has to pay for the injury, the principal must indemnify (protect from harm) the agent.

Third Party Rights and Duties   Normally, the third party has no action available against the agent. If the agent has acted according to the expressed and implied authority, only the principal and the third party are bound. Ordinarily, an agent acts in a manner that indicates two aspects: (1) who his principal is and (2) the fact that he is an agent for that principal. If, however, he agrees and signs in his own name or signs "Jones, agent" but gives no principal, the third party can probably act against the agent. Also, if the agent acts as surety (one bound with and for another) for his principal, the third party could hold the agent liable.

Termination of Agency   This can be accomplished by the parties or by law. The termination by parties can be effected in four ways: (1) by stating on the contract a time limit, (2) by the agent accomplishing the specific activity for which the agency was created, (3) by discharge of the agent by the principal, and (4) renouncing the agency. Generally, the principal and agent have the power to terminate the agency, but sometimes not the right. The right might have to be settled in court, and the loser would have to pay damages. The revocation is binding only on third parties who have been given notice.

The termination by law could result in two ways: (1) by destruction of the subject matter, such as by the wrecking of an auto which an agent was commissioned to sell, and (2) by change in the status of a party. If either the principal or the agent dies, the agency is terminated. No notice to the third party is required. If either the principal or the agent becomes insane, the agency is terminated. Notice must be given to third parties unless the agent has been legally declared insane.

Specifications

At some time in their careers, all engineers seem to be involved in writing specifications and interpreting contracts in which specifications are significant. Proper specifications help to make contracts explicit and clear.

In engineering, the word *specification* refers to the statements and paragraphs by which an individual or group tells another individual or group specifically what is desired, and how, where, and when it is to be done. The *specs,* as they are often called, give two significant advantages: (1) a permanent record is produced as a memory aid and for resolving misunderstandings, and (2) a definite plan is developed. The document usually has two parts: One is general, such as for the policies of a firm in dealing with all outside parties on quality, reliability, and safety. The other part is detailed, such as for the particular purchase of a piece of equipment or the specific objectives of a project. This second part, while being as broad as possible, might have to include a required schedule of work. This schedule could be required in order to coordinate with work done in-house or by other contractors. A CPM diagram, as discussed in Chapter 5, greatly contributes to precision.

The techniques in Chapter 4 on written communication are especially relevant here. We will review the most important ones. First is clarity and conciseness. Paraphrasing Albert Einstein, make the specifications no longer than necessary, to unequivocally convey the "what, how, when, and where." Use simple grammatical structure and simple words. Avoid jargon, abbreviations, and symbols that have any possibility of being unfamiliar to the reader. Be extremely careful in the use of the pronoun. In some situations *it* could be misunderstood as to what or to whom *it* refers. Do not hesitate to use nouns instead of pronouns to be more explicit. Bids on equipment and projects are increased to handle contingencies that might result from unclear statements. Clearness is the paramount quality of a specification. This is impossible, however, without a thorough knowledge and understanding by the writer. Seek that understanding and knowledge first. Then, write not so that you will be understood, but so that you cannot be misunderstood.

Reference specifications are those well established and generally familiar to the trade. Through use, they have been purged of errors and vague wording. The U.S. government has a vast array of published standard specifications. However, there is a very real danger in using them. There could exist some small difference between the reference specs and your project. These are difficult to recognize, and only one or two can cause expensive misunderstandings. I recommend your reading these specifications in a relaxed mood. Try to imagine that you are the contractor or bidder to whom the specs will be sent. A relaxed, undisturbed environment is of greatest importance, to help ensure that the specs say what you want them to say.

A major purpose of contracts, agencies, and specifications involves personal and real properties. The following material on *them* (I would not use a pronoun here if this was part of a specification.) is important in both the engineer's personal and professional activities.

8-3 PERSONAL PROPERTY

Real property is defined as land and anything, as by reason of its mass, that is firmly attached to it. Personal property, also called *chattel,* is all other property, as can be removed without damage to the real property. This includes tangibles such as goods, money, accounts receivable, and other evidence of debt. Intangibles include rights-of-way, patents, goodwill, and franchise rights. Combinations of real and personal property are called *mixed property.*

Personal property can be acquired by (a) original acquisition, (b) procedure of law, and (c) acts of other persons.

Original acquisition does not occur often today since it involves being the first person to find an unowned object and claim it. Engineers, however, often acquire intellectual property by creating it. Examples are inventions and patents through mental and physical efforts.

Procedures of law involve the transfer of property. This usually occurs legally in four ways: from intestate death, mortgage foreclosure, judicial sale, and bankruptcy.

Intestate is the term which indicates that a person died without a will. In such instances, the deceased's property will be distributed as directed by state statutes. If no relatives of the deceased can be found, the property *escheats* to the state. The details of wills are examined in Section 8-4.

Title to personal property may be acquired lawfully by the acts of other persons: will, gift, contract, or abandonment.

Bailment

This relationship usually benefits both parties involved, such as your leaving your auto at a garage for repair. Similar to property ownership, bailment occurs when personal property is left by the bailor (the owner) with a second party (the bailee). There are three requirements: (1) Title of the property remains with the bailor, (2) possession of the property is completely surrendered by the bailor to the bailee, and (3) the intent by both is to return the property from the bailee to the bailor at the end of the bailment.

Property being held for a bailor by a bailee cannot successfully be taken for the bailee's debt. For a judgment against a bailor, his bailed property can be obtained from the bailee.

Intellectual Property*

This type of personal property can be classified as (1) copyrights, (2) trademarks, (3) trade secrets, and (4) patents. The last two are most important to the engineer and will be considered in more detail.

Copyrights   Today they are controlled exclusively by federal law; before 1978 copyrights were governed by state laws and federal statutes. A copyright grants certain rights to the originator of copyrightable work. This category includes anything that can be copied: printed material such as books (even maps and tables); lectures, including addresses and sermons; dramatic or music compositions; photographs; motion pictures; and software. When an individual commissions in writing another to do the work, it is referred to in the statute as "work for hire." In such a case, the copyright is owned by the one who commissioned it.

The owner of the copyright is granted the exclusive rights to reproduce and distribute the work; to prepare derivatives such as condensations, abridgments, and trans-

*Material on pages 321–328 has been adapted from *What Every Engineer Should Know About Patents* by Konold, Tittel, Frei, and Stallard, published by Marcel Dekker, Inc., N.Y. 1979. Reprinted from pp. 3, 12, 17, 19, 20, 26–27, 79, 83–85. By courtesy of Marcel Dekker, Inc.

lations; and to perform or display it publicly. Any of these rights can individually or collectively be licensed, sold, or assigned.

The protection gained by the copyright covers only the manner in which the copy-righted idea is expressed, not the idea itself. No more than 250 words (even scattered) or 5 percent of the original work can be duplicated without written permission. To a limited extent, however, others can copy the material without infringing the copyright: under the "doctrine of fair use," libraries and teachers can make reproductions for classroom use or the material can be quoted in reviews. Generally, material can be copied if the act does not unreasonably deprive the owner of profits otherwise forth-coming.

Under the new law of 1978, works created after January 1, 1978, have protection when they are created, and this protection extends for the originator's life plus 50 yr. When the works are distributed, one must be sure that a copyright notice, such as a symbol © or the word *copyright* or its abbreviation *copr,* is on each copy in a location reasonably sure to be seen, like the title page. Also, along with the symbol, the year of first publication and the name of the copyright owner must be indicated.

Trademark   This can be a word, name, symbol, device, or combination of these. Trademarks are used by manufacturers, organizations, or merchants. The purpose is to uniquely identify the source of goods or merchandise. A company such as IBM may use the same trademark on a variety of its products (computers, monitors, printers). As for Sears, the trademark *Craftsman* is used on many products sold by them but made by various manufacturers to Sears' specifications.

Trade Secrets   For many years our courts have respected a firm's rights regarding trade secrets. The protection prevented disclosure or unauthorized use of this knowl-edge. The trade secret right was reaffirmed by the Supreme Court in 1974. It protects the competitive advantage of a firm, derived from its private (i.e., confidential) business infor-mation. The information could be a special design, formula, or method of production.

The trade secret right depends significantly on whether the owner can establish that adequate security precautions had been taken; for instance, the information was in only a few laboratory notebooks which were properly safeguarded, and employees signed contracts agreeing not to divulge the information.

Trade Secret versus Patents   As reviewed below, patents give protection for only 17 yr. Trade secrets last as long as they are not known by others. Sometimes another firm attempts a reverse-analysis, that is, tries to determine the information from the end product. Recently, Advanced Micro Devices reverse-engineered Intel's 386 microprocessor. (It seems that competitors are highly motivated.)* Also, the infor-mation could be independently developed and patented. In Chapter 2 we considered decisions under uncertainty. The technique using the estimated costs and rewards is appropriate here. Recall that the cost and rewards are multiplied by estimates of their respective probabilities to indicate the better decision.

---

*After considerable legal costs. Intel was able to reestablish its right (1992).

Patents You are in a fortunate position to be creative. Your engineering education and environment are conducive to inventiveness. Since creativity is part of your work, it is highly probable that some of your creations will be worthy of being patented. Beyond suggestions made below, your first move in patenting a new invention is to contact an attorney or a patent agent with special knowledge in patent law. A *patent attorney* is one who is officially registered to practice (i.e., assist you or your company) before the U.S. Patent and Trademark Office. A *patent agent* is one who is so registered to practice but is not an attorney. The material in this section will enable you to understand the work of both and will improve your interactions and communications with them.

What Is a Patent? It is a right. The word *patent* is from Latin; it means "to be open (to the public)." In exchange for the inventor's disclosure, the U.S. Patent and Trademark Office grants to the inventor or her assignee the *right* to exclude others, for a limited term, from making, using, or selling the invention other than for experimental purposes. The term is for 17 yr, except for design patents discussed below. In this way, the goal of the U.S. Constitution is fulfilled: "to promote the progress of . . . useful arts, by securing for limited times to . . . inventors the exclusive right to their . . . discoveries." Generally an inventor has always had the right to manufacture and sell any product made possible by the invention, as long as it is not covered by another's patent. Without a patent, however, so could everybody else. The patent must be filed in the name of the original inventor(s). This means that a person cannot patent subject matter created by another, even if no effort was made to patent it. It also means that even though a corporation is a legal person, only "natural" persons may apply for a patent.

What Can Be Patented The latest revision of the patent law is a 1952 Act of Congress (effective January 1, 1953). The right to patent and what may be patented, however, is well stated in section 31 of the old U.S. Patent Act: "Any person who has invented or discovered any new and useful art, machine, manufacture, or composition of matter, or any new and useful improvements thereof, or who has invented or discovered and asexually reproduced any distinct and new variety of plant, other than a tuberpropagated plant, not known or used by others in this country, before his invention or discovery thereof, and not patented or described in any printed publication in this or any foreign country, before his invention or discovery thereof, or more than one year prior to his application, and not in public use or on sale in this country for more than one year prior to his application, unless the same is proved to have been abandoned, may, upon payment of the fees required by law, and other due proceeding had, obtain a patent therefor." In short, this quotation states that the Patent Office grants patents for processes, machines, articles of manufacture and compositions of matter, and improvements thereof (also asexually produced plants), provided they are new, useful, and unobvious.

The *newness* requirement is not met if any of the following exist:

a The subject matter was publicly known, used, or on sale in this country before you made your invention, or for more than a year before you filed a patent application.

b The subject matter was published anywhere (worldwide) before you made your invention, or for more than a year before your patent application was filed.

c Any other person, who did not suppress or conceal it, made the invention before you did.

From (a) and (b) you recognize the importance of filing for the patent application within 12 months after any commercial operation or publication. An unrestricted disclosure of your invention to any outside party can constitute publication. It would initiate the beginning of the 12-months period. This period is also called the 1-yr "grace period." Furthermore, (b) indicates the significance of properly dated, well-kept research notebooks.

All of the foregoing relates to U.S. patent rights only. If your company is possibly involved in foreign patents, stricter rules might apply. For example, the U.S. 1-yr grace period might not be available. Naturally, review the specific requirements with your patent attorney or supervisor.

As reviewed later, item (c) could be the basis by which the Patent Office could decide who (if anybody) gets the patent when there are two or more competing applicants.

The *utility* requirement simply means that some purpose or usefulness must be apparent or demonstrated.

Unobviousness indicates that the invention is not obvious to persons having ordinary skills in the art of the invention. This standard works on a hypothetical-type person and is clearly a "floating" standard. In application, it causes different results in different areas of technology.

Types of Patents   There are two types of concern to engineers: the design and the utility.

The design type is solely for ornamental purposes. It can be copyrighted. The fees were determined by the term of these patents. The terms of new design patents are now all 14 yr.

The utility type is generally more difficult to obtain and is potentially much more valuable because of "broader" coverage. This type constitutes the subject matter presented in the above quotation from section 31 of the old U.S. Patent Act and includes processes, machines, etc.

Your Invention and Its Documentation   Recall that all projects have two phases: exploratory and implementation. So it is with inventions. For them, the phases are officially referred to, respectively, as "conception" and "reduction to practice." The dates of each are very important in establishing your rights against alleged patentees. Furthermore, after conception, progress toward reducing it to practice must be done diligently. *Diligence* means nearly continuous work by you or by those working with you toward producing "a physical implementation." Except for illnesses or factors such as component availability, even a few weeks' delay could cause you to lose the patent to a challenger.

Your task is to establish three things: the conception date, the date of reduction to practice, and the diligence in between. This is best done by well-kept records in a bound (not loose-leaf) notebook with numbered pages. Be sure not to tear out any pages or leave any pages blank. Use a pen, erase nothing, but delete material by using one neat line. As you remember from college laboratories, you might need the information later. Firms usually supply the bound notebooks to their engineers. If doing personal investigation, get one from a bookstore. Records should be *contemporary* (a

Latin word meaning "in time") with your work. Any graphs, photos, etc., should be taped in, signed, and dated. For experimental work, record the setup, the model, and preferably the serial numbers of equipment used. Keep accurate records of efforts, successes, and failures; recall that you may need to establish diligence. That means continuous efforts over time, successes and mistakes. Lack of success with diligent efforts can help establish the unobviousness of your invention. Let us review the overall process.

When attempting to prepare adequate entries, bear in mind that many years could go by before any controversy might develop involving your notes. Also, the notes should be self-explanatory, i.e., not requiring your presence for a full understanding by others generally knowledgeable in your area of expertise.

At the conception of your invention, explain it in your notebook. Have two persons who are capable of understanding it sign and date your entry with a note: "Read and understood." These witnesses should not be those who might now or later be considered coinventors. Witnessing records by a coinventor might be considered inadequate by the Patent Office to prove diligence or confirm any dates. As with all engineering documentation, these records will justify your making the effort to be well-organized and to present the information as briefly as clarity permits.

Shortly after conception and after you have reflected for a while, start your tentative plans for patenting your invention. You will notice that I used the word *invention* here, instead of the word *idea.* An idea cannot be patented, only the "reduction of the idea to practice," that is, the physical machine, process, etc. The first step in the plan is to discuss your concept with a patent agent or a qualified patent attorney. As with any legal work, she will advise you on the best approach if further pursuit would be worthwhile. Although the patent procedure is designed so that an inventor can patent his own work without help from others, the aid of a qualified person can prevent future complications, such as technicalities concerning special legal wording. Most important, the claims should be written with optimum limits. This is referred to as the "scope" of the patent. If the limits are too broad, the patent will be rejected; if too narrow, the value of the patent will be greatly reduced. Your attorney has expertise in this area. Assuming that your attorney's response has been positive so far, let us continue with the plan.

Patent Procedure   This frequently starts with various types of searches, as prudence indicates, and then the preparation of the application. There is no legal obligation for a search to be performed, and sometimes in the interest of saving time (or money) it is omitted. Either way it is critical that you discuss with the patent attorney any "prior art" of which you are aware, that is, anything similar that you know anything about. This is true because of the statutory date of disclosure to which you must execute an oath or declaration (see below) when filing a patent application. It is important because failure to satisfy the "duty of disclosure" can result in later unenforceability of any issued patent (even if the patent is held valid and infringed). Examples of such "prior art" which you would want to tell your patent attorney about might be (1) prior patents of your employer, (2) pending applications of your employer, (3) patents you know about which belong to competitors (or the competitors' products which you knew to exist before your invention), and (4) pertinent magazine articles or publications. The above list is not all-inclusive; talk with your attorney about them.

There are four types of searches: state-of-the-art, patentability, infringement, and validity. Each can be conveniently preformed at the Patent Office in Arlington, Virginia (Crystal City Complex). There, patents are retrievable by classes and subclasses. The searches are done by attorneys who travel there or by their associates who live nearby. The reports are written and give details relevant to the type of search requested, as well as normally identifying patents, etc., noted during the search. The attorney's knowledge and (equally important) the information you give her are of special value in retrieving the needed information.

State-of-the-Art Search   This is also called a *collection search*. It is particularly valuable early in a project assigned to you. The project could be to improve the productivity of a machine or perhaps the response time of a VLSI (very large-scale integrated) circuit. Goal-oriented projects are conducive to patentable developments. In state-of-the-art searches, patents are sought which give information pertinent to your project, i.e., "background." Typically, the search will include several reports on relevant material. The concepts gained from them will probably suggest original ideas to you. Reviewing these state-of-the-art reports will be effective in your exploratory phase. Personally, I have gained many design ideas from reviewing what others have done. Dissemination of technical knowledge is one of the founding principles behind the patent system (which was 200 yr old in 1990).

Assume now that independently or with suggestions gained by reviewing the state-of-the-art search, you have an invention. Furthermore, assume that you have had it properly documented in your notebook and have a positive response following a discussion with your patent attorney. The next search is for patentability.

Patentability Search   This is to determine two aspects: (1) whether the invention can be patented and (2) what would be the likely scope of its protection. If a state-of-the-art search was made, it could be the basis for the patentability search. Possibly no further search would be needed. Conceivably, either search might show that pertinent prior work will eliminate or sufficiently limit your rights and that the patent would be economically unwise. More likely, however, you have an invention truly new and worth patenting. A thorough patentability search will give two advantages in preparing the application: (1) By knowing about similar patents, your lawyer will be able to stress their differences and highlight the patentable features of yours; (2) also the knowledge about similar patents will permit the lawyer to develop optimum claims, specifically, those as broad as possible but still patentable over other patents. Remember that the broader the claims are the greater the potential value of the patent will be. As in every purchase, it is wise to at least approximately know its cost and weigh that against the probable benefits. If seeking a patent on your own, a discussion about the cost with your lawyer could encourage her to help you keep it low. Generally, the searches are "good investments."

Infringement Search   This search gives information regarding your making, using, or selling products of your invention without infringing an existing, unexpired patent. The information will indicate if "all is clear" or if you will need to negotiate a license to use an existing patent. A *KISS* (keep it simple, senior) story goes like this: The Stool Company had a patent on stools. Mr. Black had a patent on backs (for stools). Neither could make chairs without infringing on the other. Of course these prudent people negotiated to their mutual advantage.

Any prior searches will be helpful in initiating the infringement search. The attorney will first study patents classified in the area of your potential patent. She will also study patents cited by the examiner (who examines patent applications). She will be seeking any material which could limit your patent's claims. The attorney will also order and study a copy of the file wrapper history* of any patent closely questioned, to study its prosecution history. This is important to determine the precise meaning of the patent claims.

On completion, the lawyer will give an opinion. It will consider the range of products that can be made using your patent without incurring infringement liabilities. If needed, she might be able to suggest alternative approaches to avoid infringement. If small, the infringement risk might be a desirable business risk. This study is in greater depth than the others. Thus, from it you should gain substantial information by which to judge these probabilities.

Validity Search   Perhaps your invention could give high promise of great utility and profit, except for a probable infringement. In this situation a validity search might remove the difficulty. The patent lawyer in this search will seek weaknesses in the restraining patent. She will check past literature by which the offending patent can be declared "obvious." Accordingly, she would seek information which could be used to invalidate the problem patent on any of the other aforementioned criteria: novelty and usefulness. The validity search could possibly give an absolute defense, by putting the owner of the problem patent in an uncertain position.

The Patent Application   This has two major parts: the specification, including claims and drawings as needed, and an oath. In addition, the proper fee must be paid: in the early 1990s, it is usually $370. Under the law, the Patent Office Commissioner can (and does!) increase fees every 3 yr to account for inflation.

Specifications   Their purpose is to clearly describe the invention, how to make it, and how to use it. From them, a person skilled in the patent's subject area should be able to build and use the invention. Such a disclosure is termed *adequate.* Your assistance to ensure an adequate disclosure is crucial. Also, the person should be able to employ the best mode of operation as contemplated by the inventor. Specifications should be as clear, concise, and exact as you and your attorney can express them. These characteristics are so important that it is wise to reflect over several days, and for you and her to "brainstorm" the plan and the wording. In these preparations, your attorney will continuously stress the need for the patent's enforcement and wording that strengthens it against challenges. Accordingly, it is best to start by giving her a full understanding of your invention, its technical field, and the relevant terms. Your contribution is particularly important in describing the invention, its implementation, and its operation. This material will be read by the patent examiner. From it, he can more easily establish your invention as being novel, useful, and unobvious. Building on the understanding you give her, the attorney will use her expertise in writing the claims.

---

*The *file wrapper history* includes arguments and challenges to a patent. It also includes a review of applications for amendments and arguments in reference to prior art.

Claims Except for the proper format, claims are the most important aspect of the application. Simply, they indicate what has been invented and what is sought to be reserved for the patent holder. Naturally, it is best to establish as many claims as you can justify. Again, the claims must be sufficient for the patent to be granted and to be of value. The value includes the broadness of its utility; and whether the patent is granted or challenged depends on the present art and whether the patent infringes on existing ones.

Drawings These are frequently used to make clearer the invention's description and operation. They are not normal engineering drawings. For that reason your attorney will probably advise their being prepared by a patent drafter. He is familiar with the drawing formality required by the Patent Office. Drawings are usually advantageous in patent applications. Exceptions could be in those concerning things such as chemical compositions.

The Oath This is an essential of every application. In it, the inventor must attest to a variety of items.

a. He must state that he is the first inventor of the subject matter for which a patent is sought.

b. He denies any knowledge of specific items which would render the application or the sought-after patent invalid or improper. Also, he identifies any corresponding application filed in a foreign country.

c. He must acknowledge, in oath or declaration, a duty to disclose information of which he is aware that is material to the examination of the application.

Execute the oath only after careful review of the application. The care should be for accuracy and completeness of details, especially regarding claims. Discuss with your attorney any deletions, additions, or corrections. Make those you and she feel advisable before executing the oath. If any changes or blanks are filled in after signing, a new oath must be executed. An altered oath would invalidate any patent issued.

Once the application is filed, you are normally not permitted to add new material. Occasionally, at the time of filing, information needed to fully define a claim, or describe subsequent but highly related developments, is not known. In that case, it could be possible to better protect the claim by filing a continuation in part (*CIP*).

In this chapter, the material on patents is presented with three objectives: (1) to give you the confidence to seek the thrill of being creative and to patent the offspring of that creativity, your invention; (2) to put you in an informed position when communicating and working with your patent attorney; and (3) to minimize mistakes and efforts in obtaining a patent.

While our daily lives are usually more concerned with personal property, real property is more likely to require legal action and knowledge. We consider those aspects next.

8-4 REAL PROPERTY

An engineer will probably be involved with real property in two ways: personally when either buying a lot on which to build a home, buying or leasing a house and lot, or (professionally as a manager) building or renovating a plant. We consider details of property transfer (or conveyance) and concurrent rights and duties.

Transfers

Real property and the attendant rights are transferred in four ways: by (a) will or inheritance, (b) sale, (c) gift, or (d) legal action.

Wills   Interestingly, the will was first used only for disposal of real property, and a testament was used to dispose of personal property, hence the common usage of "last will and testament" for one document to dispose of both.

Anyone of sound mind and 21 yr of age may make a valid will in the United States. For women, the age is 18 in some states.

Sanity (or sound mind) by law has three requirements: (1) sufficient mental capacity to comprehend her property, (2) capacity to consider all persons to whom she might desire to leave her property, and (3) understanding that she is making a will.

Inheritance is not limited to relatives of the deceased. Almost anyone or any legal organization may inherit. By law, the murderer of the testator is prevented from the inheritance.

Types of Wills   Besides those ordinarily written, with witnesses described below, there is the nuncupative (or oral) will and the holographic* will (one written entirely in longhand by the testator). Usually no witnesses are required for the holographic. Conditions for validity vary from state to state.

A Valid Will, the Essentials   There are four:

1 The will must be written, that is, typed or in longhand.

2 It must be signed by the testator and sealed in states which require it. *Sealed* originally meant use of a wax to seal the document; the wax was impressed with the coat of arms. A contract so sealed was considered adequately settled as not to require other consideration. Now, as in some states, the written word *seal,* a scratch of the pen, or the letters *L.S.* are sufficient.

3 All wills, except holographic (in some states), must be witnessed. Two or three, depending on the state, are required to attest the signing of the will. The will must be signed by the testator in the presence and sight of the witnesses, and they must sign in the sight and presence of each other and the testator.

4 The will must be published. Publication of a will means that the testator must declare that it is her last will and testament. The witnesses must know that it is a will being signed.

Codicils

After making a will, moderate changes are easily made by means of a codicil. It is used to explain, change, add to, or revoke part of an existing will. Requirements for making a codicil are the same as for a will.

---

*Holo is Greek for "whole"; add *graph* ("write") and you get a word that means "wholly handwritten."

Probate

After the death of the testator, the will is presented to the probate court for official proof that it is valid. State statutes determine the pattern; usually an opportunity is given for challenge of its validity. If it is unchallenged or no challenge is successful, the will is admitted to probate; i.e., it is received by the court as a valid statement of the testator's intent. When the will is drawn up, someone is named as the executor or executrix. After the will is probated, the executor or executrix carries out the terms of the will. This must be done under bond unless, as is usual, the will states that the executor or executrix is specifically exempted from bond. If there is no will or no executor was named, the court will appoint an administrator or administratrix. Their functions are the same as an executor. In the case of no will the administrator must follow state laws for distribution of the property. We now return to the other methods for transfer of real property.

Sales or Gifts   Transfer of real property by sale or gift requires a deed. There are two types: warranty and quit-claim. Usually, only the former is adequate.

A warranty deed warrants that the title transferred to the buyer is good. For both types of deeds, the buyer gets only what the seller has to give. In a warranty deed, however, if the buyer's title under warranty is ever successfully attacked, the buyer may recover any damage suffered from the seller.

A quit-claim deed transfers title but does not warrant it. It transfers to the buyer whatever title to the property the seller may have. Such a deed might be used when the title to the property is clouded because of inheritance by several members of a family.

Elements of a Deed   To be valid a deed must have several essential elements. It must name grantor (seller, if not a gift) and grantee (buyer or receiver). Also the consideration involved must be given, and the property must be described or otherwise identified. Expression of conveyance must be used; it must be signed, sealed (in some states), witnessed, delivered, and then accepted by the grantee.

Any instrument involving real property must be recorded. This is essential for if the grantor were to make a second, fraudulent deed to another, the first grantee to record the deed would have the only valid claim.

Legal Action   The property right of possession and use is lesser than than the right of a public necessity. The legal right to take private property for public need is known as *eminent domain* (or condemnation). This right may be exercised by the state, municipality, or other public entity. Quasi-public private enterprises which mainly serve the public, such as water and utility companies, may petition for and be granted this right.

Only private rights as necessary will be taken, and compensation is made according to an assessment of the market value of the right lost. Either side may appeal.

Landlord's and Tenant's Rights

It is probable that during your first job or after a job transfer, you might rent for short or long periods of time. Making a home does not require owning the house. The wisdom concerning the purchase of a house can be analyzed economically by the techniques in Chapter 6; intangibles can be weighted by the decision theory approach presented in Chapter 2.

The landlord naturally has a right to the rent set in your mutual agreement. Furthermore, she has the right to come pleasantly on the property to collect rent due. In certain states, she has the statutory rights to exercise a lien against the tenant's personal property if other efforts to collect the rent fail. As last resort, she may obtain an order of eviction against the tenant.

The leased property must be returned to the landlord at the end of the lease, in substantially the same condition as when leased. She has available an action of waste against the tenant if waste (disposal of leased assets, even a tree or top soil) can be shown.

The tenant has rights to the leased property free from interference by the landlord. Unless stated otherwise, the lease might also include rights for use of appurtenances such as a small building, tool house, or a separate garage. The tenant has the right to use the property in any way which does not violate the lease. However, if the property is unsuitable for the tenant's purposes, this is not relevant to the lease's legality, unless fraud can be established. If the property will not meet the tenant's needs, he should not have signed the lease.

The landlord might consider certain improvements depending on the terms of the lease. Those done by the tenant's labor or at the tenant's expense should be agreed upon before signing the lease. Furthermore, in all cases, details of the improvements should be expressed in writing.

Unless stated otherwise in the lease, the landlord is responsible for all taxes and assessments.

The tenant has the same liabilities to guests and other third parties as he would if he were the owner. He has a duty to keep the property clean and generally in a safe condition. Only unsafe conditions known to the landlord and not obvious to the tenant might make the landlord liable.

8-5 CRIMINAL AND CIVIL LAW—TORTS

Major divisions in the field of law are

1 Criminal law
2 Civil law
    a Torts
    b Breach of contracts

Society is injured when a crime is committed. Accordingly, the state undertakes prosecution of a crime. Its purpose, if the person is found guilty, is to prevent recurrences of the crime. Of course, the penal system attempts rehabilitation. In a few but growing number of instances, compensation to the victim is sought.

As noted earlier, except in the few states in the United States which follow Roman law, *civil law* means law of private rights. It is divided into torts and breach of contracts. In these cases the plaintiffs must bring action, usually to recover damages.

Criminal Law

An act prohibited by common law or by a statute is a crime. The criminal statutes must present three questions: (1) What act (or omission) is prohibited? (2) Who can or cannot commit the crime? (3) What is the punishment for committing the crime?

Punishment for a criminal offense could be one of the following: (1) death; (2) imprisonment; (3) removal from office; (4) fine; (5) disqualification to hold and enjoy any office of trust, honor, or profit under the Constitution.

There are three degrees of crime: treason, felony, and misdemeanor.

Treason is the highest crime. It is defined in the Constitution: "Treason against the United States shall consist only in levying war against them, or in adhering to their enemies, giving them aid and comfort. No person shall be convicted of treason unless on the testimony of two witnesses to the same overt act, or on confession in open court."

Felony, the second level of crime, is generally defined as an act punishable by death or imprisonment in a penitentiary for 1 yr or more.

Misdemeanor (Latin for "bad behavior") is the lowest level. It consists of all prohibited acts less than felonies. Typical misdemeanors are traffic violations, violations of zoning laws, and breach of the peace. Punishment usually consists of a fine or jail sentence less than 1 yr, generally, anything less than death or imprisonment in a penitentiary.

Civil Law—Torts

Civil law provides procedures to redress wrongs to an individual. The wrongs fall into two categories: torts and breach of contract. Details of contracts are treated above. A tort is any wrong, not arising from breach of contract, for which private action for damages may be maintained. Note that for torts, infringement of an individual or personal right, or a wrong against an individual will go unpunished unless the offended person chooses to bring action. We review torts next.

Torts are infringements of absolute or relative rights of an individual. There are three absolute rights: (1) personal security, (2) personal liberty, and (3) private property. Respectively, examples of infringement of these rights are assault and battery, unjust imprisonment, and trespassing. There are two relative rights: (1) public and (2) private. Examples against these rights, respectively, are violation of official duty and abduction of a family member by a third party. The appropriate remedy for a tort is civil action for damages brought by the injured party.

Wrongs can result from acts of commission, acts of omission, or doing something which in itself is not wrong but which is done with insufficient caution. These three are considered by law to make an individual liable for tort through (1) performing a wrong act which causes injury to another, such as destroying another's property, (2) failure to carry out a legal duty and causing damage to an individual (simple examples might be inadequate signs warning of dangerous high voltages, or improper guards on dangerous machines), (3) doing something which itself is not illegal but which is done in a manner or at a time or place that it causes damage to another. An example could be keeping a large or aggressive pet and not adequately restraining it from damaging a neighbor's property or intimidating her safety. Torts of special importance to engineers involve the reliability and safety of product designs. Product liability is reviewed next. Because of its importance, the presentation is in detail.

8-6 PRODUCT LIABILITY [2]

As in other cases of tort law, the plaintiff makes his case by showing that the defendant neglected a duty owed to the plaintiff. Furthermore, the plaintiff must show that the negligence was the proximate (Latin meaning "nearest") cause of the plaintiff's injury. Until early in this century, *caveat emptor* (buyer beware) was the legal response concerning product liability. However, a decision written by Judge Benjamin Cardozo in *McPherson v. Buick Motor Co.* [217 N.Y. 382 (1916)] changed all that; his reasoning underpins product liability law today. Especially important are two legal philosophies: negligence and warranty.

Negligence

As stated above, the plaintiff's approach is to show that the producer was negligent. It could be in one or both aspects: the producer neglected the duties of careful design or careful production. The plaintiff must prove either one, especially if the defendant (producer) can indicate contributory negligence. Nonetheless, any producer negligence is effective in winning cases for the plaintiff. The degree of neglect is heavily weighed in determining damages.

Warranty

This term concerns the implication (by the producer) that the product will preform its intended job and can be used safely by the operator. If the operator incurs injury using the product as intended, the plaintiff has cause to act against the seller. The implied warranty was breached; this is equivalent to a breach in contract, discussed later.

Strict Liability

Most states' laws of product liability follow section 402A of the Restatement (Second) of Torts. This section is often called *strict liability.* Because of its importance, let us give attention to its wording: (1) "One who sells any product in a defective condition unreasonably dangerous to the user or consumer or to his property is subject to liability for physical harm thereby caused to the ultimate user or consumer, or to his property, if (a) the seller is engaged in the business of selling such a product, and (b) it is expected to and does reach the user or consumer in the condition in which it is sold. (2) The rule stated in subsection (1) applies although (a) the seller has exercised all possible care in preparation and sale of his product, and (b) the user or consumer has not bought the product from or entered into any contractual relation with the seller."

Although seemingly explicit, interpretations and extensions are the current "arena" of product liability.

Plaintiff's Case

In making a case of recovery for injury against the producer-defendant, the plaintiff must prove that the product was defective when it left the producer's control. Furthermore, evidence must prove that the defect was the cause of injury. Usually juries decide on these cases. Experts are frequently used to give opinions to the jury as to whether or not the product was defective when it left the producer, as alleged. Use of an obviously defective product readily establishes contributory negligence. A hidden defect, of course, significantly helps the plaintiff's case. Labels warning the prospective users, as well as instructions and warnings, weigh for the defendant; if they were closely followed by the user, defense is difficult.

Damages are usually to compensate for the plaintiff's loss. Occasionally, as alluded to above, extremely hazardous or careless behavior by the defendant will influence the court to go further in punishment.

Defendant's Case

Beyond those factors mentioned in the paragraph on the plaintiff's case, we will consider others. First, it is well to recognize that if faulty design or manufacturing is evident to the defendant's expert, it will be so to the plaintiff's expert as well. In turn, the best approach is to settle out of court. The only reason to incur the lawyer's fees and the court expenses is that the probability is high for either winning or reducing the damages by a greater amount. In some cases, however, companies might "invest" in establishing a particular position to help prevent future multiple liabilities. In other words, sometimes there is more involved with a company's litigation strategy than simply defending the present matter. There are several possible defenses:

1 *Product alteration.* Any change, such as by repairs, could indicate the the original manufacturer was not at fault.

2 *No proximate cause.* If this condition can be shown, the defendant is cleared. [*Caveat:* In certain instances this is "presumed" in light of the facts. For example, if a plane crashes after its wing falls off, it might be said that "the thing speaks for itself" (*res ipsa loquitur*). A rebuttal presumption would be established that those controlling and managing the plane, as opposed to the passengers, are liable.]

3 *Obviousness.* If the fault is obvious, then the plaintiff has no case.

4 *Abuse.* Although the producer should expect considerable abuse by the user, evidence of abuse will influence the jury to reduce damages, if it does not rule for the defendant.

5 *Functional necessity and state of the art.* Progress in modern technology requires risk. Possibly nothing can be produced completely risk-free. Although recognizing this situation, the manufacturer will be expected to produce the product as safe as the "state of the art" will permit. For products manufactured years past, a study can be advantageous to the defendant. Review of the literature of that time, comparing the manufacturer's product with the safety of others of that era, could be effective. The study could show that the safety of the product was superior for its time. The patent system is often employed during such searches, as well as your own experience and knowledge.

Other factors such as the producer's close adherence to standards and careful attention to design and production will weigh in favor of the defendant.

Finally, we consider state acts. The "overkill" of product liability cases has been recognized by the states. Thus, many product liability acts have included (a) a statute of repose indicating that after a certain period (e.g., 10 yr) the producer's responsibility for the product ends, (b) that the "state of the art" is a usable defense, (c) that compliance with recognized standards (e.g., military standards) is a reasonable defense, (d) that alteration or abuse by the plaintiff, or both, is a defense, and (e) that failure to warn is not a component of strict liability. This new orientation in judging product liability promises more equitable damages and, in turn, a long-term savings for the consumer.

8-7 THE JUDICIAL SYSTEM

Often the engineer will encounter honest misunderstandings and occasionally unethical acts that can be resolved only in court. The following material presents the elements of that system. It will greatly facilitate your interchange with attorneys. There are 51 court systems in the United States: the federal court and 50 state systems. We first review the general court system and then give specifics for the federal and state courts. There are two basic types of courts: trial and appellate (or appeal) courts.

Trial Courts

They are the original jurisdiction. Trial courts hear and decide arguments by determining facts and applying appropriate rules. The process may be carried out by a judge, with or without a jury. Of course, trial by jury is guaranteed by both the federal and the state constitutions, unless waived by all parties. In a trial by jury, in reaching a decision, the judge decides points of law and the jury, points of fact.

Appellate Courts

They review the decisions of trial courts. If either party in a civil suit is dissatisfied, the case may be appealed to a higher appeal court. The power of appellate courts is confined to a review of potential or alleged errors made in the lower court. No new arguments or proof is admissible, for the appellate courts do not retry cases. There are no witnesses or jury. Instead, both sides present *briefs* (legal and fact-based arguments addressing the issues on appeal) and then engage in a relatively brief (perhaps 20 min per side) oral argument before the judges (usually three or more) who will decide the case. The decision concerns only whether the lower court acted according to the law, or whether there was clear error in the finding of the facts.

The ultimate court of appeals is the United States Supreme Court. As reviewed later, it exercises both appellate and original jurisdiction. Only in a few circumstances can a decision be appealed to the Supreme Court as a matter of right. It controls its docket and reserves its time for the most importance cases, those deemed to have far-reaching impact. We now consider the federal and state court systems.

Federal Courts

They consist of the Supreme Court, courts of appeal, district courts, and others.

The United States Supreme Court   This is our highest court. It consists of a chief justice and eight associate justices. Again, it is the final court of appeal. The only cases that may originate in the Supreme Court are those which involve ambassadors, public ministers, and consuls, or those in which a state is a party. All others go to the Court by appeal.

Cases are appealed to the Supreme Court from either U.S. Courts of Appeal or state supreme courts. However, cases may go to it directly from any court if the question involves the U.S. Constitution or is of very great public interest. As noted above, in only a few circumstances can a decision be appealed to the Supreme Court as a matter of right.

Courts of Appeal   There are now 13 federal courts of appeal. They were established (1891) by Congress to ease the burden on the Supreme Court. They only handle appeals and do not hold trials. The issues of appeal are only on the basis of conflict in the law. One of the circuits (the "federal circuit," established in 1982) has exclusive jurisdiction over patent appeals. This has resulted in a greatly renewed and strengthened interest in (and value of) patents in the 1980s, which will probably continue in the future.

District Courts   These are the trial courts of the federal system. They are the most numerous of federal courts. The nation is divided into more than 90 geographical territories (districts). Each state has at least one district and, in turn, at least one district court. In these courts, suits are started and issues of fact are determined before a judge or before a judge and jury. District courts handle most of the cases which come to federal courts.

In civil action, the district courts have jurisdiction in two major areas. The most obvious involves strictly federal matters as might arise under the Constitution or a controversy under federal statutes. The other is important to the engineer and is relevant to product liability. It involves amounts in excess of $50,000 and "diversity of citizenship," citizens of different states. A *citizen* is an individual, corporation, or similar entity. A corporation is considered a citizen in the state in which it is incorporated and the state of its principal place of activity. It cannot sue or be sued in a federal court in either of these states if the other party is also a citizen of either of these states. Such a situation requires trial in a state court. As previously stated, the federal district court may not try a case involving less than $50,000. For amounts in excess, and when both parties are of the same state, the option is for trial in either state or federal district court.

State Courts

These systems differ considerably from state to state, but all are formed similar to the federal court. Each has a final court of appeal, a state supreme court; lower courts of original jurisdiction; and in-between courts of appeal. The supreme court is confined mainly to appeals from the lower courts.

Below the courts of appeal are usually superior courts, and below them are the municipal and justice courts. These include courts of common plea, county courts, small claims courts, and many others including the probate court. The probate courts handle cases of wills, trust, etc.

Trial Procedure

This section deals with the procedure a plaintiff pursues to initiate and carry out court action in our judicial system. Consider that you have been significantly injured and wish to seek damages. You would see an attorney and review with her the facts relevant to your case. As with other services, you would probably choose an attorney on the basis of recommendations from friends. If after hearing your information she feels that you have a reasonable case, she will prepare and file a complaint.

Complaint   It is also called a *petition.* The complaint has two purposes:

1 State as clearly and concisely as possible the cause of action.
2 Demand the remedy sought by the plaintiff.

This document must be filed in the court in which the remedy is sought.

Summons   On receipt of the complaint, the court issues a summons to the defendant. It simply notifies the defendant that action is going to be brought against him. Depending on the jurisdiction, a copy of the complaint accompanies the summons or is available from the clerk of the court.

Response to Summons   If the defendant does not answer within the statutory time, the result is a judgment against him by default. He usually responds by one of three ways:

1 *Demurrer.* This admits the facts in the complaint; however, it is contended that these facts are insufficient to support legal action. Furthermore, it then raises the question of law: Do the facts give the plaintiff cause for action at law? Demurrer is abolished in the federal system and in many state systems, although the same sort of defensive approach is accommodated in other forms.

2 *Motion to dismiss.* It disagrees with the facts. Specific defects might be noted.

3 *Counterclaim.* This attack recognizes the facts presented, but alleges others. On the basis of them, the defendant demands remedy from the plaintiff (that is, he countersues the plaintiff) who, in turn, has an opportunity to reply.

The pleadings set forth plaintiff's cause and claim, and defendant's defense(s). They are not complete until an issue has been framed. Furthermore, the case may go to trial only when the issue (statement of opposing arguments) is clear.

The Jury   A trial jury is called a *petit* (*French* meaning "small," the opposite of *grand*) jury. There are usually 12 members. By statute, in several states, there may be fewer for civil cases or for crimes less than capital offenses. In choosing members, both attorneys and the judge are involved. The basis for challenging prospective jurors is set by statute. This generally concerns a financial or blood connection between the prospective jurors and the litigants. Also, each attorney may exercise the right of pre-emptory challenge to disqualify a limited number. After the jury has been impaneled (selected and sworn), the trial may begin. Normally, one juror is selected by the judge as the foreperson and assumes a leadership role during deliberations and actions of the jury.

Courtroom Procedure   When the pleadings are complete, the scheduled time on the court docket has come, and the jury has been selected, then the trial may begin. First the attorneys give opening statements, then the witnesses are sworn in, and subsequently the evidence is examined. Each attorney sums up his and her case to the jury, the judge charges the jury (on the principles of law involved), and the jury retires to reach a verdict. In the charge to the jury, the judge also sums the case and instructs them concerning the issues to be decided. On reaching a verdict, the jury returns and the foreperson announces the decision. The judge then gives the judgment for the case.

New Trial—Appeal   Within a statutory time, a party displeased with the outcome of the trial may seek a new trial (in that court) or may appeal the case to a higher court. Either party could be dissatisfied; the party in whose favor the court ruled could feel that the damages paid are not enough, or the other party might feel that they are too much.

A new trial is usually sought first. Grounds might be that a member of the jury was unfairly selected, that financial or blood relations exist between a member and a litigant, or that there was misconduct, such as sleeping or gambling. Occasionally, error by the judge is cited, for instance, allowing or disallowing, respectively, improper or proper evidence.

An appeal may be sought by either party in a civil court. In a criminal case, however, only the defendant may appeal an adverse judgment.

The appeal brief presents the reasons for the appeal to the appeal court. It gives the objections or exceptions taken to the rulings of the trial court. The objections usually concern three protests: (a) those involving the admissibility of evidence, (b) errors in the conduct of the trial, and (c) instructions by the judge. The judge's instructions to the jury include evidence presented, testimony taken, and the relevant law to be considered in deciding the case.

The appellant (the one appealing), when presenting a brief of the case to the appellate court, lists errors and information on the case from the appellant's viewpoint. The most important points are normally reviewed during oral argument; judges also ask questions during oral argument concerning law and facts of the case, which must be answered then by the attorneys.

Evidence   This is used to prove questions of fact, the means of establishing proof. A civil case is won or lost on the comparative weight of the proof. Evidence is classified as real or testimonial. Real evidence can be determined by inspection, such as a broken cable or rung on a ladder. Testimony consists of statements by witnesses. It regards information a witness received by one or more of her senses, such as the unsafe operation of a piece of equipment.

Witnesses   Competence of a witness to testify is decided by the judge. The determination is generally based on the person's mental capacity and mental ability. Counsel (one of the attorneys) might object to testimony, claiming that the witness is incompetent or that the testimony is irrelevant or immaterial.

Generally, opinion evidence is excluded as improper when offered by the average lay witness. Then the court looks to the expert regarding matters concerning, for example, the adequacy of a design or the capabilities of equipment. An expert is a person who, because of technical training and experience, possesses special knowledge in a particular field.

8-8 ENGINEER: EXPERT ASSISTANCE

Attorneys and judges are learned persons. Law requires a broad general knowledge in many specialized areas in order to understand the factual material presented in cases. Actor and humorist Will Rogers said, "We are all ignorant, just about different things." Accordingly, the court seeks special information in technical areas. In this matter, the engineer can help in three important ways: advice, expert testimony, and assistance in analyzing the other experts' testimony. You immediately recognize the importance of your in-depth knowledge and thorough understanding of the specific fundamentals and technology. Of equal importance is your ability to communicate clearly.

Advice

By your thorough knowledge of the technical subject, you will be able to explain it in the terms of nonprofessionals. This will be of great help to your attorney as she prepares her brief and later examines and cross-examines witnesses. Also, as reviewed next, as a witness, during the trial you will be able to clearly and simply explain the technical questions of the case to the judge and jury.

Testimony

In your area of expertise, you can consider everyone in the court to be a layperson. Your challenge is to present the fundamentals in simple language, thus ensuring that they are understood by the judge and jury. It is not easy. Lay a simple foundation of the fundamentals, relating them to common experiences. Then, based on them, develop the technicalities of the case. Use of visual aids, reviewed in Chapter 4, greatly enhances and facilitates your presentation.

The effectiveness of your testimony will depend mostly on how well you prepare. This includes three parts: (a) your thorough study of the details of the case, as well as the technical issues, (b) your knowledge of the fundamentals, and (c) a presentation easy for laypeople to understand, complemented by visual aids that are easy to follow. This will be a strong basis for your answers to unexpected questions. Remember, however, that the key to a quick mind is anticipation.

Analysis during Trial

The trial is dynamic. In addition to helping your attorney in preparing the case, you will need to assist her during the trial. Be sensitive to technical errors in the testimony of the opponent's expert witness. Also be quick to react, from fundamentals, on unexpected evidence. As indicated above, anticipation is the best approach. However, do not try to outguess your own attorney. Both attorneys may be attempting to question you in an area not previously discussed with you. Do your best to provide straightforward and honest answers to both attorneys.

The Legal System and the Engineer

The engineer interacts with the legal system in three areas: professionally, personally, and as a responsible citizen.

Professionally the engineer relates to other engineers, the employer, and the clients. Harmony is quickly lost when the rights of others are infringed. The infringement is often the result of negligence more than malice. In either case, it is most undesirable and can easily lead to costly court or out-of-court settlements. This chapter gives the legal information needed to avoid most cases and, if there is a lawsuit, to minimize the damages.

Many of the professional legalities are applicable to your personal life (contracts, purchases, etc.). In addition, other information of personal importance is presented, relevant to wills, rights and responsibilities of landlords and tenants, and the purchase of land and houses.

Equally important is the legal knowledge which is part of the background required of the responsible citizen. The material in this chapter is a first step.

DISCUSSION QUESTIONS AND PROBLEMS

8-1 LEGAL ASPECTS OF ENGINEERING

1 How does the phrase "preparing ahead of need" relate to the subject matter of this chapter?
2 Give two codes of law which influenced Western civilization.
3 In present times, to what aspect of law does civil law refer?
4 Give a brief written review of the four basic types of law.
5 Is common law statutory? Explain.
6 What is the hierarchy of common law?
7 What remedies are available under common law?
8 What are the most significant equity remedies?
9 Define an *injunction.*
10 Is a jury required for a preliminary or permanent injunction?
11 Define a *consent order* aka (also known as) a *consent decree.*
12 Define a *specific performance.*
13 Is the remedy of a specific performance available in courts of common law? Give reasons.
14 Is a jury required for cases in a court of equity?

8-2 CONTRACTS

15 Discuss the five elements of a contract.
16 What types of contracts must be written?
17 In the English Statute of Fraud:
   a In section 4, what type of enforcements are probably of interest to engineers?
   b What constraints are put on contracts by section 17?
   c Under what conditions can a contract be enforceable over the limits of section 17?
18 **a** What three relationships may be involved in "delegation"?
   b In legal terms, what people (parties) are involved in an "agency"?
19 Name the four ways an agency is established.
20 **a** What characteristics of the principal and agent are basic to implementing the agent's duties?
   b How many principals should an agent represent?
21 Name a few of the ways parties may terminate the agency.
22 What are the significant advantages of "specs"?

8-3 PERSONAL PROPERTY

23 Define the terms:
   a *Chattel*
   b *Bailment*
24 Explain briefly the various ways personal property can be acquired.
25 What does it mean when a person dies *intestate ?*
26 What constitutes "intellectual property"?
27 How are copyrights controlled?
28 Besides the rights to reproduce and distribute the work, what other rights are granted by a copyright?
29 How do trade secrets differ from patents?
30 Make a brief outline of the steps used in applying for a patent.

8-4 REAL PROPERTY

31 What are the essentials for a valid will?
32 How may changes be made in a will?
33 What are the elements of a valid deed?

8-5 CRIMINAL AND CIVIL LAW—TORTS

34 Discuss the three degrees of crime.

35 Review the two categories of individual wrongs redressed under civil law.

36 Define and give an example of a *tort*.

8-6 PRODUCT LIABILITY

37 Using legal terms, what recourse does a customer have when injured by defective merchandise?

38 a Define *strict liability*.

   b Under what conditions is it probable that the court will require damages in excess of the plaintiff's loss?

8-7 THE JUDICIAL SYSTEM

39 Name the two types of courts, and review their functions.

40 Give a brief review of the trial procedure.

8-8 ENGINEER: EXPERT ASSISTANCE

41 In what ways does an engineer interact with the legal system?

REFERENCES

1 Vaughn, R. C. *Legal aspects of engineering* (3rd ed.). Dubuque, IA: Kendall/Hunt, 1983.

2 Thorpe, J. F., & Middendorf, W. H. *What every engineer should know about product liability*. New York: Marcel Dekker, 1979.

Comstock

ETHICS AND ENGINEERING

CHAPTER OUTLINE

The word *ethics* comes from *ethos,* the Greek word for "character." As a subject, ethics concerns concepts of the individual or group by which actions are judged "right" or "wrong." The subject involves the deliberate, free actions of a person or group. These actions alone are in our power and, concerning these alone, rules can be prescribed. We cannot and should not try to control our feelings (the Child*), but we can and must control our behavior (responses permitted by the Adult from the Child or Parent). Ethics is the discipline which can act as the "performance index or reference" for our control system.

Let us model ourselves as an adaptive control system. Our muscles, including those of the tongue, are the motors of our body. Our Adult acts as the controller, comparing the body's output with the references stored in our Parent ("Do what's right because it's right") and our Child ("Because of my position, I deserve special privileges"). Between them, our Adult continually tries to establish moral virtues. These, as defined by Aristotle, are discussed below. Self-interest is the primary controller of humans; ethical self-interest is the optimum controller.

Engineering is generally understood to be the application of scientific principles, through the use of models and technology. The purpose is to design new systems and improve existing ones, all for the benefit of the human race. Ethics is the discipline which examines the moral significance of that creative activity. After reviewing the fundamentals of ethics, we proceed in using the engineering approach for the practical application of ethics to moral conflicts. The main objective of this chapter is to improve your ability to reflect logically on moral issues.

9-1 FUNDAMENTALS

These involve the information our Adult needs in making rational comparisons, with minimum emotional influence from the creative but often selfish Child. We first review three levels in moral development. Next, we consider theories from which we can recognize why certain acts are wrong. Finally, we present three phases in evaluating moral conduct.

Moral Development

Extending concepts pioneered by Jean Piaget, Lawrence Kohlberg suggested three main levels in moral development. Generally they parallel the growth stages of a normal person [1].

Preconventional Level   This is the most primitive. Right conduct is judged by actions that directly benefit the person. Motivation is dominated by the desire to avoid punishment, by acquiescence to power, and by seeking self-gratification. This is naturally apparent in young children, but can still be significant in some adults.

---

*Transactional analysis is presented in Chapter 1.

Conventional Level   In it, the norms of one's family, group, or society are accepted as the standards of morality. Typical of our Parent ego, they are accepted uncritically. At this level we are motivated by the desire to please others and to meet the expectations of the social group. Self-interest is suppressed. Loyalty and identification with others are most important. Kohlberg believes that most adults never go beyond this level. Morality is equated with legality. Our engineering thought processes make it easier for us to move to a higher level, one of personal responsibility.

Postconventional Level   At this level, an individual is autonomous. These persons think for themselves. They do not, without analysis, accept customs and precepts. Their analyses are reasoned from general principles, which universally apply to all people independent of race, religion, or culture. Of course, the rule "Expect no privileges not accorded others" is such a principle. Their motivation is to act morally, above self-interest. Moral integrity, self-respect, and respect for others are their guides for all activities.

This chapter and the entire book are directed toward aiding students and engineers in achieving this last level of autonomous morality. As an engineer I have found it easy to be considerate of and generous to others. The credit is not mine, but is rather the result of three good fortunes: (1) self-confidence gained from having technical competence, even in dealing with daily life; (2) the financial security gained by using the principles of Chapter 7 on financial planning; and (3) my having started early to accumulate resources. Returning to Kohlberg's concepts on moral development, let us now consider a variation.

Carol Gilligan presents a different basis for a person's moral development. In her book *In a Different Voice,* she charges Kohlberg with a male bias. She suggests that males tend to resolve dilemmas, moral and others, by determining and applying the relevant rules (a limited model). In contrast, women tend to try harder, by studying the personal relationships of the people involved. Accordingly, they focus more on the context of the situation (the total system) in which the dilemma evolves. She calls this the "ethics of caring." As a result, Gilligan offers a different perspective on the levels of development. Her first level is the same as Kohlberg's. In the second level, however, she contends that women, while tending to nurture others, fall prey to a cultural stereotype. It pressures them to give up personal interest to serve the needs of others. Thus, their virtue of caring for others makes it seem that women stay at the conventional level. In reality, most caring individuals, male and female, do achieve the postconventional level. They can evolve a reasoned balance between caring for others and exercising one's rights in pursuing self-interest. Further perspective on the levels of normal development is given by the three phases of moral inquiry, considered next.

Three Phases of Moral Inquiry (in Evaluation)

These phases are instrumental in making moral evaluations. They are (1) gathering factual information; (2) analyzing it through understanding based on concepts, principles, and related problems; and (3) evaluating the total complex in light of ethical norms or standards. Frequently, we tend to operate in reverse order. Our Parent ego state often jumps immediately to the evaluation phase, using only prescriptive guides such as an organization's code of ethics. It judges a person or action without the benefits of the other phases. Early engineering codes gave primary emphasis to an employee's allegiance to the employer and the client. The most recent codes reflect the influence of phases (1) and (2). These codes recognize the rights to safety by the public. Moral theories are helpful in identifying the factors needed for the analyses of phase (2). They also give guidance for phase (1). All three phases are used in our case studies presented later.

Moral Theories [2]

There are four types; they differ in their stress on which is the most fundamental concept: (1) good consequences for all (defined as *utility*), (2) duties, (3) human rights, or (4) virtue. They are complementary in making moral inquiries.

Utility Theory This can be defined as the balance of good over bad consequences. The theory, often called *utilitarianism,* indicates that we should always act to maximize utility. John Stuart Mill (1806–1873) contended that acts and their consequences should have more importance than rules. He argued that while maxims such as "Your word is your bond" and other rules based on past experience are generally best, but they are not always true. We immediately recognize that this theory could easily become dangerous without the support of the three phases of inquiry presented above. The bad reputation of Machiavelli (1469–1527), passed down over centuries, resulted from his theory: any means, however unscrupulous, may be justified to maintain a strong central government. He claimed it to be "the most good for the most people." Notice that "the most good for the most people" is clearly the desired utilitarian outcome. However, many of the actions which Machiavelli used to achieve this goal violated human rights. Today, the term *machiavellian* implies political cunning or bad faith. Later in this section we present Locke's tenets on human rights.

Returning to John Stuart Mill, he believed that happiness is the only intrinsic good, that is, something good or desirable only for its own sake. Other goods are instrumental goods. Their value is that they can produce good, such as a visit to a physician by a sick person. The guidance gained will probably lead the patient back to health and permit normal activities. If happiness is the only good, what is happiness? Mill offered that happiness is a life comprising many pleasures in great variety, mixed with the inevitable brief pains. The happiest life consists mostly of the higher pleasures. He contended that these are derived through intellectual inquiry, creative accomplishment, appreciation of beauty, friendship, and love. Lesser goods are the physical pleasures such as eating and sex. We have recognized that abstinence greatly enhances them.

Considering Mill's concepts as act-utilitarianism, Richard Brandt, a contemporary, offers a second version: rule-utilitarianism. It regards moral rules as primary. If these rules are generally followed, the resulting acts would produce the most good for the most people. A set of rules would be an optimal code, if on being adopted and followed by all, these rules would produce more public good than any alternative code.

Of course, this is the goal of the code of ethics of our professional societies. Often rules clearly identify improper acts. Their use alone, however, has the failing of not considering the overall situation in making the final judgment.

Duty Theory   Immanuel Kant (1724–1804) regarded duty as most fundamental in ethics. He valued goodwill, the intention to do one's duties, as a "good" greater than happiness. Right actions are required by duties to others and duties to ourselves. The first duties include maxims similar to the rules of action offered by Mill: be honest, keep your promises, do not inflict suffering on others, be fair, make reparations if ever unfair, and show gratitude for the kindnesses of others. Second are the duties to ourselves: strive to improve our intelligence and character, continually work toward developing our capabilities, and avoid dangers of self-harm.

Kant gave three criteria for duties: They must show respect for others, express an unconditional imperative for moral acts, and be a universal principle.

Respect   The honest and conscientious effort to fulfill duties is a great value. It is "goodwill." Furthermore, people are rational beings and have the capacity for that goodwill. Accordingly, they deserve respect. Respecting others (their rights and dignity) is fulfilling our duty to them, and respecting ourselves fulfills our duty to ourselves.

Moral Imperatives   These differ from nonmoral imperatives in that the nonmoral are conditioned. For example, a nonmoral imperative is the charge that if you wish to lose weight, significantly reduce your intake of fats (the body stores about 90 to 95 percent of fats, but converts to fats only about 25 percent of proteins and 3 to 5 percent of carbohydrates; fats have approximately twice as many calories per ounce as the other foods) [3]. This is a nonmoral imperative. The results of this imperative are conditioned on a change in behavior. Moral imperatives are unconditioned. We should follow them regardless of the conditions, whether we want to or not. They are our duty because of our autonomous commitment to morality. An example of this type is "Do not steal." The fact that you are being paid an unfair wage cannot morally justify stealing to compensate for the low pay.

Universality   This criterion means that the duty is applicable to everyone; there are no privileged individuals. Moral principles are for everyone to use as guides for action. Most rules meet this criterion: For example, "Do not deceive," and "Protect innocent life." These rules, however, could be in conflict. This situation regards *prima facie* duties. A weakness of Kant's duty-based ethics is that he thought duties were absolute.

Prima Facie Duties   The first two words are Latin for "first appearance." The term is used to indicate laws and principles which could have exceptions because of facts not first apparent. As alluded to above, "Do not deceive" might not be the best moral action if deception could "Protect innocent life."

John Rawl, a contemporary, has developed Kant's ideas in a new way. He suggests that valid principles of duty are those upon which all rational people would voluntarily agree. He presents two principles agreeable to all rational people: (1) Each person is entitled to the most extensive liberties compatible with equal liberties for others; and (2) differences in social power and economic benefits are justified only when they are likely to benefit everyone, including the most disadvantaged. The first principle should be satisfied first. The second, of course, is a premise of free enterprise. These principles are a bridge to the next theory, based on human rights.

Human Rights Theory   Ethicists of this school contend that rights do not derive from duties, but that duties exist because of the rights of others. The duty "Do not kill" arises because of a human's right to life. John Locke (1632–1704) argued that a person's human rights are to life, liberty, and property gained by one's labor. His views were significant at the time of the French and American revolutions and probably influenced the wording of our Declaration of Independence. Contemporary libertarian ideology reflects his views that human rights place a duty on others to not interfere with one's life.

A. I. Melden* argues that having moral rights presupposes the capacity to have concern for others and to be accountable within a community. This concern gives the engineer the right to warn a client or the public about the safety of a product. Melden extends his concepts to "welfare rights": the right to community benefits needed for living a minimal human life.

Virtues Theory   Virtues are defined by Aristotle as certain habits acquired by human beings. There are two types. One is intellectual, that is, virtues which enable us to engage effectively in rational activities. These virtues include foresight, mental discipline, perseverance, and creativity. The other type is moral. These virtues are reflected by our tendency to balance between the extremes of excess and deficiency in conduct, emotion, desire, and attitude. Aristotle calls this balance the Golden Mean (average). For example, truthfulness is the mean between extremes. Excess would be telling everything in violation of tact and confidentiality. Deficiency would be secretiveness and lack of candor regarding information relating to a client's welfare. Also, generosity is a virtue lying between two extremes. At one extreme you would be giving away all of your goods and resources. In cases of misfortune, you would become a burden on society. At the other extreme you would become a miser, keeping everything for your benefit only. You would not contribute your share to society (for those less fortunate). Moral virtues lead us to pursue social goods within a community.

The Essential Fundamental: The Golden Rule

From early societies to present times, a golden rule—"Treat others as you would have them treat you"—has been the main guide for social (ethical) conduct. A study of comparative ethics shows that only slight variations of this rule were stressed by Siddhartha Guatama (ca 563–483 B.C.), founder of Buddhism; Confucius (551–479 B.C.); Aristotle (348–322 B.C.); Jesus Christ (ca 6 B.C.–ca 30 A.D.); Muhammad (ca 570–632 A.D.), Arab prophet and founder of Islam; and the Old Testament (Lev. xix, 17), similarly interpreted in the Talmud [4].

The constant guide for the complete person, professionally and personally, is *love:*

---

*A. I. Melden is a professor in the Department of Philosophy, University of California at Irvine.

the willingness to help others in ways often unpleasant and at times inconvenient, all with no expectation of getting anything in return. Note that these actions show love, but genuine love is probably a state (of willingness to act), not the act.

Practical Ethics

Alasdair MacIntyre* extended virtue ethics to apply to professional practices as they relate to cooperative activities. The aim of these practices is toward social goods that otherwise could not be achieved. These goods are internal and are the objectives of the practices. External goods, on the other hand, are those such as fame and public recognition. They would be sought by means not specified by (but guided by) these practices.

The primary internal good of engineering is developing safe, useful economic products, while respecting the rights of clients and the public. It is probable that these high expectations will be reached only by the completely professional engineer: one who is conscientious, safety-conscious, and imaginative. These virtues give the integrity by which the engineers can morally unite their professional and personal lives. Integrity is a bridge between them.

As noted above, virtues are generally patterns of action, emotion, and attitude that permeate all actions in life. Moral integrity (inner unity based on moral commitments) is sustained when these virtues strongly influence our actions in both personal and professional life. This integrity is the peace professionals constantly seek.

Moral problems are those that arise in situations calling for decisions based on moral reasons. Significant in them is the virtue of taking responsibility for one's actions. This encompasses trustworthiness and benevolence, two qualities that are especially important in the engineering profession.

Trustworthiness is a fundamental virtue for engineers engaged in relationships with employers and clients. These relationships can be effective only when there is mutual trust: that the engineer will responsibly and effectively perform the services contracted, and that the client will make proper payment on completion.

Benevolence means the desire to foster good. It is important in the interchange of all people, especially between engineers and clients.

The four ethical theories reviewed above give complementary perspectives for making moral judgments. All give guidance for the special safety obligations engineers have to a technologically uniformed public. This responsibility derives from the way the particular expertise and functions of engineers relate to the rights of the people affected by their work. This interplay links engineering and moral philosophy.

9-2 CODES OF ETHICS [5]

These are standards of conduct for an organization's members. They are not enforceable at law. The Civil Engineers of Great Britain adopted a Code of Ethics in 1910. Shortly afterward, the engineering societies of this country adopted codes. They were

*Alasdair MacIntyre is a professor in the Department of Philosophy, Notre Dame University, Indiana.

modeled after the British code, which emphasized responsibility to the client. The American Institute of Electrical Engineers (AIEE) was the first in this country to adopt a code (1912). That institute has since joined with the Institute of Radio Engineers (IRE) to form the Institute of Electrical and Electronics Engineers (IEEE). The original AIEE code served as a model for the codes of most of the other engineering societies. In them, the principal stipulation was that "engineers should consider the protection of a client's or employer's interest as their first professional obligation, and therefore should avoid every act contrary to that duty."

In 1947 there was a major reformulation of the codes of the various societies. They followed the Canons (Latin for "rules") of Ethics for Engineers of the Engineers' Council for Professional Development (ECPD). The reformulations began to show the societies' concern for the public welfare. That concern should be weighed against responsibilities to employers and clients. In 1974, the ECPD (now ABET, Accreditation Board for Engineering and Technology) canons were revised: "Engineers shall hold paramount the safety, health and welfare of the public in performance of their duties." In a recently proposed IEEE code, as well as in an earlier one, the public interest is not specified as being paramount, but it does indicate that engineers' responsibilities to employers and clients are *limited* by their responsibilities to "protect the safety, health and welfare of the public. . . ." The proposed code is reprinted below.

Significant is that none of the codes consider the *rights* (versus responsibilities) of the engineer. Most of the engineering societies have promoted "Guidelines to Professional Employment for Engineers and Scientists." These include rights (primarily economic) of the engineers employed by large organizations. Regretfully, however, no consideration is presently given to protection of engineers from reprisals taken because of actions in conformity with the codes (see Case Number 1 below).

IEEE SIMPLIFIED CODE OF ETHICS [6]

(Approved August 1990 by the Board of Directors.)
We, the members of the IEEE, in recognition of the importance of our technologies in improving the quality of life throughout the world, and in accepting a personal obligation to our profession, its members and the communities we serve, do hereby commit ourselves to conduct of the highest ethical and professional order. We further agree and covenant:

1 to make engineering decisions consistent with the safety, health, and welfare of the public, and to disclose promptly factors that might endanger the public;

2 to avoid real or perceived conflicts of interest whenever possible, and to disclose them to affected parties when they do exist;

3 to help improve understanding of technology and of its proper use;

4 to maintain and improve our technical competence and to undertake technological tasks for others only if qualified by training or experience, or after full disclosure of pertinent limitations;

5 to be honest and realistic in stating claims or estimates based on available data;

6 to seek, accept, and offer honest criticism of technical work, to acknowledge and correct errors, and to credit properly the contributions of others;

7 to neither offer nor accept bribes;

8 to treat fairly all persons regardless of such factors as race, religion, gender, disability, age, or national origin;

9 to never maliciously or falsely attempt to injure the person, property, reputation, or employment of others;

10 to assist colleagues and coworkers in their professional development and to support them in following this code of ethics.

It appears that through moral reasoning, these codes reflect the societies' progress into the postconventional level. This progress was the result of the societies' leaders using the three phases of moral evaluation presented earlier: gathering information, analyzing it through understanding based on moral concepts, and evaluating the total complex. Nonetheless, all codes are static. Like a firm's year-end financial statement, the codes give the respective organization's moral status at the time of their proposal.

The progress of the engineering societies focuses on an important condition: Ethical systems must be dynamic. This attribute is needed to guide a person's actions in a dynamic world, especially in areas of technology. As fundamentals, generally, the ethical principles presented in this chapter are virtually unchanging. By using them and building on them, engineers can develop their own system of ethics. This ethical system must be based on moral principles, similar to the way a terrestrial guidance system is based on the unchanging cardinal references of north, east, south, and west. To be effective, however, the controller's guidance signals must continuously reflect consideration of the changing environment. By their education and technical experience, engineers are particularly capable of making "on-line" assessments and decisions of this type. This behavior is at a higher level of moral development than simply following rules. It is at the level of postconventional personal responsibility.

9-3 CASE STUDIES

Next, we will consider two ethical problems. One is almost a classic. It reflects the economic aspects which have had great influence in the industrial-defense complex. Former president Dwight Eisenhower warned of this national weakness in 1962. The second case is fictitious, but is in the recent literature (*Spectrum,* April 1990, page 19) and will probably be discussed in future years.

CASE NUMBER 1 [7]

This case concerns a successful bid to LTV, Ling-Tenco-Vought Co., by the B. F. Goodrich Co. The bid was to supply wheels and brakes for the A7D light attack aircraft. The aircraft was being built for the U.S. Air Force by LTV of Dallas, Texas. A decade earlier, a landing gear assembly built by Goodrich for LTV was a complete failure. As a result, Goodrich had been unsuccessful in subsequent bids to LTV. Thus, Goodrich was especially jubilant about getting this bid. The word circulated early that the brake must be successful.

Although no weight penalty was imposed in the request for bid, the low weight and the very low price influenced LTV to award the contract to Goodrich. Their proposal was for a four-rotor brake weighing only 105 lb. The design was made by a senior design engineer. A new member of the engineering staff was assigned to assist him. A data analyst, who was also a technical writer, was also assigned to the project.

From the very beginning, the brake failed in dynamometer tests. The four lining carriers were insufficient. Temperatures within the brake reached 2000°F, and excessive torque caused the linings to disintegrate. The junior engineer conducted his own experiments using a five-rotor brake. They were successful. The junior engineer urged redesign, but was ignored. Failures on the four-rotor brake continued for a year. Despite these failures, formal attempts were made to qualify the brake. Good engineering practice and common ethics were ignored. Six weeks before flight test, Goodrich issued false qualifications. The official qualifications contained about 200 pages of elaborate curves and graphic displays; nearly all were falsified.

At the flight test, the brake welded and the pilot was just able to control the skidding aircraft. The wheel had to be removed and the brake pried apart before the plane could be towed. Within a week after the test, the data analyst reported everything to the FBI. Military officials called a halt to the flight tests. Senator William Proxmire (D-Wis.) learned of the project and requested an investigation by the General Accounting Office. In response, the Department of Defense ordered sweeping changes in inspection and reporting procedures.

Why would a reputable firm engage in such devious and fraudulent practices? The data analyst had reported the brake falsification to a manager of technical services. Later, the manager said that he only did what he was told. The chief engineer called the analyst irresponsible and disloyal.

Let us review the ethical aspects of this case. We start with the three steps for inquiry:

1. Gather all relevant facts. Most of these are given above; however, the analyst later commented that "not an excuse, but a reason for not acting earlier, was fear of financial loss." He felt that he had no choice except to follow instructions. He had six children to support. *A comment:* Building a small estate early in your career reduces the pressure to act other than as you feel is right; it also reduces the need to prematurely accept a job. But, we probably agree, finances should not be our guide on moral issues.

2. Study the facts and review the ethical fundamentals. In this case, the foundations of all four of the moral theories were broken.

   a. Utility, the maximum good for the most people, was spurned.

   b. Duties, the respect for the safety and rights of others, were completely disregarded, especially concerning the safety and rights of the test pilot.

   c. Human rights, the concern for others' property and self-respect, were ignored. These others included the engineers, those associated with the project, and the employees and stockholders of the B. F. Goodrich Company.

d. Virtues of trustworthiness and accountability for one's actions were completely rejected.

3. Evaluate. It is easy for us to make an impressive analysis and condemn those involved with this project. Fortunately, we do not have to make a formal evaluation of them. To do so ethically would require much more information and a more extensive study than we have done. Nonetheless, our method is correct, and for this case or one involving us personally, we would only need to further pursue the data and study.

CASE NUMBER 2

This one is hypothetical. It is based on a case presented by the National Society of Professional Engineers' (NSPE) Board of Ethical Review (BER). It was published in the *IEEE Spectrum* [8].

A small town advertised a request for proposals for a townwide emergency communication system. A large consulting firm assigned the proposal to three staff engineers. Shortly after the proposal was submitted to the town, a municipal official approached the engineers. The official sought to deal directly with them, bypassing the consulting firm. The engineers informed their department head of the solicitation, resigned from the firm, and negotiated with the town officials.

The NSPE Board of Ethical Review (BER) posed the question: Was the action of the engineers unethical? As before, we will use the three phases of inquiry, guided by the four moral theories.

1. Gather all relevant facts. (a) Did the engineers gain any "particular and specialized information" while employed by the consulting firm? (b) How much "company time" was used developing the proposal? (c) In that time the engineers could have been assigned to an ongoing project. What lost opportunity did the firm experience?

2. Study the information. For this we can make only probable answers to the questions above.

a. Since the consulting firm probably had previous projects in communications, the engineers had most likely improved their expertise while working for the firm.

b. Consider their time used to generate the proposal. Chapter 5 shows that effective proposals require time and deep involvement in both the exploratory and the implementation phases.

c. The nation's economy could have been in a slump; however, it seems probable that the firm could have used the engineers profitably on other projects.

3. Evaluate. We will consider the fundamentals for each of the moral theories: utility, duties, rights, and virtues.

a. *Utility.* A story seems appropriate. A gentleman came upon several youths by a stream. They were throwing rocks at frogs. The gentleman asked what they were doing. They replied, "We are having fun." The gentleman replied, "It's not fun for the frogs."

The engineers and the town were "having fun." From the conditions we think most probable, the fun was at the expense of the consulting firm. Thus, the situation did not meet the condition of the most good for some without reducing the good for others.

b. *Duties.* Since 1974, the code of ethics of the engineering societies has noted that the engineers' responsibility to employers and clients should be limited by their responsibility to the public's safety and welfare. Here it seems that no safety aspects were relevant. Perhaps we might consider that allegiance to the employer was offset by the monetary gain of the town's population. This assumes that the total cost to the town of the communication system was lower without including the consulting firm's higher overhead. This is probable, but is not definite or given.

c. *Rights.* The expenses incurred and the efforts exerted by an organization to develop an experienced staff establish rights of the firm. It pays taxes and other overhead which contributes to the economy of the community. It seems that the right for a reasonable profit on this investment was denied.

d. *Virtues.* The intellectual ones such as foresight, creativity, etc., were not infringed. The moral ones of generosity and benevolence, however, did not seem to have been in the character of any party involved.

I suggest you make an assessment of this case, and then review the responses published by NSPE or IEEE.

9-4 SUMMARY

Ethics transcends the legal system. The two overlap only when you believe that the probability of "being caught" on a legal wrong is nearly zero. Then, you are your own judge. A system of ethics gives the engineer integrity and consistency of behavior in all matters: personal and professional. This integrity and consistency are the essence of the peace all complete professionals seek.

This chapter presents the fundamental thinking of a number of renowned ethicists. These virtually unchanging fundamentals can serve you well in developing a personal system of ethical behavior. Few characteristics exceed personal ethics in establishing a feeling of self-worth. My colleague Jack McCormac (well-published in the field of civil engineering) advises that "the most important possession a person owes himself or herself is a spotless reputation. No amount of money, fame, or knowledge is an adequate substitute. A person who can be depended on and who will work hard and learn as much about a profession as she or he can is going to be successful" [9].

DISCUSSION QUESTIONS AND PROBLEMS

1 On the basis of your reflections, define *ethics.*
2 A *purpose* is an unending endeavor: a life mission. Give your present status in moral development, or that of a friend, with the anticipated final level.
3 Review the acronyms in Chapter 4, and give a brief written review of the four ethical theories. You are encouraged to include your personal opinions.
4 Regarding codes of ethics, what inferences can be made from Brandt's rule-utilitarianism?
5 "Do not steal" is offered as a moral imperative. Offer another, and explain why it is a moral imperative.
6 Define *prima facie* duties.
7 How are A. I. Melden's concepts (on human rights) relevant to engineers?

8 Considering virtue theory, discuss proper balance for a generous person.

9 From your reflections, how do you think the Golden Rule, love, and ethics are related?

10 Congress, on the premise that home ownership promotes responsible and stable family life, legislated that interest on home mortgages is deductible from taxable income. Using the three phases of moral inquiry, give your judgment on an individual who uses a home equity loan to buy a boat.

11 a Are codes of ethics enforceable at law?

b In 1947 how was the Engineers' Council for Professional Development (ECPD) code of ethics significantly changed from the orientation of earlier codes?

ENHANCEMENT EXERCISES

12 With explanations, match each of the 10 covenants in the IEEE Simplified Code of Ethics with the corresponding type of moral theory.

13 The *Engineering Times* is a monthly publication of the National Society of Professional Engineers. Review a hypothetical case from the section titled "Engineering Ethics" (usually on page 3) in a current issue and correlate the Board of Ethical Review (BER) with the moral theories presented in this chapter. As a graduate engineer, it will be advantageous to frequently study this section and its complement, the "Legal Corner."

14 In the 1960s, Congress passed the Brooks Act requiring qualifications-based selection (QBS) for federally funded A/E (architect/engineer) services. In 1992, about 35 states have a type of mini-Brooks statute or procedure for nonbid A/E procurement. Nonetheless, QBS advocates say that with tight municipal budgets it is still a constant struggle defending those mandates in court. The press is too happy to report how a city picked an engineering firm without getting competitive bids. Use your judgment and the three phases of moral inquiry to answer the challenge: "Is the press promoting ethics in engineering?"

REFERENCES

1 Martin, M. W., & Schinzinger, R. *Ethics in engineering.* New York: McGraw-Hill, 1989, pp. 17–21.

2 Martin. *Ethics.* pp. 33–42.

3 Christian, J. L., & Gregor, J. L. *Nutrition for living.* Menlo Park, CA: Benjamin/Cummings, 1985, pp. 226–261.

4 The golden rule. *Encyclopedia Britannica,* 1988, *16,* 656:1a.

5 Wear, K., & Callinan, J. P. The engineer, professionalism, and ethics. Knoxville: *The Bent of Tau Beta Pi,* Spring 1984, pp. 24–29.

6 Proposed new code of ethics. *The Institute,* New York: Institute of Electrical and Electronics Engineers, March 1990, p. 5.

7 *Conference on engineering ethics.* New York: American Society of Civil Engineers, 1975, pp. 20–24.

8 Christiansen, D. Ethical judgement. *IEEE Spectrum,* Institute of Electrical and Electronics Engineers, April 1990, p. 19.

9 McCormac, J. C. *Surveying fundamentals* (2nd ed.). Englewood Cliffs, NJ: Prentice-Hall, 1991, pp. 528–531.

EPILOGUE

A NEW BEGINNING

This subtitle may seem overly dramatic. But with a better knowledge of the fundamentals which influence you, your colleagues, and society, you have an opportunity for a new, more satisfying professional and personal life. This book offers many of those fundamentals. Two threads reinforce its fabric: (1) the engineering approach, and (2) the technique of accomplishing tasks and enjoying life, little by little. Sure and steady wins the race. Control of one's life is the sovereign factor in achieving success. That control depends on understanding the fundamentals of the situation. Quoting Malcolm Forbes: "With all thy getting, get understanding." With understanding you will recognize that the ever-present randomness (entropy) need not make life an ongoing struggle. Rather, it can be a challenge and opportunity for continuous growth. While maintaining control, your Adult can orchestrate the creativity of the Child. The guidelines below focus on the more important concepts of this text. They are presented in three sections: those relevant to your self-esteem, interaction with others, and engineering economic systems.

THE GUIDELINES

Self-Esteem

This important, reflective evaluation is highly dependent on your achievements and your receiving positive recognition.

1. We all need "strokes" of appreciation and recognition. In childhood, we received them from loving parents. In adult life, we can earn them by our capabilities and actions. These "strokes" can be started by simply enunciating your words well and using proper English. With a little more effort and greater reward, you might enjoy becoming a gourmet chef. When reading, look up unfamiliar words and always note their roots. These will act as memory aids to encode the meanings. Also, the roots act as building elements to understand other words.

Be especially competent in a few areas. Everyone needs to have a few special niches in which she or he feels secure. There is no security like being competent in a needed service. An example is CPR (cardiopulmonary resuscitation: heart, lungs, revive). These courses are frequently available at universities or firms, or from the Red Cross. My Child enjoys having learned a few phrases in foreign languages, such as *après vous* and *despues usted* ("after you" in French and Spanish). "Hello" in Chinese (Mandarin) and Japanese is frequently fun to say and useful as well.

2. Anticipate change. The essence of modern life is being adaptive to changes, but only after proper analyses. Anticipation is the key to a quick mind. It gives the advantage of considered judgment. That judgment is best when based on full knowledge of relevent fundamentals and their understanding.

3. It is well documented that physical and mental exercise enhances our mental attitude and general well-being. Furthermore, exercise retards aging: "We do not wear out, we rust out" [1].

4. Take advantage of compound interest. Financial security strongly boosts our self-esteem. The results of the fundamentals and techniques in Chapter 7 are multiplied manyfold by the magic of compound interest over long periods. Start now!

5. Be ethical. The peak of self-esteem is ethically analyzing a situation and acting according to objective judgment. However, do not feel self-righteous; we are doing only what is proper. But that feels great!

Interaction with Others

More important than technical competence, an engineer's success depends on the ability to

- Pleasantly interact with others.
- Communicate clearly and pleasantly.
- Plan activities; use discipline and assertiveness to follow the plan.

The guidelines below will be helpful in continuing your development of these capabilities.

1. Reward others for their help. We tend to do what (and only what) we perceive as rewarding. Recognition from those important in our lives is a top reward. These "important people" are usually our first- or second-line managers and our spouse. In the text, we considered the Praise Principle: When someone is especially helpful or competent, or both, advise his or her "important person," preferably by a handwritten note. It is often inconvenient, but highly rewarding.

2. Avoid emotional power struggles. Take advantage of the fact that with time, emotions die. Try to anticipate the situation, and be prepared with a "change the subject" comment. Humor (Child to Child) is unmatched for relieving tensions.

3. Strive to give criticism with humility. A colleague, former vice president of Sangamo Electric, observed that criticism is accepted in proportion to the humility of the giver. Use expressions such as "I'm not sure, but it seems that. . . ." or "I wonder if. . . ."

4. Use models to analyze problems. All rational decisions are based on models. When you and a colleague differ on approaches to a problem solution, you are probably using different models or have different objectives, or both. The different models may be due to a lack of knowledge of the situation's fundamentals. No problem: Actor and humorist Will Rogers once said that "we are all ignorant, only about different things." A noncriticizing discussion could reveal where and in what areas the lack of information exists. The same discussion could reveal any differences in objectives. In a similar vein, quickly admit mistakes. It is refreshing to others. They then feel less threatened by their mistakes, and everyone can learn from the experience.

5. Speak with care. Oral communication is the primary way in which we inter-change with peers and first-line management. In all speaking, practice enunciating well. Do not be mentally lazy and use catchall phrases such as "those things" and "you know." Rather, take a few seconds to organize your thoughts. Daily, practice being explicit. The clarity of your communication will be pleasing to your colleagues and enhance your self-respect.

6. Try keeping a telephone file. In it, note the date and time of all calls. Include the essence of the conversation. Your colleagues will think you have a "terrific memory," and the notes can be used to document your agreements. Making notes before a tele-phone call is effective in presenting your thoughts clearly.

7. Organize your oral and written communications by first using *PWRR:* planning, writing, relaxing, and revising. The time given to the exercise should be proportional to the importance of the message.

8. Plan. As implied above, planning is surely of great significance in all achieve-ments. By following a comprehensive plan we can achieve our goals with minimum resources: time, cost, and emotional and physical effort. Follow the advice in Chapter 1 concerning self-discipline and assertiveness. It will help you minimize the effort required to carry out your plans.

9. Use self-discipline. Everybody knows it's easier to plan something than to carry it out. The necessary discipline, however, is developed easiest with the "little-by-little" technique. The satisfaction of the smallest step encourages the next.

10. Control interruptions and distractive influences from others, especially when you bring work home. This is a serious problem for achievements. Assertiveness seems to be the only solution.

Finally, in the Chapter 1 Appendix, How to Succeed in Life through and in Engineering, we note the importance of our associates. These include our neighbors. Most of us are at home a considerable part of our lives, and the influence of neighbors on our whole family can be great. Try to choose such that it will be a positive experi-ence. Like marriage, similar values and attitudes are very important. Your efforts will be rewarded for your total life and for the lives of your children.

Engineering Economic Systems

These deal mostly with your professional life.

1 Make every effort to learn the fundamentals regarding your project.

2 These fundamentals make possible the full understanding which Malcolm Forbes exhorts our getting.

3 From fundamentals with understanding, you will be able to make a reasoned approach to any difficulties in life, professional and personal. It is truly wonderful. This text is a first step.

Reference

1 Miller, S. S. *Family medical guide.* New York: Funk & Wagnalls, 1983, p. 612.

ANSWERS TO SELECTED PROBLEMS

I took great care in triple checking these answers, but please let me know of any errors I missed. Write in care of the College Division, McGraw-Hill.

Chapter 2

4a Three categories of a decision maker's (DM's) environment are

(1) Certainty—Although certainty never really exists, the outcome of the decision is thought to be nearly certain.

(2) Uncertainty—The DM does not know all the alternatives, the possible results, or the probabilities of the results of the known alternatives.

(3) Under conditions of risk—The possible alternatives are generally known, and the probabilities of their outcomes are known. This category is amenable to an analytic approach.

b For decisions under a state of uncertainty, if the DM feels that all possible outcomes are known, the state can be converted to one "under risk." This state can be achieved by the DM's assigning subjective or inferred probability to each outcome.

5a Often the expected value for the roll of dice is calculated by summing the products of each possible sum of the dice and its probability. There is an easier way (for fair dice): The rolls and outcome of each die can be considered statistically independent. Thus, the expected value of their sum is the sum of the expected value of each:

$$\begin{aligned} \text{EV (roll of dice)} &= 2 \text{ EV(roll of a die)} \\ &= 2 (1 + 2 + 3 + 4 + 5 + 6)/6 \\ &= 7 \end{aligned}$$

Recall that the sum of $1 + \cdots + n = \frac{1}{2} n (n + 1) = 21$ for $n = 6$.

   b $10

   c $27.50

6a EUV(A) = 73.4 (only 11 percent greater for 100 percent greater reward)
  b EUV(A) = 77
7a E(lost) = $1023.75. Your risk-averse value: $5118.75. It makes the premium cost of $105 desirable.
  b Case B: EV ($500 deductible, $250 greater than the $250 required) $23.50; CME = $118, EMV = $118 × 0.094 = $11.09, long-term cost of additional deduction; Net saving = $22 (premium deduction) − $11 paid over long-term. This $11 is 4.4 percent on the extra $250.
  c Case C: EV ($1000 deductible, $750 greater than $250 required) = $52.50; CME = $382.4; EMV = $382.4 × 0.07 = $26.8, Net savings = $70 − $26.8 = $43.2, 5.76 percent on the extra $750

Chapter 5

3a

Activity	A	B	C	D	E	F	G	H
EAT	7	10	6	14	—	24	16	12
LAT	14	10	22	14	—	24	24	24

  b BDF: 24 days

4a

Activity	A	C	E	G	H
Slack: free	7	4	—	8	12
Total	7	16	—	8	12

  b Free < total slack

5a 0
  b 16
8a Critical path ACF, 11 days

Activity	B	D	E
Slack: free	0	2	1
Total	1	1	1

  b $3600   **c** $3700
9b Critical path: BCF, 12 days
10 $3750
11a .0013
  b $6.5
  c Crash for any cost < $4743.50 (.95 × 5000 − 6.5)
12b

Event	1	2	3	4	5	6	7	8	9
EAT	0	3	4	8	7	14	14	19	21
LAT	0	9	4	14	7	14	15	19	21

  c

Activity	A	C	D	E	F	J
Slack: Free	0	0	2	5	5	1
Total	6	6	8	6	6	1

  d $379

Chapter 6

2a 414 $M,

 b 138 $M

 c $3.74

3a 343 $M,

 b $2.23

4a 80,000 pencils

 b $42,000

 c Yes, earnings from special order would be $25,000.

6a $2676.45

 b $2041.75

 c $5143.17

7a 4.60247 percent

 b 4.60278 percent

8 8.2 percent per month or 98 APR

9 $2427

10a $649.32

 b $454.92, savings by purchasing machine A = $194.40

11a BV(3) = $1050

 b D_{SD} (2) = $233.33

12a SL: machine A, $1814.77, $292.25

      machine B, $1598.17, $379.40

 b SD: machine A, $1793.32, $288.80

      machine B, $1587.06, $376.76

13 $400.29

14 1.67 years

15 1.95 years

16 23.9 percent

17 41 percent

18 $288,426

19 34.7 percent

Chapter 7

16 $1116.90

17 $111.11

18a 8

 b 8 < 14 → conservative, income stock

19a ROI = 3.53 percent

 b ROE = 12.98 percent

 c Debt ratio = 72.8 percent; ratio greater than 50 percent indicates undesirable financial risk.

INDEX